CORN CRUSADE

CORN CRUSADE

Khrushchev's Farming Revolution in the Post-Stalin Soviet Union

Aaron Hale-Dorrell

OXFORD
UNIVERSITY PRESS

OXFORD
UNIVERSITY PRESS

Oxford University Press is a department of the University of Oxford. It furthers
the University's objective of excellence in research, scholarship, and education
by publishing worldwide. Oxford is a registered trade mark of Oxford University
Press in the UK and certain other countries.

Published in the United States of America by Oxford University Press
198 Madison Avenue, New York, NY 10016, United States of America.

© Oxford University Press 2019

All rights reserved. No part of this publication may be reproduced, stored in
a retrieval system, or transmitted, in any form or by any means, without the
prior permission in writing of Oxford University Press, or as expressly permitted
by law, by license, or under terms agreed with the appropriate reproduction
rights organization. Inquiries concerning reproduction outside the scope of the
above should be sent to the Rights Department, Oxford University Press, at the
address above.

You must not circulate this work in any other form
and you must impose this same condition on any acquirer.

Library of Congress Cataloging-in-Publication Data
Names: Hale-Dorrell, Aaron Todd, author.
Title: Corn crusade : Khrushchev's farming revolution in the post-Stalin
Soviet Union / Aaron Hale-Dorrell.
Description: New York, NY : Oxford University Press, 2018. |
Includes bibliographical references and index.
Identifiers: LCCN 2018019524 (print) | LCCN 2018028937 (ebook) |
ISBN 9780190644680 (updf) | ISBN 9780190644697 (epub) |
ISBN 9780190644703 (online component) | ISBN 9780190644673 (hardcover)
Subjects: LCSH: Khrushchev, Nikita Sergeevich, 1894–1971. |
Corn industry—Soviet Union—History—20th century. |
Corn—Government policy—Soviet Union—History—20th century. |
Agriculture and state—Soviet Union—History—20th century.
Classification: LCC HD9049.C8 (ebook) | LCC HD9049.C8 S738 2018 (print) |
DDC 338.1/7315094709045—dc23
LC record available at https://lccn.loc.gov/2018019524

CONTENTS

Prologue	vii
Acknowledgments	ix
Introduction: Khrushchev Crusades for Corn	1
1. Scarcely Making Ends Meet	10
2. Industrial Agriculture, the Logic of Corn	27
3. Corn Politics	55
4. Better Living through Corn	87
5. Growing Corn, Raising Citizens	109
6. From Kolkhoznik to Wage Earner	137
7. American Technology, Soviet Practice	167
8. Battles over Corn	197
Conclusion	226
Notes	235
Glossary and Notes on Transliteration	295
Bibliography	299
Index	317

PROLOGUE

From 1953 to 1964, Nikita Khrushchev campaigned to improve the Soviet Union's food supply by compelling farms to use the latest technologies to plant corn, a crop hitherto unfamiliar in most of his vast, environmentally diverse country.[1] Meant to feed the livestock necessary to provide citizens a rich and varied diet, corn was to contribute to domestic stability, the abundance required by Khrushchev's concept of communism, and success in the Cold War contest against the United States, which he reimagined as a "peaceful competition" to provide higher living standards. Scholars have long dismissed this corngrowing strategy as irresponsible because the crop demanded warm, humid summers that much of the Soviet Union lacked. Yet a story of climate alone captures only part of a history of modest successes, policy blunders, and the everyday dysfunction inherited from Iosif Stalin.

Long before Khrushchev, rulers of Russia had imposed potentially useful innovations, only to see them frustrated by the realities of governing. The satirist Mikhail Saltykov-Shchedrin caricatured overbearing sovereigns' attempts to reform an empire and people so apparently intractable. His *History of a Town*, published in 1870, takes the form of a fanciful chronicle of Glupov—a name derived from the Russian *glupyi*, or foolish—that serves as an allegory for the whole of Russia. Many town governors benignly neglected their duties, but the dangerous ones aimed to improve the people, culture, and economy.

One of the latter, Vasilisk Borodavkin, resurrected the reforms of Dvoekurov, a predecessor who had been so progressive as to institute discriminant flogging in place of indiscriminant *and* to compel the townspeople to cultivate mustard. The latter initiative had nonetheless provoked disturbances. "Wherever there's flogging, you can bet your life there's a rebellion. . . . A lot even went to Siberia over that business!" the citizens informed Borodavkin when he inquired about Dvoekurov's mustard experiment. "Speaking in some detail concerning the basis of society in general, and of mustard as the basis of society in particular," Borodavkin tried to convince the inhabitants to grow it again. The chronicle records that at first he did not succeed. "Because, as was his habit, he did not speak but shouted," it notes, "his speech resulted only in the Glupovites being seized by terror."² Saltykov-Shchedrin's narrator, a later commentator on the chronicle, clarifies that the citizens resisted such an obviously beneficial policy only because they misread Borodavkin's confusingly worded proclamation. In response, the governor waged a "war of enlightenment" against the townspeople. After a nine-day campaign shedding no blood, the inhabitants accepted mustard "in all places and for all time," but they soon grew so much of the condiment that a market crash ruined the town. Reaction to the French Revolution then forced Borodavkin to add to the chaos by waging new campaigns "against enlightenment"—and mustard.³

Saltykov-Shchedrin drew inspiration for the mustard-planting scheme from efforts during the reign of Catherine the Great in the eighteenth century to coerce peasants to plant potatoes, another unfamiliar New World crop. In a 2011 interview, Sergei Khrushchev linked his father's program to that episode. He contended that historians misrepresent the policies by claiming that "Khrushchev brought corn from America to plant it beyond the Arctic Circle." Defending his father's legacy, the younger Khrushchev continued, "Of course, this was not the case. Father simply ascertained that ... corn contained the maximum amount of feed, and said, 'Let's adopt that.' Corn became a joke, but there were no corn rebellions like the potato rebellions of Catherine's time."⁴ Or, at least, no one overtly revolted.

In hindsight, my decision to write first a doctoral dissertation and then this book about Khrushchev's devotion to corn seems only logical. Fascinated by the Soviet Union, I noticed that many historians offered only a cursory explanation for the seemingly quixotic scheme. Yet the episode spoke to my experience growing up in America's rural Midwest, where corn is king. In June and July of the summer I turned fourteen, I spent several weeks detasseling corn, a job description likely familiar only to readers raised in the Corn Belt, home to commercial production of hybrid seeds. The work meant getting up before sunrise to ride a school bus to a field. There, from dewy morning to sweltering midday, a crew of teenagers manually completed a job begun by a machine designed to chop the pollen-producing top, the tassel, from corn plants in rows each composed of one genetically uniform line. This ensured that they were pollinated by plants of a different line in a neighboring row, which kept their tassels. This produced seeds sharing desirable characteristics of both lines. Earning only minimum wage, the job functioned as a rite of passage teaching the value of manual labor. Although I returned each working day that summer, by its end I had my fill of tassels. Little did I know that I would recall that work years later while sitting in a Moscow archive reading reports to the Central Committee of the Young Communist League. It turns out that beginning in the late 1950s, teenagers across the Soviet Union's southwestern reaches did the same job as part of Khrushchev's corngrowing program.

ACKNOWLEDGMENTS

I could never have succeeded in the field of history without the many people who helped me along the way and thereby made this book possible. Like all products of academic labor, it would have been impossible without the help, advice, support, and encouragement of colleagues, friends, and family. This book began as my doctoral dissertation at the University of North Carolina at Chapel Hill, which I completed under the expert guidance of Donald J. Raleigh. I cannot sufficiently acknowledge all that Don taught me about the craft of history and the hard work of writing it, both of which were substantial, time-consuming endeavors. Moreover, he consistently demonstrated what it meant to be a scholar and mentor.

In the department of history, I also benefited from the mentorship of Louise McReynolds and Chad Bryant, who each readily offered advice and encouragement through every phase of the project. Other faculty members contributed to my graduate training in teaching and scholarship in ways that, while perhaps not obvious, shaped my approach to the topic and to the work that went into the project. Christopher Browning, Donald Reid, Michael Tsin, Lisa Lindsay, Christopher Lee, Miles Fletcher, Marcus Bull, and Theda Perdue all belong on this long list, which is nonetheless not exhaustive.

All members of the *kollektiv* of UNC Russianists offered advice and encouragement, and set examples of how to go about all of this. I must offer special thanks to those who shared part of my years there: Jen Parks, Marko Dumančic, Mike Paulauskas, Emily Baran, Gleb Tsipursky, Edward Geist, Adrianne Jacobs, Andrew Ringlee, Audra Yoder, Dan Giblin, Kathleen Conti, Gary Guadagnolo, Stephen Reigg, Trevor Erlacher, Louis Porter, Virginia Olmstead, and Dakota Irvin.

In addition, my graduate student colleagues across all fields contributed to making Chapel Hill home to both a rigorous intellectual life and a collegial departmental culture. Chapel Hill and Carrboro felt like home through those years, moreover, thanks to good friends. Jonathan Hancock, David Williard, and Julie Ault each turned the necessity of stretching a graduate student's budget by sharing a house on Todd Street into a real pleasure. They also—mostly—turned a blind eye to my penchant for laughing a little too loudly over a cup of coffee and the morning's news before 8 A.M. The friends who lived on Laurel Avenue for many of those graduate school years each individually and as a group helped me make it through. I cannot express how much I appreciate the willingness to commiserate that Kim Kutz Elliott, Friederike Brühöfner, Nora Doyle, Anna Krome-Leukens, and Rachel Henson offered during those years. Derek Holmgren and Sarah Lowry offered many evenings of excellent company and outstanding food, making sure that I never forgot to stop and taste whatever was fresh from the farmers' market. Mark Slagle made sure our team of historians dominated the local pub quiz but, more important, has offered years of friendship. He was the one, naturally, who showed up for dinner with a bottle of whiskey on November 9, 2016.

During the two academic years and two summers I spent researching this project in Moscow and other parts of the former Soviet Union, many colleagues and friends made the Russian capital and other cities feel a bit like home. I offer a hearty thank you to Markian Dobczansky, Anna Whittington, Masha Kirasirova, Krista Goff, Ben Sawyer, Kelly Kolar Sawyer, Robin Brooks, Jared McBride, Lindsay Martin, Adrianne Jacobs, Sara Brinegar, Brendan McGeever, Sylvain Dufraisse, Maya Holzman, Lucy Dunlop, Alan Roe, Patryk Reid, Andy Bruno, Kristy Ironside, Betty Banks, Marilyn Campeau, Katie McElvanney, Siobhan Hearne, Ben Bamberger, Yana Skorobogatov, Abigail Bratcher, Cassandra Hartblay, and many others. Other colleagues in the field who have made Moscow an inviting place to do Soviet history include Andrew Sloin, Stephen Bittner, Alan Barenberg, and Saulius Grybkauskas. I must particularly thank Saulius for encouraging me to delve into the archives in Vilnius and then going a long way to making it possible for me to do so. Oscar Sanchez provided insightful feedback on the project and always proved willing to talk about the research, but also made sure I took time out to have a beer and mull over the World Cup, the Euros, the Champions League, or whatever other football was on the horizon.

In the second of those academic years in Moscow, I was fortunate enough to benefit from a postdoctoral research fellowship at the International Center for the History and Sociology of World War II and Its Consequences at the National Research University–Higher School of Economics. I must thank Oleg Budnitskii and Liudmila Novikova, who made the center tick. Oleg Khlevniuk, Michael David-Fox, and the other researchers helped to make the center a welcoming environment, which I found enriching and conducive to immense productivity. Fellow postdocs Natalie Belsky, Seth Bernstein, and Angelina Lucento provided friendship, a sounding board, and the encouragement to soldier through the long Moscow winter.

I should also mention Anna Krylova, who graciously agreed to serve on my dissertation committee, encouraged me to expand the project's ambition as it became a book manuscript, and offered the opportunity to give papers as far from home as a workshop, "Twentieth Century Socialism: Ideas and Practices in Soviet Russia and China," held at Tsinghua University in Beijing. During the course of this project, I was fortunate to share panels at a range of conferences with a number of colleagues, whom I thank for their comments and questions. These include the Fifth European Congress on World and Global History in Budapest; the Association for Slavic, East European, and Eurasian Studies Annual Convention; the Southern Conference on Slavic Studies Annual Meeting; and the Southern Forum on Agricultural, Rural, and Environmental History in Columbia, South Carolina.

During the final phases of the project, an unexpected move took me to Rhode Island, where I was fortunate enough to have a ready-made group of colleagues in the field. Ethan Pollack and Bathsheba Demuth graciously welcomed me to the state. Moreover, I was fortunate enough to join a small but lively group of young historians in the field based in Boston.

Since I conceived this project in early 2009, I have accrued more than a few debts to institutions. In Chapel Hill, this includes the Department of History and its Departmental Research Colloquium, but also the Carolina Seminar "Russia and Its Empires, East and West"; the Center for Slavic, Eurasian, and East European Studies, and the helpful folks at Davis Library. The Graduate School provided a dissertation completion fellowship for 2013–2014 that enabled me to power through, finish, defend, and graduate during that academic year. Support and funding also came from external organizations, including the University of Illinois Center for Russian, East European, and Eurasian Center's Summer Research Lab; the Summer Institute for Conducting Archival Research at George Washington University; Andrew W. Mellon Foundation and the Institute for International Education, Moscow; and the American Councils Research Scholar Program. The Duke University Center for Slavic, Eurasian, and East European Studies provided Foreign Language and Area Studies fellowships, which funded several years of my graduate training. I received a great deal of help from the archivists at the many archives in Russia where I worked over the course of several years. In particular, Galina Mikhailovna Tokareva of RGASPI and Liudmila Ivanovna Stepanich of RGANI stood out for their warmth.

I must of course thank the staff of Oxford University Press and, in particular, Nancy Toff, who shepherded this book through the multiyear process of transforming a partially revised dissertation into a fully formed scholarly work. I also would like to thank the anonymous reviewers of the manuscript, who offered insightful and encouraging feedback on the details, but, more important, on the larger significance of the project.

The odyssey that has brought this book into being would have proven vastly more difficult, if not impossible, without my family: Mom, Dad, Jessa, Doug, and Madilyn. Their quiet, consistent support made it possible to keep plugging away. Above all, Mary provided love and understanding. She put up with me packing up to go to Russia not once, but twice—the second time for many months at a stretch—within the first three years of our relationship. And she married me anyway. I dedicate this book to her, with love.

INTRODUCTION

KHRUSHCHEV CRUSADES FOR CORN

A revolution in agriculture transformed the Soviet Union's kolkhozes and sovkhozes during the leadership of Nikita Khrushchev, ensuring that in 1964 they bore only a passing resemblance to their counterparts of 1953. Throughout that decade, Khrushchev championed corn with a zeal that citizens mocked by deriding him as a *kukuruznik*, or corn-man.[1] They told jokes about the venture, including one alluding to Vladimir Lenin's 1920 maxim that the power of elected soviets and electrification of the revolutionary state promised to create technological modernity, a prerequisite to communism. "What is communism?" went the question. The punchline followed: "It is Soviet power plus the corn-ification of the entire country."[2]

To Khrushchev, corn was not a joke, but rather part of a serious effort to fundamentally reform agriculture. Since the end of 1920s, Iosif Stalin's version of socialism had imposed spartan living conditions on all but a privileged few. Striving to provide everyone the abundance required for the anticipated higher stage, communism, Khrushchev viewed corn as feed for the livestock necessary to enrich diets hitherto poor in meat and dairy products. He therefore imported from the United States advanced technologies designed to efficiently produce corn and other crops. Animating his policies, these ideas permeated his voluminous speeches on agriculture, an anthology of which filled eight volumes.[3]

Pursuing his goals, Khrushchev insisted that millions of people work tens of millions of hectares in far-flung corners of the Soviet Union. Corn plantings increased fivefold from only 3.5 million hectares in 1953, or 3.3 percent of all crops, to 17.9 million in 1955. Setting a target of 30 million hectares by 1960, Khrushchev pushed for matching the United States by devoting 30 percent of croplands to corn. In 1962, plantings in the Soviet Union peaked at an astounding 37.2 million hectares. At nearly 20 percent of the total area, this figure fell short of his goal only because that area had nearly doubled since 1953.[4] Khrushchev repeatedly promised that corn was facilitating efforts to win admirers around the world by building a society equal to the United States in material abundance, but distinguished by egalitarianism. He habitually claimed that his country was soon to outdo its Cold War rival by "catching up to and surpassing America" in food output, as he did in the 1959 Kitchen Debate with US vice president Richard Nixon.

When ousting Khrushchev in October 1964, his former followers savaged corn as a component of the agrarian program they denounced as "harebrained scheming."[5] They blamed it for the 1963 harvest failure that had forced the Soviet Union to make unprecedented grain purchases abroad. In the late 1980s, critics echoed that verdict by counting corn as one of Khrushchev's many apparent policy "failures," making them proxies for Mikhail Gorbachev's reforms, which they opposed.[6] Influenced by conceptions of Communist Party rule as totalitarian, foreign analysts during Khrushchev's time presumed that his power over party and government guaranteed authority to make any policy a reality. They therefore blamed all failures on him. One agricultural economist consequently wrote of the "superficiality" and "lack of seriousness" evidenced by shifts in the direction of Khrushchev's policy initiatives.[7] Using words such as "incoherent" and "contradictory," historians have described his reforms as a quixotic search for a "miracle."[8] Amid all this, scholars have viewed corn as inexplicable and as evidence of Khrushchev's imprudence. As William Taubman's Pulitzer Prize–winning biography concludes, this "crusade" for corn "turned into an irrational obsession."[9]

Each of these mistaken conclusions has shaped subsequent perspectives on Khrushchev and his initiatives. In fact, he inherited from Stalin a Soviet Union burdened by crises in rural society and agrarian production. Determined to address them, Khrushchev transformed the idea of encouraging farms to plant corn into a crusade sustained by an almost religious belief that science and technology would improve on conventional agricultural practices. Instead of a debacle following inevitably from outlandish schemes imposed using unconstrained power, the corn policy seems much more sensible if considered as part of Khrushchev's quite rational decision to adopt a model of industrial agriculture originating in the United States. Formidable party and government mechanisms gave Khrushchev license to attempt reforms to the Soviet Union's agrarian political economy: that is, the relationship among labor, production processes, custom, regulation, and state policy. But the bureaucracies Khrushchev nominally controlled concurrently limited his capacity to transform policies into practices. Despite doggedly pursuing a vision of abundance based on corn, Khrushchev achieved yields only fractions of those needed to make the crop a cost-effective improvement on oats, hay, and other proven alternatives.

Misinterpretations of Khrushchev's corn crusade and far-reaching agrarian reforms have shaped conventional understandings of his leadership and policies. Observers at the time saw that the Soviet Union had begun to act like something other than a Stalinist dictatorship, although precisely what remained unclear. Most scholars considered Khrushchev's agricultural programs unpromising, although a few gave him credit for devoting some resources and attention to the lagging agrarian sector.[10] Even though the Soviet Union's archives opened in the 1990s, interpretations of agriculture under Khrushchev have hardly adapted. The most influential history of the era, Taubman's biography persuaded scholars to reimagine Khrushchev as a complex, serious historical figure, rather than an unpredictable enigma. Concentrating on personality and politics, however, the biography considers agrarian reform within the bounds of conventions defining Khrushchev the reformer as, in Taubman's words, "daring but bumbling."[11] Since 2000, historians have begun to study the postwar and post-Stalin years in earnest, redefining conceptions of Khrushchev's approaches to international trade, relations with newly independent former colonies of European overseas empires, the battle against Stalin's legacies of repression, the urban housing shortage, and the fraught relationship with the Communist Party of China.[12] As Khrushchev's signature domestic policy area, agriculture and food require a similar reappraisal, without which any re-evaluation of the postwar years will remain unfruitful.

Rigorous scholarly re-examination of the era indicates the need for a revived inquiry into rural society and agriculture, which have hitherto been relegated to the margins.[13] Promoting corn, Khrushchev intensified an existing commitment to a transformative, industrial vision for agriculture, a fundamental part of human societies' complex relationships with environmental processes. From the 1920s, the Bolsheviks' doctrines and early experiences ingrained ways of thinking about technology and power. Their new country adopted farming methods that became common across the globe, evidence that they shared a vision of technology, the environment, and rural development with other modernizers and industrial societies. Even a program as repressive as Stalin's collectivization applied technologies of rule analogous to those employed by other high modernists, albeit in radical forms inflected by recurring agrarian crises, Bolshevik teachings, inherited political culture, and experiences of World War I and the Russian Civil War. Collectivization was an archetypical example of a common high modernism, but one taken to authoritarian extremes against a peasant society too weakened to successfully resist.[14]

The Bolsheviks founded a state committed to transforming the countryside, a mission familiar to many industrializing societies.[15] Although self-identifying as communist, the Soviet Union behaved in ways comparable to other states founded to modernize the former empires remade by the catastrophic effects of World War I.[16] To build the future, the Soviet Union had to solve problems related to peasants, agriculture, rural development, and the environment. The preoccupation with such transformations predisposed Khrushchev to see the joining of modern farming and corn as a panacea for the chronically low productivity afflicting kolkhozes and sovkhozes. For its part, corn was poised for the role. Since 1500, corn had gradually spread from the New World and across the globe. By around 1900, the crop had become unequaled as a commodity and integral to the industrial farming systems subsequently invented in North America, which spread rapidly after World War II.[17]

Khrushchev's agrarian reforms testify to efforts to redefine rural economic and social relations, which continued to evolve long after collectivization. Turning preexisting principles of industrial agriculture into widespread practices, Khrushchev's agrarian reforms coincided with a burst of rural modernization. Rather than imagining a fixed state, modernity, I have chosen rural modernization as the term for historical processes intertwined with mass urbanization, education, consumption, and other transformations.[18] Adapting to these trends, the Soviet Union borrowed from kindred industrial societies those technologies needed to get the most from corn and, more broadly, the potential of industrial agriculture. Much as in other contexts, these technologies altered the ways farmers in the Soviet Union produced food and how citizens consumed it.[19]

Progressing furthest in North America and Western Europe, these developments produced urban majorities fed by a tiny number of farmers who increased per-hectare productivity by utilizing technologies that allowed them to concentrate energy in their fields by applying various forms of petroleum. Each society's unique experience of these transformations influenced the speed and trajectory of those individuals pushed from villages and pulled by cities into migrating in search of non-agricultural livings. Many societies witnessed a decrease in the number of farmers. Each of those who remained on the farm produced more food, and many benefited from subsidies designed to offset chronically depressed prices.

These processes were ongoing everywhere during the Khrushchev era. Even in 1945, societies in Western Europe retained substantial rural populations, leaving large numbers to move to cities only during the following thirty years.[20] Because the Soviet Union's rural majority lasted until about 1960, it more closely resembled societies of Eastern and Southern Europe. After 1953, the country's new leaders faced a daunting challenge. Exploiting peasants to finance breakneck industrialization from the 1920s onward, Stalin had altered processes of rural modernization already underway before 1917, leaving his successors the gargantuan task of seeking a new balance between farm and factory.

First in the United States, then in Europe, and later in parts of Asia, Africa, and Latin America, farmers came to apply new technologies: not only machines, but also irrigation, chemicals, fertilizers, agronomy, accounting, and technical education.[21] The Soviet Union also committed itself to what historian of technology Deborah Fitzgerald terms "the industrial ideal in American agriculture." After World War I, American entrepreneurs and engineers had refined a combination of machines, transportation infrastructure, communications, commodity markets, migrant labor, bank loans, technical expertise, and the capital needed to set them all to producing value. To coordinate this, they applied methods pioneered in industrial management.[22] Peasant societies often prioritize local self-sufficiency and maintaining social institutions. Modern alternatives to industrial farming favor conservation of soil, water, and biological diversity. Discounting all of these, the industrial ideal encourages pursuit of the most marketable produce at the lowest cost, which measures only monetary value of inputs without acknowledging resulting social or ecological harm.[23]

In the late 1920s, the Americans who developed these technologies and practices transplanted them to the Soviet Union, which embraced this approach to agriculture almost from its invention. Returning to the United States a few years later, the Americans brought back new experience to apply at home, which they did on a constantly expanding scale. The whole cycle demonstrates the transfer mentality that encouraged the belief that these methods were applicable in almost any social, cultural, or environmental context.[24] In the Soviet Union, however, industrial farming principles began to define practice widely only under Khrushchev. His Virgin Lands program, which converted millions of hectares of sparsely populated steppe into mechanized, industrial-scale wheat farms, joined with initiatives to put genetics, chemistry, engineering, and accounting to work down on the farm. He thus endeavored to integrate industrial agriculture into everyday activities.

Khrushchev's embrace of corn and related technologies emphasizes common processes of rural modernization and technology transfers through the Iron Curtain, calling into question binaries resulting from the Cold War. In these spheres, the world did not divide neatly into equivalent competing blocs separated by capitalists' commitment to markets and the communists' espousal of the opposite of markets. After 1991, scholars came to see how the Cold War had distorted their understanding by encouraging all sides to imagine two separate worlds of ideology, culture, domestic economics, and international trade.[25] The Soviet Union was distinctive for its statist practices, repressiveness, and restrictions on movement, yet it shared modernist conceptions of technological agriculture with other modernizing, industrial societies. The country's isolation was not as complete as the metaphor of the Iron Curtain suggested. Except for during some rare periods of heightened tension, technology and trade crossed back and forth with the encouragement of the Soviet Union's leaders.

Moreover, the Cold War encouraged all sides to compare and compete. In the face of American economic and military preponderance, the Soviet Union was always at a disadvantage. Its ambitions and capacities could hardly match those of the United States because, despite the Soviet Union's much larger population, its national income was approximately equal to that of France or the United Kingdom.[26]

Rather than an isolated, separate system, the Soviet Union was an integral part of larger histories of science and technology. Its leaders declared their country different and, when they sensed external threats, portrayed themselves as walled off from the world of capitalism. Nonetheless, they often borrowed from the industrial societies they planned to defeat and remained sensitive to external pressures posed by technological advancement.

After 1953, Khrushchev's agrarian reforms provided impetus for full-scale adoption of farming technologies he and his experts learned of abroad. He thus revived a dialogue about such technology with foreign—especially American—interlocutors. The history of corn therefore forms part of a crossed history of technology. Building on comparative histories and histories of transfers, the concept of crossed history emphasizes multilateral, multidirectional interactions and processes, rather than static systems. The term evokes the meeting of two or more roads, an interaction in which systems and other entities cannot remain intrinsically separate. Instead, processes of crossing alter the object of the transfer and all historical subjects involved, albeit often in asymmetrical ways.[27] Rather than a distinct non-capitalist system embodying the antithesis of the modernity defined by liberal capitalism, the Soviet Union was a complex entity embedded in a larger world involving many actors and ideas. Khrushchev's corn crusade therefore forms part of larger currents of rural modernization and developments in farming technology operating in parallel across the world.[28]

The first post-Stalin decade was an era of intense economic and social change, symbolized by processes of the Thaw in cultural matters and de-Stalinization in politics. The Khrushchev years did witness official and popular grappling with Stalin's legacies of crash industrialization and violent repression. The relatively relaxed atmosphere of the Thaw fostered the critical discussion of agriculture that encouraged reform. Yet popular and scholarly conceptions of these better-known political and cultural currents have never critically examined Khrushchev's agrarian policies. Often defined by concepts of the Thaw and de-Stalinization, analyses of the period typically do not take into account developments in rural society, agrarian policy, and agricultural production, which did not remain static backwaters.[29]

Although continuities across the 1953 divide are evident, an appreciation of the era must acknowledge the post-1953 dynamism of the Soviet Union in economic and technological spheres. At the time, both critics and admirers noted the stunning economic growth rates and technological prowess demonstrated by the launching of *Sputnik*, the first satellite, in 1957. Subsequent observers, however, have been influenced by knowledge of later slowdowns. In the late 1980s, Gorbachev popularized an influential narrative of economic and social stagnation during the late 1970s and early 1980s under Leonid Brezhnev. This allowed Gorbachev to draw a contrast between his reformist leadership and the relative hostility to change under Brezhnev. During the post-Soviet era, the postwar years have acquired a reputation for stasis and dysfunction, a preconceived notion that dwells on obvious failings while missing signs of dynamism.[30] During the 1950s and 1960s, the Soviet Union exhibited considerable sophistication that became evident when its experts applied chemical engineering, nuclear research, jet propulsion, and other inventions of what became known as the Scientific-Technical Revolution, which aligned its experiences with those of other industrial societies.[31]

The Soviet Union's agrarian sector witnessed the application of similar technological advances, albeit with results limited by everyday practices. Evident to contemporaries, significant changes in agrarian society have disappeared from view, overwhelmed by presumptions about the invariability of defining features of the Soviet Union. Understandings of the industrializing society of 1930s relied on categories of analysis appropriate to the violence, expropriation, resistance, and accommodation of collectivization, categories presumed to have remained applicable to later periods. It therefore followed that these characteristics and the Soviet Union's ideology bound the country to remain the mirror image of its capitalist counterparts throughout the postwar era.[32] This conventional thinking about the postwar era maintains a binary that defines the Soviet Union only in relation to those supposed polar opposites: dynamic capitalist societies based on markets and private property.[33]

Historians of the Soviet Union therefore have a great deal of learning ahead of them, which requires a questioning of apparently stable economic, social, and administrative processes fundamental to postwar agrarian society.[34] Instead of merely tinkering with a failed agrarian system, Khrushchev initiated an ambitious program that only partially succeeded. This book therefore inquires into the basics of how agrarian economics worked, jettisoning conceptions rooted in the Soviet Union's differences, which both sides of the Cold War divide defined in opposition to the capitalist other. Put simply, the country's farms and other enterprises were not controlled by a static set of command structures antithetical to market-like mechanisms. Even the capitalist systems from which earlier conceptions drew their contrasts changed constantly, being themselves products of fierce contests among competing social and economic interests. In much the same way, the antimarket and illiberal principles long considered fixed in the Soviet context during the 1930s and stable thereafter were in fact subject to struggle and change. Doctrine prompted leaders to reject private capital in agriculture, as in other sectors, but they concurrently intensified wage labor, the commodification of the basic human capacity to expend labor power and a vital element of Karl Marx's critique of capitalism and capitalist political economy.[35] Far from relying on enthusiasm and mobilization alone, Khrushchev responded to the entrenched but unacknowledged labor market by intensifying the influence of wage labor over kolkhozniks' lives. He emphasized prices, production costs, bonuses for exceeding the plan, guaranteed wages, and financial accounting, all while expanding money relationships in the countryside.

The thematically organized chapters that follow examine ideals, goals, technology, organization, management, and wages, all to understand what the Soviet Union *did*, rather than what it *was not*.[36] This inquiry relies on declassified documents from the Moscow archives of the Communist Party and the government, which defined policy and oversaw its implementation, as well as from the archives of local administrations in Vilnius, Kyiv, and Stavropol tasked with turning policy into practice. These documents, combined with contemporary newspapers and periodicals, disclose important information in what they say, what they leave unspoken, and what they assume, all of which sheds light on the goals, ideals, values, and worldviews not only of top leaders, but also individuals beyond Moscow. The "habits of mind" of these local officials and the kolkhozniks during the years after Stalin have received comparatively little attention from scholars.[37]

Subjected to inquiry on these terms, the Soviet Union sheds its isolation as the Soviet Experiment. Instead, it becomes one of a set of industrial societies that also includes liberal and capitalist ones. Although peculiar due to its history and its leaders' doctrines, the Soviet Union was a participant in global trends relating to industrial agriculture, a fact placing that sphere alongside media, culture, education, urbanization, consumption, and trade, all sites of well-documented interactions.[38]

Having endured extreme privation during the century's first half, the Soviet Union became a comparative haven for the postwar generation. The tranquility of its members' youth and early adulthood contrasted with parents' and grandparents' experiences of revolution, civil war, collectivization, terror, and wartime sacrifice. These Soviet Baby Boomers valued education, social mobility, love, Beatles records, and family life, pursuits in both private and professional spheres familiar to contemporaries abroad.[39] Society continued to evolve after the war, becoming more educated, urbanized, comfortable, and cosmopolitan, even as its utopian impulses became less palpable after Khrushchev. Rising mass consumption meant that more citizens aspired to an apartment, a personal car, or a summer cottage.[40] Into the bargain, they expected a rich and varied diet befitting their newfound status. Khrushchev expected corn to provide just that.

1 SCARCELY MAKING ENDS MEET

"Later on I found out that . . . it was a long time since Matryona Vasilyevna had earned a single ruble," recounts the narrator of Alexander Solzhenitsyn's novella *Matryona's Home*, referring to the aged kolkhoznik with whom he had lodged. In the tradition of nineteenth-century Russian realism, the story captures the severity of everyday life in the early 1950s. "On the kolkhoz [Matryona] had worked not for money," the narrator explains, "but for . . . the marks recording labor-days in her well-thumbed workbook."[1] Labor-days entitled her only to a share of the kolkhoz harvest—if any—left over after taxes and other obligations to the state. Personifying the rural heart of Russia, Solzhenitsyn's protagonist had to toil, despite her infirmity, to secure food and fuel for the long, harsh winter.

Documents corroborate Solzhenitsyn's sense that Iosif Stalin left his successors an agrarian crisis to when he died in March 1953. The next month, Nikita Khrushchev and his competitors for power, Georgii Malenkov and secret-police chief Lavrenti Beria, received a survey of intercepted statements intended to illustrate popular feeling.[2] Resenting the kolkhozes compulsorily established in 1947 and 1948, writers in restive, recently annexed Western Ukraine penned letters complaining of overwork, miserly earnings, and cheating bosses. "Dearest sister," began one, "I have still not received so much as a single potato from the kolkhoz so long as it has existed." Sardonically mimicking Stalin's 1935 statement touting purported prosperity, another correspondent concluded, "Life has become 'better and more joyous.'"[3]

Facing an agrarian emergency upon assuming power, Stalin's successors had little choice but to reform because his policies had not transformed the countryside in the manner forecast by Bolshevik visions of rural modernization. The revolutionaries had considered industrialization the herald of large-scale modern farming and of the end of the peasant class whose existence they considered archaic. Hastily collectivizing individual allotments into kolkhozes, Stalin and his followers employed coercion to bring peasants' grain under state control, but this nationalized agriculture failed to realize the Bolshevik ideal of farming.[4] Enforced by threats and violence, this arrangement robbed agriculture of capital and caused social dislocation, imposing constraints that grew only more pronounced under the strain of World War II and during a postwar food crisis. Stalin's assurances notwithstanding, life had become neither "better" nor "more joyous" for the rural majority of the population. The failings of the agrarian sector meant that after the war, even allegedly favored factory workers struggled to put food on the table. Stalin's successors concluded that only modern industrial farming could provide the rich and varied foods needed to improve living standards and stabilize society, each essential to contesting the Cold War, which they reimagined as a competition between economic systems. Within months of interring Stalin, his heirs responded to pressures from below, giving Khrushchev an opening to place a bet on a crop he had come to value while in Ukraine: corn.

By the time of the Russian Revolution of 1917, modernizing forces had unsettled society and frustrated two agrarian reform programs. Pressure to end serfdom came to a head in the late 1850s, after the Russian Empire had lost the Crimean War on its own soil to technologically advanced rivals. In 1861, emancipation ended the bondage tying the peasant majority to land they did not own. One of the Great Reforms designed to reinforce the tsarist order and strengthen military preparedness, the settlement required peasants to pay for the land and subordinated to the existing village communes their individual allotments and right to migrate. Over subsequent decades, population growth exhausted the communes' limited supply of land, encouraging the industrious to clamor for more. In the 1880s and 1890s, peasants began to relocate to growing cities and industrial settlements. During the Revolution of 1905, urban strikers responded to economic downturn and the battlefield disasters of the Russo-Japanese War (1904–1905). In the countryside, peasants rebelled against weakened civil authority, protesting the land shortage and legal restrictions. Employing repression tempered by a few concessions, the government launched new reforms that reduced the influence of the commune by empowering individual peasants to separate allotments from it. The final years before World War I saw measurable increases in agrarian productivity and incomes, but pressure continued to build.[5]

The Bolsheviks seized power amid wartime chaos, faced new crises, and then constructed a market-based agrarian system. In 1917, peasants exploited the collapse of authority to confiscate and reapportion large estates, the farms most oriented to the market. During the savage Russian Civil War (1917–1922), the countryside endured grain seizures, violence, crop failures, famine, and disease. Added to revolutionary egalitarianism, these pressures strengthened communes at the expense of the incipient individualism. Poor peasants received expropriated lands while disorder disproportionately harmed the rich, compressing the socioeconomic scale. Peasants fought requisitioning, causing shortages in the cities. In late 1921, these compelled Vladimir Lenin to propose the New Economic Policy (NEP) to aid reconstruction. Making concessions to the peasants, it replaced confiscation with taxes and legalized markets for after-tax surpluses. Enterprising farmers expanded within legal limits, including by hiring farmhands. To some Bolshevik critics, markets and wage labor heralded rural capitalism. Before Lenin died in 1924, he cautioned that NEP might have to endure for decades while the countryside evolved toward socialism.[6] Even a revolutionary workers' state could not hope to remake overnight a country where the 100 million peasants comprised 80 percent of the population.

Within a few seasons, NEP returned harvests to the 1913 level, but then fell victim to economic and political pressures. It permitted the state to acquire grain cheaply, sell it abroad, and use the foreign currency to import capital goods required to rebuild ruined factories. This caused a scissors crisis, so named for the appearance of a graph tracking the declining prices for grain and rising ones for industrial manufactures. Responding, peasants withheld and consumed surpluses. By the mid-1920s, policy debates raged within the Communist Party about how to facilitate development in the face of adverse international economic conditions.[7] A group headed by Nikolai Bukharin favored a gradual approach premised on stability bought with concessions to the peasants, to whom many communists remained hostile. Those to their left, led by Lev Trotskii, argued for treating the peasants as an internal colony by raising taxes and tilting the balance further toward the state to finance breakneck industrialization.

NEP reached a crossroads in 1927 when Stalin and his followers took a series of fateful decisions. The United Kingdom broke off diplomatic relations and then communist policy in China suffered a sharp reversal, heightening fears of war. Having aligned in 1925 with Bukharin and others to defeat Trotskii, Stalin reversed course, arguing that his erstwhile allies' cautious path left the country vulnerable. He took up the left faction's argument for a crash program to build up the heavy industry required for defense. The First Five-Year Plan handed the bill for imported machinery to the peasantry, whose grain would join timber, petroleum, and other raw materials among the exports. Already low in the 1920s, global commodity prices declined as capitalist economies sank into the Great Depression, meaning that each unit exported earned less foreign currency. Stalin responded by exporting more to make up the difference and maintain planned imports.

Needing to export grain and feed the swelling industrial workforce, Stalin turned to authoritarian methods hearkening back to the civil war. After the 1927 harvest, peasants responded to the price scissors by slashing sales of grain just as industrialization began to gather steam. Considering this sabotage, Stalin and his comrades declared war on well-off peasants, labeling them class enemies. Receiving reports of unrest, they sponsored the Ural-Siberian method entailing forced delivery quotas, sparking a multilateral conflict engulfing peasants, local officials, urban party militants dispatched to the scene, and leaders in Moscow. Branding any resisters kulaks, a Russian word meaning "fists" and connoting grasping avarice, zealots marked many for dispossession, internal exile, and even death. By 1930, activists were compelling peasants to sign land, livestock, and tools over to the new kolkhozes. A multiyear series of violent outbursts consolidated most of the 25 million peasant farms into 250,000 kolkhozes.[8]

Bolshevik modernizers touted economies of scale, mechanization, and agronomy. Stalin extolled the kolkhozes as necessary to secure the marketable surpluses required by industrialization. Celebrating the 1929 procurement campaign in his article proclaiming the Year of the Great Breakthrough, he declared that automobiles and tractors were proof that socialist development was overcoming backwardness and proving superior to capitalism.[9] The five-year plan promised tractors providing 9.5 million horsepower by 1933, but actual imports and manufactures totaled only 3.2 million by that year and 8 million by 1936.[10]

Rejecting market mechanisms and gradual evolution, Stalin prioritized economic and social control. A nationalized system almost colonial in nature, the kolkhozes allowed the government to extract grain, taxes, labor, and military recruits, much as its imperial predecessor had done.[11] The farms ensured little more than crude control over the harvest. The five-year plan mandated planting 20 percent more crops and adopting technologies to increase yields by 20 percent. In reality, capital stocks declined by approximately 25 percent as peasants slaughtered livestock rather than sign over property under compulsion, leaving the kolkhozes short on draft animals and organic fertilizer. The country's tractors, chemical fertilizers, harvesters, and other resources were insufficient to rectify the shortage.[12] Milking, cutting hay, and other jobs required much labor, while kolkhozniks sowed and harvested without enthusiasm, dragged their feet, and stole produce. Far from increasing, grain output declined by about 25 percent even as government purchases escalated to cover exports, which soared from 200,000 metric tons to 5 million. In 1932 and 1933, state requisitions caused famine in breadbasket regions across Ukraine and southern Russia. Famine also struck Kazakhstan, where authorities forced nomadic herders into sedentary life. Estimates of total deaths range from 5.5 million to 8 million.[13]

Even as these conditions pushed peasants from the village, factory jobs pulled them into the industrializing cities. Much of the population seemed to be on the move. By the 1930s, a system of internal passports prohibited relocating to urban areas, but millions did so anyway to avoid kolkhoz membership.

Accused kulaks had to chose between exile to remote regions and flight to the cities. The number of factory workers increased from 28.1 million in 1928 to 54.7 million in 1931, while the rural population declined from 121.9 million to 114.8 million.[14]

By controlling the kolkhozes and their produce, Stalin's government impeded realization of the potential of industrial farming. Nominally democratic, the kolkhozes faced constant interference from local bosses imposing distant ministries' decrees, imprudent and ill adapted to local conditions.[15] Kolkhozes faced escalating quotas for grain, meat, milk, and raw materials such as cotton, even as they paid the MTS, or machine-tractor station, in kind. Representing government and party power otherwise limited in rural communities, these state-owned enterprises provided few tractors, harvesters, and specialists. Far weaker in rural areas than in urban centers, the new state resembled the Russian Empire, which had undergoverned its vast expanses.[16]

After collectivization, however, the state intervened in peasant life primarily during annual sowing and procurement campaigns. Urbanites negotiated propaganda barrages, parades, and other relatively benign intrusions. Some individuals sought to live up to the official ideals, seeking to become properly Soviet. Stalin's rule exposed them to terror and other malevolent sides of the police state. Kolkhozniks, however, primarily interested themselves in work, local bosses' demands, and the weather. Religions remained strongest in the villages, ensuring that peasants skeptically viewed occasional rural campaigns promoting socialist ideals and state priorities.[17]

From 1933 to the eve of World War II, uneasy equilibrium prevailed. Harvests remained at the 1928 level even as the population grew and the government took an increasing share. By 1939, that share reached 189 percent of the 1928 level, while small rises in prices granted some relief to the kolkhozes.[18] Offering little benefit to kolkhozniks, labor on the kolkhoz produced commodities supporting continued investment in industry, transportation, and defense. Stalin's dictums precluded a return to private farming, or even to NEP, but the kolkhozniks remembered better times while toiling under conditions they decried as a second serfdom. Each family devoted intensive labor to its personal plot and the few personal livestock allowed, obtaining the most fruit, vegetables, potatoes, and other produce possible. This provided subsistence and a surplus to sell in urban markets for otherwise scarce money. In the late 1930s, these plots encompassed less than 5 percent of cropland, but yielded at least 25 percent of all produce.[19] Most kolkhozniks conserved time by working for the kolkhoz only the minimum required to avoid sanction. In 1940, 8.7 percent of the 42.7 million kolkhozniks did not tally the required number of labor-days, a figure including 500,000 who refused to work at all on the kolkhoz.[20]

Nikita Khrushchev earned his education in politics and policy in this era under Stalin. Born into a peasant family, Khrushchev was formed by time spent in mine, factory, Red Army detachment, and party committee. He returned to agriculture only in 1938 when Stalin dispatched him to govern Ukraine, where he gained experience that shaped the policies he pursued after 1953. A Stalin supporter, Khrushchev retained a streak of independence from his master. He masked a faculty for political maneuvering by portraying himself as a country bumpkin. A skilled machinist, he received only rudimentary formal schooling. As an adult, he dreamed of further education that might carry him to a factory director's office, but party duties kept him from both and forced him to become an autodidact in modern farming.[21]

Khrushchev was born in 1894 into an impoverished peasant family in Kalinovka, a village in Kursk province of the Russian Empire.[22] There he earned a few years of primary schooling before moving in 1908 with his father to the Donbas, in today's Ukraine. Khrushchev later considered himself a worker because from that time, excepting brief returns to Kalinovka, he labored in the region's factories and mines. A dapper youth with a talent for machine trades, Khrushchev became an activist. In 1912, he came to the attention of the police for collecting aid for victims of an infamous massacre of striking miners in the distant Lena goldfields.[23] Engaging in local politics after the February Revolution, he joined the Bolsheviks, whose presence in the Donbas had initially been weak, only in late 1918. During the Civil War, he enlisted in the Red Army and became a commissar responsible for educating soldiers in political matters. Having served dutifully but with no great distinction, he returned to the Donbas to manage a mine. Soon, he entered a technical training program for adult workers, the first of two attempts to further his education and earn a position as an industrial manager. Secretary of the technical-college party cell, Khrushchev soon left to head a nearby district committee.

Thus Khrushchev began a rise that gathered momentum every year and established him as an exemplary Stalinist. Under his patron, Lazar Kaganovich, Khrushchev took administrative positions first in Kharkiv and then Kyiv, each time demonstrating ambition even as he hinted at insecurity about limited formal education.[24] When rephrasing Marx and Lenin into plain slogans, Stalin spoke to an initial wave of working-class officials, which included Khrushchev. In 1929, Kaganovich, a stalwart of the Stalin faction, helped Khrushchev secure admission to the Industrial Academy in Moscow. He soon abandoned his studies, however, to plunge into duties as secretary of the academy's party cell and then of the committee for the Bauman District where it was located. Proving his loyalty in the struggle against opponents of Stalin, he soon moved to the prestigious Krasnopresnensk District committee and then the city committee, where he served as Kaganovich's deputy. By the mid-1930s, Khrushchev was in charge of the everyday operations of Moscow, a metropolis in the making that swarmed with workers building factories and the first lines of the monumental Metro. Having fought to expel Stalin's opponents from the party, Khrushchev likely assisted in the Terror by signing death warrants produced by the dreaded secret police.

Yet actions recalled by others suggest that Khrushchev never lost his humanity. Building on self-directed practical learning, he enjoyed the tutelage of his wife, Nina Kukharchuk, an instructor of theory and party history. Maintaining popular connections, Khrushchev conceived of socialism as a promise to provide justice and necessities to the least in society. Between 1938 and 1949, he served as Stalin's satrap in Ukraine, developing a certain independence from a master who was demanding but distant. Khrushchev seems to have believed in the innocence of General Iona Iakir and other friends who perished between 1936 and 1938 amid unconvincing accusations of treason. Khrushchev even violated the need for silence by expressing misgivings and fears about his own arrest. Once, when speaking to a former comrade from the Donbas, he swore to "settle" with Stalin when he could. Referring to the dictator by combining a Russian obscenity with his Georgian family name, Jughashvili, Khrushchev demonstrated his penchants for stunning indiscretion and vulgarity.[25]

In Ukraine, Khrushchev helped direct efforts to win the Great Patriotic War, victory in which became the crowning achievement of Stalin's Soviet Union. In the meantime, however, the war rained blows on the people and farms. From 1941 to 1943, the Germans and their allies occupied Ukraine, whose people and farms endured atrocities and the effects of combat. Even those on the home front stretched themselves to the limit. Wartime triumph came at the price of millions dead, millions more displaced, and devastation in rural areas that combined with the unsoundness of the kolkhozes to cause a postwar famine.

Even behind Red Army lines, the war drained the kolkhozes and kolkhozniks. Already short on labor and horsepower, the kolkhozes exhausted capital to feed the war effort. Peasants flooded into the industrial workforce and comprised at least 60 percent of the Red Army. The number of male kolkhozniks fell by nearly 6 million, as 32 percent of the 1941 total left. A million women joined in departing. Destroyed or converted to producing war matériel, factories did not manufacture spare parts for tractors and harvesters, let alone replace those worn out or destroyed. In the 1930s, reaping grain was one of the few processes partially mechanized. By 1942, the kolkhozes carried out 79 percent of the job using horse-drawn machines and even manual labor. Kolkhozniks harnessed cattle to the plow and, in extreme cases, pulled it themselves. Already low fertilizer stocks declined further. Wartime demands disrupted crop rotations. In 1942, yields fell from 0.7 or 0.8 metric tons per hectare to a mere 0.46 tons. Yet the state procured the needed grain, 90 percent of it from the kolkhozes.[26]

Living and working conditions deteriorated. Each kolkhoznik faced a quota of labor-days that rose during the war. Juveniles and the aged received smaller assignments. Local authorities had at their disposal harsh punishments for any who failed to fulfill the quota, although they enforced them only to flagrant or repeat violators.[27] Kolkhozniks earned few material rewards. In 1940, kolkhozes in the Russian Soviet Federative Socialist Republic (RSFSR) had allocated 58 million rubles of their 10.3 billion in earnings, as well as 9.15 million tons of grain, for distribution in proportion to the number of labor-days each kolkhoznik earned. By 1946, those figures had fallen to 15 million rubles of 9.8 billion in earnings, and a mere 1.98 million tons of grain.[28] More than 75 percent of kolkhozes distributed less than 1 kilogram per labor-day, and 7.7 percent paid nothing at all.[29]

Kolkhozniks went to work anyway. Setting aside talk of socialism, the authorities gave out medals and appealed to concern about husbands and sons at the front.[30] Local bosses looked the other way when kolkhozniks enlarged their personal plots beyond the legal maximum, permitting them to feed themselves, pay taxes, and sell produce in legal markets.[31] These markets augmented the government supply system, which fully fed the army while providing workers with a bread ration and facilitating their efforts to produce or purchase other necessities.[32]

Life was worse in occupied areas. Millions of noncombatants perished from Nazi starvation policies and other atrocities, joining the millions of military casualties in the Soviet Union's total of at least 27 million war dead. Millions more found themselves maimed, homeless, displaced, or forced to labor. The invasion destroyed machines, livestock, buildings, and other capital. An official survey reckoned the ruin of 98,000 kolkhozes, 1,875 sovkhozes, and 2,890 MTSs, as well as the loss of 20 million horses, cattle, and other livestock. In 1945, a special commission calculated the damage to the whole country at 679 billion rubles, some 30 percent of which had been inflicted on the kolkhozes.[33]

In 1943 and 1944, the advancing Red Army brought Ukraine back under control, relieving Khrushchev of his army post and restoring him to duties in the republic. He became responsible for imposing control and jumpstarting production in the face of a deepening humanitarian crisis. Although exhibiting considerable continuity with the prewar years, postwar life reflected wartime experiences, forming a phase distinct from both eras.[34] As the front passed westward toward Berlin, life behind the lines saw little relief. Official accounts credited the party with guiding everyone through a short phase of so-called reconstruction. "Having victoriously concluded the Great Patriotic War," a later history of rural development triumphantly declared, "the Soviet people embarked on completing the tasks and realizing the plans for peaceful construction established by the party before the war."[35] This tale fabricates a sense of unity by suggesting that postwar challenges required the Soviet Union to simply restore that prewar system.[36] Instead, circumstances continued to evolve.

Victory fostered hopes founded on wartime latitude that Stalin had allowed the party, government, and military. Some were optimistic that, having triumphed on the battlefield, the people had earned rewards.[37] Predictably, the kolkhozniks favored changes, albeit in forms incompatible with Stalin's plans. Rumors abounded that Moscow was soon to dissolve the kolkhozes, perhaps under pressure from the Allies. Others supposed that Stalin had already given the order, but that corrupt local officials concealed the concession to benefit themselves.[38]

Amid anxieties about the Cold War, Stalin responded to optimism by ending wartime expedients and redoubling repression. Innovation in culture, policy, and economic management waned. Already visible in the 1930s, a conservative trend made Russian nationalism acceptable, a reversal the policy of the 1920s. Waves of antiforeign sentiment culminated in campaigns against disloyal "cosmopolitanism" used to defame scientific and cultural figures with ties abroad, especially those of Jewish extraction. Courting elite loyalty by offering benefits unavailable to most, Stalin reinforced hierarchy and enhanced enterprise managers' authority.[39] The atmosphere in Stalin's inner circle duplicated the mood darkening society. In his memoirs, Khrushchev recalled returning to Moscow from Kyiv in 1949. He found that the aging dictator humiliated even loyal aides by "behaving... as though he were God and had created them. His attitude was at once patronizing and contemptuous."[40] Persevering against declining health, a suspicious Stalin manipulated his underlings. Isolated by his cult of personality, as it was later labeled, the nocturnal dictator required their attendance at tense, alcohol-fueled soirées lasting into the early morning hours. For the political elite as for society as a whole, these were grim years.[41]

The narrative of reconstruction also minimizes the agrarian crisis that reached its tragic peak in the famine of 1946 and 1947. Compounding the chronic inadequacy of the kolkhozes, the weather depressed harvests. Typically, in one year out of every three, one of the Soviet Union's grain-producing regions experienced extreme weather. During the war, the weather had been uncharacteristically favorable. Exacerbating the exhaustion of farmworkers and capital, a rainless spell began in 1945 and culminated in a 1946 drought, which drove the harvest toward its nadir.[42] As many as 2 million starved or succumbed to accompanying epidemic diseases. Many more suffered from malnutrition, especially in the rich farming regions of Ukraine and Moldova facing Stalin's demands for grain. In letters, inhabitants despaired at the "nightmarish conditions" and feared that they would "die of hunger." While authorities euphemistically informed superiors about "provisioning difficulties," peasants experienced the reality: "There is no bread and we know not how we'll live through the hunger."[43] Indifferent to the weather and the effects of war, Stalin expected farms to provide grain to feed the cities and support clients abroad, exacerbating the calamity and ensuring that those who suffered most were kolkhozniks forced to relinquish their grain. In 1940, the government had procured 36.3 million tons from a harvest of 95.5 million.[44] In 1945, the harvest hit rock bottom, registering only 39.6 million tons.

In charge of Ukraine, Khrushchev faced intense pressure and bore some responsibility for the famine. Receiving rosy reports about the expected harvest in the spring and early summer of 1946, Stalin raised his demands in July, disregarding new word of looming harvest failure. As always, quotas reflected the government's requirements, rather than the kolkhozes' capacity to sell. To Khrushchev's credit, when evidence of crop failure and human suffering streamed in, he took the considerable risk of appealing for aid. Citing reports of death and even cannibalism, Khrushchev wrote to Stalin at least three times during the autumn. The latter countered that his subordinate had been fooled by local officials and Ukrainian nationalists, who hoped to keep hidden grain for themselves. In time, Ukraine received limited aid, which prevented the calamity from becoming still worse.

The political consequences took shape only in early 1947, when Stalin ordered Khrushchev removed from his post as secretary of the republic's central committee, leaving him only to head the government. Sent from Moscow to head the party, Kaganovich seemed poised to exchange his role as Khrushchev's patron for that of hangman. Khrushchev suddenly fell ill with a case of pneumonia, one likely exacerbated by chronic stress and the acute danger of repression. Just as swiftly, he recovered and his political fortunes followed. Late the same year, Stalin recalled Kaganovich and reinstalled Khrushchev.[45]

The famine exacerbated average citizens' already poor diet, a condition Khrushchev had to address once he came to power. Even outside the drought zone, shortages diminished workers' and peasants' calorie intake, compounding chronic malnourishment brought on by war. Beginning in September 1946, workers, clerical personnel, and their dependents faced stricter rationing and rising prices. During the war and for several years afterward, most subsisted on a poor diet based on bread and potatoes. Demanding workplace conditions, pitiable healthcare, and failing sanitation infrastructure contributed to rising mortality.[46]

Even after the famine passed, grain supplies remained at best sufficient. Dry years followed in 1948, 1950, 1951, and 1954. After bottoming out, grain procurements rose to 30.1 million metric tons in 1948, but that quantity did not satisfy rising demand. Subsequent annual growth measured 5.6 percent in 1949, only 0.6 percent in 1950, 4 percent in 1951, and 3.3 percent in 1952, a year that gave way to a 12.7 percent decline in 1953. Even the successful harvests did not satisfy increased requirements. At 91.4 million tons, the 1952 harvest was short of the 1940 total, giving the lie to Malenkov's declaration that it met needs, which he made with Stalin's approval to the October 1952 Nineteenth Party Congress. Procurements of 34.6 million tons failed to keep pace with a 6.7 percent increase in demands on government stocks that year alone. As Khrushchev phrased it eighteen months later when disparaging Malenkov's claim, "In terms of staples, we scarcely make ends meet."[47]

The problem shaped the average diet even after the famine relented by the end of 1948, when workers' food intake began to rebound. At the end of 1947, the government made a political statement by ending wartime rationing, a bid by Stalin for social stability and a propaganda victory. In the long run, the kolkhozes and kolkhozniks could not shoulder the burden of exploitive pricing policies, which paid them less than production costs for meat, milk, eggs, and other foodstuffs. Predictably, output of these foods stagnated. By 1950, household surveys showed that basic food security had returned, but nutrition remained poor, as it would well into the 1950s. Calorie consumption rose, but the carbohydrate-heavy diet had little of the officially cheap but actually rare meat, milk, eggs, and animal fats.[48] In 1953, average citizens consumed less meat and milk than they had before collectivization, or even before 1914.[49] A 1949 policy to remedy the shortfall by expanding livestock herds and growing more of traditional fodder crops achieved little because the prices kolkhozes received continued to make those endeavors financially ruinous.

During the postwar years in Ukraine, Khrushchev affirmed his convictions about agriculture. In some instances, he modified his approach after trials achieved little. In the case of corn, experience confirmed his opinions, strengthening his enthusiasm for the crop. In his memoirs, he recalled that his grandmother and other Kalinovka peasants had served ears of corn grown in kitchen gardens as a summertime treat.[50] Of the Donbas years, he noted only that corn had grown near a metallurgy plant where he had worked.[51] While Khrushchev was away in Moscow in the 1930s, the chaos of collectivization and famine cut short halting propaganda campaigns and research programs designed to encourage corngrowing in southern regions. The Soviet Union did not annex the former imperial province of Bessarabia—renamed Moldova—until World War II. Western Georgia and southwestern Ukraine, other locales where the crop was common, accounted for only a small percentage of total cropland. Then, in the late 1930s, Anastas Mikoyan relayed to Khrushchev his observation that American farmers used corn as animal feed, encouraging the latter to spread knowledge garnered from foreign sources.[52]

In 1949, Ukraine averted a renewed famine by planting corn, strengthening Khrushchev's faith in the crop. When frost killed nearly 2 million hectares of winter wheat, kolkhozes and sovkhozes replanted them with corn, permitting the republic to fill Stalin's crushing grain quota.[53] The crop served its traditional role as Ukrainian farmers' last resort. Its life cycle poorly suited crop rotations based on winter wheat and barley, which were sown in fall and left to take root, lie dormant during winter, and ripen the following summer. When an exceptionally hard winter or dry spring ruined these crops, however, peasants planted corn in late spring. Becoming edible sooner than other spring grains, the cobs allowed them to feed themselves and their animals.[54]

In late 1949, Khrushchev returned to Moscow as a devoted advocate for corn. He badgered the oblast's kolkhozes to plant the crop, previously unfamiliar to the area, and to apply the knowledge he had gained in Ukraine. After he left that republic, corn plantings there declined toward their postwar nadir in 1953. During Stalin's final years, farms planted less corn than was economical and lacked technologies developed in the United States.[55]

During his tenure in Ukraine, Khrushchev absorbed information about corn—and much else—by observing researchers and farmworkers in the fields, a style of hands-on leadership increasingly rare in the 1940s. Limited in competence with the written word, Khrushchev refused to retreat behind a desk and loved visiting the fields, factories, and mines, getting his hands dirty and seeing for himself. Short on formal training, he peppered experts with questions, and often took the advice of the self-taught and the formally educated in equal measure. In some cases, this had disastrous results. At key moments, he supported Trofim Lysenko, whose pseudoscientific ideas about breeding plants and animals so often promised Stalin and then Khrushchev exactly what each wanted. Although never unquestioning, Khrushchev's support earned condemnations from contemporaries and historians alike.[56] To his credit, Khrushchev also backed experts who introduced valuable innovations in fields such as metalworking.[57]

In Ukraine, Khrushchev attempted to reform the vital agrarian sector. Amalgamating small kolkhozes into larger units, he began a process that soon spread across the country and continued after 1953. During collectivization, kolkhozes often included the residents of one village, resulting in many very small units. By merging several kolkhozes, Khrushchev hoped to save on administrative costs, ensure economies of scale, and strengthen farms' ability to invest in their own production. Having begun in 1950, this policy, later combined with limited conversions of kolkhozes into sovkhozes, reduced the number of the former to only 17,900 by 1963.[58]

Ukrainian experiences also spurred Khrushchev's 1951 proposal to modernize village life. To remedy wartime destruction, the leader thought to transform amalgamated kolkhozes into town-like settlements. In Ukraine, he had a model village built with apartment blocks and social services demonstrating care for living conditions. He sustained these concerns after returning to Moscow to combine duties as a Central Committee secretary and as chief of the city and oblast party organizations. Amalgamating the local kolkhozes, he called for creating agrotowns similar to the model village. In March 1951, *Pravda* published the text of a speech advocating demolishing ramshackle houses to make way for multifamily apartment buildings outfitted with modern conveniences. Promising the smart physical appearance and cultural resources of urban life, Khrushchev did not account for the enormous prospective cost or the ideological implications of privileging consumption over production, especially in rural areas viewed so suspiciously under Stalin. The speech summoned swift attacks from Malenkov and others. Because these criticisms had Stalin's blessing, they threatened Khrushchev's political fortunes and forced him to write a groveling letter to the leader. The danger soon passed, but the experience stung.[59]

After the war as much as before it, the government compelled the kolkhozes to sell grain and other produce cheaply while paying dearly for MTS services. As harvests rose after 1946, Stalin continued to extract disproportionate financial and material contributions to rebuilding, much as during the initial industrialization drive. Kolkhozes' payments for MTS services were high, even as machines and trained specialists remained in short supply. By 1950, the numbers of tractors, harvesters, and other machines had reached only the insufficient prewar levels. Tax and procurement policies shifted part of the financial burden of the Cold War military buildup onto the agrarian sector. As wartime official policy intended, kolkhozniks had amassed considerable savings by selling produce at market. The 1947 currency reform wiped out these savings by converting old banknotes to new ones at low rates.[60] The government further alienated the fruits of peasant labor by exacting obligatory deliveries and high taxes.

Mass migration denuded the kolkhozes of farmworkers who might have helped the farms meet Stalin's demands. Millions of discharged Red Army soldiers left to escape famine, taxes, harsh working conditions, poverty, and hopelessness.[61] Between 1948 and 1950, a further 7.6 million kolkhozniks and veterans followed, approximately 40 percent from Russia and the remainder from the other union republics.[62] In 1950, rural residents remained 60 percent of the population, a proportion that continued to shrink thereafter as the rural population declined only slightly in absolute terms, but urban areas expanded rapidly.[63] The war had killed more men than women, while military service opened more opportunities for male veterans to leave the village. This imbalance meant that in rural Russia the ratio of women to men rose to an average of two to one. In some locales, it reached three to one.[64]

Authorities demanded more intense manual labor from the kolkhozniks, who responded by demonstrating their dissatisfaction. Because the kolkhozes paid little or even nothing, kolkhozniks refused to work. Pay reached rock bottom in 1946, causing nearly 20 percent of kolkhozniks to not tally the obligatory minimum number of labor-days. Variations between regions show that lower pay provoked greater reticence to work on the kolkhoz. As a result, the kolkhozniks struggled to make a living.[65] One lamented, "So beautiful a spring has come. But what can be done if there is nothing to live on?" Explaining things to his daughter, the letter writer continued, "Whatever [the kolkhoz] has is not ours. What's more, there is no way to earn anything in the village. People work the whole summer and gather in the harvest. Then the bosses divide it among themselves, leaving nothing [for the kolkhozniks]."[66] Conditions caused kolkhozniks to write 92,795 complaints to a central commission between 1947 and 1950.[67]

Stalin's government replied with coercion. In 1948, new policies increased annual labor-day quotas, norms defining the amount of work constituting a labor-day, and taxes on personal plot production. Kolkhozniks replied by slaughtering livestock and chopping down orchards to avoid the extortionate taxes.[68] Earnings from work in the kolkhoz fields declined and had no relation to the size of the harvest. In 1950, at least 70 percent of kolkhozes—in some locales more than 90 percent—paid less than 1 kilogram of grain for a labor-day. Between 4 and 8 percent of kolkhozes in each administrative region paid nothing at all.[69] When kolkhozniks refused to work, authorities imposed sterner punishments. The law mandated sentences to a labor camp for lateness, petty theft, or other violations of industrial workplace discipline. On kolkhozes, taking a potato from the field to feed a family, if prosecuted, could lead to time behind barbed wire. By 1953, the Gulag held the largest number of prisoners in all its years in existence. Some 17 million violators of the labor code were tried and convicted, although most faced docked wages and other limits on rights. Only 3.9 million received custodial sentences, comprising the largest single group of those sentenced between 1940 and 1953.[70]

An edict sanctioning stricter punishments for kolkhozniks originated with Khrushchev, marking a contrast with his later reformist policies. Proposed and adopted in early 1948, a directive granted kolkhozes power to expel members who habitually failed to meet the annual labor-day norm, stripping them of their right to a personal plot and subjecting them to higher taxes. Flagrant violators faced internal exile to remote corners of the Soviet Union. Launched on a trial basis in Ukraine, this arch-Stalinist law to combat what the text termed "antisocial and parasitical ways of life" later went into effect across the country, but ultimately failed. After authorities mounted a brief press campaign and made examples of a few violators, it lapsed.[71] By 1953, more than 4 million kolkhozniks, or 15.9 percent, failed to meet the labor-day minimum, and 600,000 did not work for the kolkhoz at all.[72]

With few incentives other than coercion, the kolkhozniks worked without enthusiasm. The state procured insufficient grain, meat, milk, wool, cotton, and other needed commodities. The agrarian sector slowed economic advance. By 1953, perceptive party members wrote to the Central Committee to describe the repressive policies' effect—or lack thereof. Some accepted the system as it was, complaining about kolkhozniks who made no effort "to engage in socially beneficial work" and "adhered to a parasitic way of life."[73] Blaming "private property-based psychology," local party authorities bemoaned how kolkhozniks expanded their personal plots beyond the legal size, cut hay from kolkhoz fields, and raided kolkhoz stocks of grain.[74] Taking a more critical stance, a journalist and party member identified only as F. N. Kirikov wrote to denounce the 1948 law. He explained that the severity of punishments dissuaded the rest of the kolkhozniks from voting to confirm them, as the formally democratic kolkhoz charter required. Violators did not care to remain kolkhozniks, moreover, making expulsion not a threat but an opportunity to work for Vologda Oblast's nearby lumber mills. "Such measures have no effect here," Kirikov reported. "Those who were expelled left the kolkhoz happily.... How could kolkhozniks consider it a punishment to leave a farm where they earned nothing?" With a little maneuvering, they might even keep a valued personal plot anyway. Even those who remained members flouted calls to the kolkhoz fields, preferring religious holidays and mushroom hunting. Flawed incentives, Kirikov concluded, were the primary reason the oblast failed to meet its quotas.[75]

A 1953 update to the law eliminated the threat of expulsion and exile, but kolkhozniks sometimes refused to vote even to confirm the higher tax on personal-plot produce. Every family had at least one member who failed to meet the minimum. "For that reason," one letter concludes, "the loafers go unpunished."[76] In one district of Stavropol Krai, between 20 and 30 percent of kolkhozniks failed to reach the minimum. Yet the chiefs of the kolkhozes declined to report violators for enforcement of the higher tax rate. Admonishing the kolkhoz bosses, tax officials implored local party authorities to intervene. On one kolkhoz, the chairman's "negligence" had allowed the kolkhoz assembly to reject the entire list of those subject to the tax penalty "on the grounds that each kolkhoznik had legitimate reasons for not fulfilling the labor-day minimum," such as caring for a dependent. This circumstance, the tax officials' petition to the krai party committee sardonically concluded, was "extremely unlikely."[77]

Central Committee personnel filed letters such as Kirikov's in the archive because they spoke to the concerns of Stalin's successors. In August 1953, Khrushchev commented specifically on the journalist's letter.[78] The leader soon denounced policies causing "low labor discipline" and "violations of the principle of material incentives," meaning failure to pay kolkhozniks.[79] Malenkov announced reforms easing the taxes paid on the output of kolkhozniks' personal plots and reducing compulsory deliveries of goods for which the kolkhozes received payment lower than production costs. As part of the personal rivalry with Malenkov then growing into a struggle for power, Khrushchev responded. Considering himself more attuned to the peasants because of his origins and tenure in Ukraine, he set out his evaluation of the predicament at the Central Committee plenum in September 1953.

With these actions, Stalin's successors first acknowledged the desperate conditions facing rural residents. Khrushchev's September speech was necessary because the leaders lacked knowledge about agriculture and rural life, hampering any attempt to remedy the crisis. Subordinates had obfuscated when reporting to Stalin, ensuring that he had known little about agriculture and kolkhozniks' lives. Khrushchev later derided Stalin for never leaving his office and therefore believing in socialist-realist films portraying singing peasants feasting at tables heavy with food and drink.[80] On other occasions, Khrushchev criticized Stalin for never questioning the deceptive statistics received from officials who likely sought to protect themselves from the dictator's displeasure.[81]

In the summer of 1953, Stalin's successors confronted the lack of data about income, labor productivity, consumption, yields, and more, all of which hindered efforts to diagnose afflictions of rural society and the agrarian economy. Preparing Khrushchev's speech, his aide Andrei Shevchenko confronted Vladimir Starovskii, head of the USSR Central Statistical Administration, who repeatedly altered data previously reported. When Shevchenko rebuked Starovskii for revising a figure for the fourth time, the statistician protested that each new one had been an improvement: that is, it had made the situation appear less dire. Shevchenko countered that he cared little for appearances and wanted only a clearer understanding. His riposte apparently had little effect because Starovskii returned the following day with new figures.[82]

Carrying out collectivization, Stalin and his supporters subjected agrarian production to control. Gaining greater power than they had possessed under NEP, authorities sought primarily to extract grain and produce. Even when kolkhozniks relinquished their produce, they resisted coercive measures and, unknowingly, the high modernist ambitions of political masters in Moscow. Lacking passports, kolkhozniks did not enjoy the right to work outside the village or other benefits of citizenship, even though they were subject to taxes, military service, and other duties of a citizen. For more than twenty years, the government suppressed their capacity to consume while alienating the produce it used to feed urbanites, provision the army, and finance industrialization.

In 1953, the new authorities recognized the need to reform Stalin's agrarian system, which did little to embody the ideals of modern socialist farming. They sought doctrinally acceptable means to reshape the way kolkhozniks lived, worked, and related to the kolkhoz and the government. In time, the kolkhozniks gained status and some room to maneuver in economic spheres. Signs of rural modernization strengthened as industrial farming moved from an ideal limited in actual use down on the farm, as it had been since the 1920s, to the principle underpinning practices across a range of production activities. Treating kolkhozniks better, these new approaches quietly superseded the coercive practices of the past. To realize their vision of modern socialist agriculture, leaders had to fully incorporate rural residents as industrious producers, active consumers, and fully fledged citizens.

On a piecemeal basis, Khrushchev and his supporters began to react to conditions down on the farm. They embarked on reforms that addressed these problems by integrating the countryside into the industrial economy. They had to act because the chronically ailing system had reached a crisis point in the years up to 1953. Yet there was no past equilibrium to which they could return. The consequences of war, famine, inadequate investment, and low output plagued the kolkhozes. Doctrine closed off a return to NEP or any move to disband the kolkhozes.

Stalin's successors, however, did have alternatives at hand. Khrushchev's bet on industrial agriculture was only one of many mutually compatible options. The modest tax and price reforms announced by Malenkov in August 1953 demarcated one course, which many of Khrushchev's policies maintained in subsequent years. Other possible strategies included Khrushchev's 1951 village modernization scheme, the 1949 livestock development plan, a proposal to grant kolkhozes authority to own machinery independent from the MTS, the idea to permit kolkhozes to plan for themselves how to meet government orders, schemes to inexpensively manage soil fertility, and the grandiose but abortive Stalinist Plan for the Transformation of Nature. Each in its own way shone light on the crisis facing the Soviet Union and the makers of agrarian policy, yet also the avenues open to reform. Of these, the high-modernism and preoccupation with America ingrained in Bolshevism in the 1920s made corn, coupled with industrial agriculture, the solution favored by Khrushchev.

2 INDUSTRIAL AGRICULTURE, THE LOGIC OF CORN

When Nikita Khrushchev visited the United States in 1959, he built goodwill by meeting with President Dwight Eisenhower at Camp David, addressing Congress, speaking with business leaders, inspecting factories, sampling the hot dogs, touring Hollywood, and captivating the press at each stage. Khrushchev's brief sojourn in Iowa, the center of the Corn Belt and emblem of American farmers' productivity, also attracted a gaggle of reporters, who documented the day he spent surveying Roswell Garst's farm from field to feedlot (Figure 2.1). Bantering with the American seed-corn magnate, the leader took in the machines, buildings, crops, and livestock. In his memoirs, Khrushchev expressed admiration for Garst and his know-how: "I walked around the farm and was delighted." "Actually, I had a dual perception of him," the former leader elaborated: "As a capitalist, he was one of my class enemies. As a man who I knew and whose guest I was, I treated him with great respect and valued him for his knowledge [and] selfless desire to share his experience with us."[1] By the time Khrushchev's borrowed powder-blue Cadillac convertible sped away, he had intensified enduring interests in corn and industrial farming.[2]

This visit aided the corn crusade by solidifying Khrushchev's commitment to the industrial farming principles that tied his agrarian reforms together. The combination of corn and these ideals deserves a place alongside mass culture, consumption, media, and other spheres in which developments in the Soviet Union intersected with counterparts in other industrial societies in ways little noted at the time, but increasingly visible to historians. In farming practices and technology no less than in these other spheres, the Soviet Union did not exist in isolation behind a physical Iron Curtain, let alone an intellectual one.

Addressing the crisis inherited from Iosif Stalin, Khrushchev strengthened existing commitments to industrial farming by browbeating his country's farms to adopt these methods for

FIGURE 2.1 Nikita Khrushchev (first row, center), chair of the USSR Council of Ministers, and his entourage visit the Coon Rapids, Iowa, farm of Roswell Garst (to Khrushchev's left) during their trip to the United States in 1959. Shadowed by crowds of reporters and curious onlookers, Khrushchev observed American industrial farming firsthand. The material in the foreground is corn that has been chopped to make silage.
Photo by V. V. Egorov, courtesy Russian State Archive of Cinema and Photo Documentation, ID#1-116190

growing corn and other crops. This process is best understood using the concept of crossed history, an approach emphasizing that transnational exchanges are not simple communications between static systems, but instead constitute multidirectional interactions potentially transforming each participant.[3] In 1955, the leader reinvigorated a long-standing but interrupted dialogue with American interlocutors by dispatching the first postwar delegation to the United States. He and his advisers sought domestic stability, vindication of promises of abundance under communism, and success in the global Cold

War struggle for hearts and minds. They viewed industrial farming as the solution to the Soviet Union's problems because they considered technology to be readily transferable from capitalist conditions to socialist ones. The United States thus served as a source of practical methods even while it was the Soviet Union's ideological opponent and fiercest competitor.

Having cooperated with Georgii Malenkov on post-Stalin agrarian reforms in 1953, Khrushchev soon seized the initiative. In early 1954, he launched a program to plow up millions of hectares of unused land, known as the Virgin Lands campaign, one of many applications of industrial farming principles characterizing the subsequent decade.[4] Competing for power and authority with rivals in Stalin's former inner circle, Khrushchev considered himself best suited to determine agrarian policy not only because of his rapport with peasants and experiences in Ukraine, but also because his proposals promised a rapid solution to the crisis. Rejecting Malenkov's cautious proposal for steady, incremental growth, Khrushchev promised a revolution.[5] From factory farms in the Virgin Lands to later enthusiasm for genetics and chemicals, he emphasized strategies facilitating efficiency and productivity, a program exemplified by plans to integrate corn and industrial farming into Soviet practice.

Khrushchev endeavored to expand and enrich food supplies, improving domestic stability and the Soviet Union's reputation abroad. Amid the otherwise spartan conditions under Stalin, consumers found only occasional luxuries, harbingers of future abundance.[6] Khrushchev sought to replace meager, monotonous fare with three square meals a day. Compelling kolkhozes and sovkhozes to plant corn, he expected bounteous harvests of feed for the livestock necessary to permit mass consumption of previously rare meat, milk, and other prized produce. "The struggle for high yields of wheat and corn, for higher production of meat, milk, wool, and other farm products," he announced, was integral to "building communism."[7] Identifying practical steps toward utopia, he explained, "We want Soviet people to eat to their hearts' content, and not just bread, but good bread, as well as sufficient meat, milk, butter, eggs, and fruits." He continued, "Living on bread alone, we might just get by. We must more than get by: we must ensure that Soviet people's lives become better and more beautiful every day. Having constructed a socialist society, we are confidently moving toward communism."[8]

Khrushchev worked to reinforce the existing order by improving living standards. Rapid urbanization burdened state food supplies, causing Stalin's successors to fear unrest sparked by shortages. "I believe only in [corn]" backed by the new technologies, Khrushchev explained to advisers in private in 1955, "otherwise no five or six years will save us. . . . We will use new means to do the job."[9] Inadequate clothing, housing, and food invited mass disturbances. In his memoirs, Khrushchev recalled the unease that made leaders cautious toward cultural innovations during the Thaw. He remembered that they feared that a "tidal wave would come along and sweep us away."[10] That flood of developments might have as easily resulted from discord sparked by empty store shelves as by the revelations about Stalin's crimes and the return of former labor-camp internees.

On occasion, such grievances spilled over into protests. Alarmed, the leaders of the Soviet Union hastened to repress those in East Germany in 1953 and in Poland in 1956.[11] Khrushchev sought to preempt similar events at home by stimulating farm output and enriching food supplies. During the 1950s, economic growth improved diets and provided material comforts. By the early 1960s, progress lagged behind Khrushchev's promises, dashing rising popular expectations. Finding less food for sale than they had come to expect, citizens grew dissatisfied. Khrushchev's apparently empty pledges tarnished his image and prompted domestic disturbances that, although they remained local, challenged official proclamations about abundance. Linked to material grievances rather than to antisystemic sentiments, the protests indicated that citizens accepted official ideals, but expected authorities to deliver the goods. Profound societal changes wrought during and after the Stalin's time had provoked stresses on the system born of its own success.[12] The most famous disorder occurred in 1962, when workers in Novocherkassk grew angry at rising food prices, falling real wages, and empty promises of abundance. Exploding in protest and attacking bosses indifferent to their grievances, those at one factory rallied sympathizers and seized the town's administrative center. Repressed in volleys of gunfire that killed many, the insurrection resulted in the arrest and trial of the leaders.[13]

Food security also shaped foreign policy. Securing the Soviet Union through nuclear deterrence, Khrushchev also sought geopolitical influence through demonstrations of communism's superiority.[14] Championing peaceful coexistence and peaceful competition, he envisioned providing living standards sufficient to impress the world. By solving seemingly quotidian problems, he attempted to validate state ownership, centralization, and other elements of the Soviet economic instruction sheet as aids to equitable development. During the 1950s, growth surged to rates above those posted by other advanced industrial economies, lending feasibility to Khrushchev's boasts that the Soviet Union was to sustain that pace and soon forge ahead of the United States in absolute terms. Reinforced by Khrushchev's habitual pledges to quickly "catch up to and surpass America," this optimism culminated in the 1961 Third Party Program, which vowed to match the United States in per-capita production and, by 1980, lay the material foundations for communism. The Soviet Union was achieving "a higher stage of development of human society," Khrushchev explained, characterized by "abundance of material and spiritual blessings . . . [ensuring] that people do not suffer from a lack of anything."[15]

Emboldened by optimism, Khrushchev accepted capitalist rivals' measurements of success. The Soviet Union needed not only to orbit satellites and eliminate unemployment, but also to provide the comforts already available to some Americans, thereby living up to its ideal and proving the superiority of communism. Consumer goods manufacturing and food processing captured some resources from privileged heavy and defense industries. Satisfying the growing numbers of consumers became integral to the survival of the Soviet Union. Citizens expected the economy to provide comforts, making the model kitchen on display at the 1959 American National Exhibition in Moscow not an object of envy, but a glimpse at the expected near future. Khrushchev insisted that socialism, with its proven formula for rapid industrialization, was also to provide an egalitarian form of the good life. Pursuing prestige in this manner, he risked his own authority and the legitimacy of the Soviet Union. The relative scarcity of consumer goods even after his improvements, however, did not destine the country to failure. Authorities learned to manage citizens' demands and consuming habits, helping their rule to survive as long as it did.[16]

The Soviet Union's leaders had long sought to enhance their prestige by convincing citizens of America's failings. Amid the Great Depression, propaganda campaigns touting Soviet supremacy had celebrated advances at home. Declaring the Soviet Union's achievements first or better, the authorities had also boasted about past Russian innovators who had supposedly mastered the telephone and airplane before American inventors. After World War II, xenophobic campaigns had curtailed exposure to foreign researchers and advances in almost every field.[17] Even as these campaigns and their counterparts, the red scares, in the United States intensified, foreign interest in advances in the Soviet Union remained. In what was called the KR Affair, Americans inquired about a reported potential cure for cancer discovered in Moscow, queries that resulted in the researchers' censure in 1947.[18]

Officially, the Soviet Union had bypassed social and economic conflicts inherent in the lower capitalist stage. Even as Khrushchev sought foreign technology, propaganda continued to spotlight racism, discrimination, and class stratification in the United States and Western Europe. Portrayals softened somewhat after 1953, but publications continued to remind readers that colonial powers oppressed subaltern peoples, former Nazis remained free in West Germany, Jim Crow laws enforced segregation in the United States, and working-class Americans struggled to make ends meet even in the citadel of consumerism.

Seeking models for industrial farming, leaders concentrated on the United States. Although at first this seems inconsistent, the rival superpower could stand for both high technology and capitalist oppression in part because these diverse images referred both to the United States itself and to the America constructed in official discourses. The latter emphasized poverty, oppression, joblessness, and racism as qualities inherent in American capitalist society and absent from the Soviet Union. Leaders valued the United States as a source of advanced technology that, despite its origins under capitalism, they might beneficially borrow. Concurrently, they measured living standards at home against those in the United States, the global benchmark.[19]

Science and technology in the Soviet Union were intertwined with counterparts abroad because its leaders embraced world-spanning developments in those areas. The search for models became urgent during transformative moments, including the decade under Khrushchev. Amid his reforms, the barrier separating the Soviet Union and its allies from capitalist economies was porous and translucent, rather than the impermeable Iron Curtain of so much Cold War imagery. People saw and goods passed through, often with official approval, a reality better captured by the image of a curtain made of *nylon*. A nylon stocking symbolizes both the divergences in access to consumer goods on each side and the presence of a porous, translucent barrier.[20]

The postwar antiforeign campaigns had curtailed long-standing dialogues with Americans and others. After 1953, former ties revived. Readers devoured Ernest Hemingway's novels in Russian translation and moviegoers flocked to see Yul Brynner on the big screen in *The Magnificent Seven*. By the 1960s, youth clamored for blue jeans and a chance to listen to the Beatles on bootleg records. Scientists, engineers, clothing designers, and many other specialists absorbed theoretical and practical advances firsthand or through specialist literature.[21]

The new contacts re-established transfers of knowledge into the Soviet Union and, in some cases, back out to friendly countries. Farming technology was no different. Endeavoring to prove communism superior to capitalism, Khrushchev tried to revive kolkhozes and sovkhozes by appropriating technologies and practices. He considered corn a "quicker, easier, and cheaper" source for the livestock feed needed to ramp up meat and dairy output and, thereby, to make good on his promises.[22] The United States was a model because Americans had long eaten more of these foods. In the years surrounding World War II, their consumption surged further because farmers adopted new technologies that increased productivity and drove down prices.[23] Khrushchev envisioned equaling that output by using the same modern methods to grow the same crop.

Passing through porous Cold War barriers, experts restarted an old conversation. Even before 1917, the Russian Empire's progressive specialists and landowners looked to American technologies and methods, while Americans reciprocated with curiosity about Russia's expanses. Between 1861 and 1914, some large landowners and a few peasant farmers slowly embraced advances.[24] The International Harvester Corporation found a growing market for its reapers and other machines and, after 1900, invested in manufacturing inside Russia.[25] American advances in corngrowing also attracted the empire's modernizers, who invited experts to consult on new seeds and methods. In fact, when in 1955 Roswell Garst prepared for his first visit to the Soviet Union, he consulted with one such visitor, Louis Michael, about his experiences working in the province of Bessarabia (today Moldova) between 1910 and 1917.[26]

After the Revolution, American practical know-how facilitated economic initiatives in the Soviet Union. This interest replicated Russians' long-term search in foreign ideologies and technologies for solutions to perceived backwardness.[27] In the 1920s, some Bolsheviks merged the factory organization principles of Frederick Winslow Taylor and the management techniques of Henry Ford into the catch-all term "Americanism." Critics charged that capitalist origins imbued resulting practices with an exploitative character. Considering technology a value-neutral good, advocates countered that conditions under socialism, a more advanced social system, precluded exploitation of the working class.[28] Believing that transferable technologies would prove more productive under socialism than under capitalism, Bolshevik authorities embraced American expertise. During the First Five-Year Plan, Americans aided construction of the steelworks at Magnitogorsk, the automobile factory in Gorkii (today Nizhnii Novgorod), and other emblematic projects.[29]

Fascinated with American industry, the Bolsheviks unsurprisingly also sought templates for modern farming in the United States. Committed to transforming their agrarian country, they lacked the capacity to build the necessary machines. In the 1920s, they imported tractors and other implements primarily from the United States.[30] While urban intellectuals embraced Americanism as a template for reformed working and living, peasants saw Ford's name on the new tractors. During the First Five-Year Plan, factories began to produce tractors using designs, machine tools, and skilled workers from the United Sates.[31]

Pursuing American levels of excellence, policy simultaneously employed recognized Russian expertise in land and crop management. Famed botanist Nikolai Vavilov built his Leningrad seed bank with samples from across the world to study the origins of domestication and develop improved cultivars. This practical research promised to allow the Soviet Union "to overtake and surpass the US Department of Agriculture in the scientific renovation of agriculture."[32]

At that time, American practitioners of industrial farming transferred their know-how to the Soviet Union. Experts in animal health taught innovative methods on model kolkhozes, albeit with limited success.[33] More consequentially, the authorities invited more than 1,000 specialists to establish demonstration farms, illustrating what scholars have called a transfer mentality characteristic of the ideal of industrial agriculture. Downplaying local differences in conditions and practices, American experts applied the technologies they had developed for growing wheat on North America's Great Plains to the sparsely populated North Caucasus because the climate and soil of each land seemed to approximate those of the other. Having gained only ten growing seasons' experience at home, these experts brought machinery, training, and advice to the Soviet Union. Motivated by profit and curiosity rather than sympathy for socialism, they quit the Soviet Union for good when contracts expired in 1932, returning home with new expertise in applying their methods.[34] Over the coming decades, Americans implemented industrial technologies in increasingly wide areas of production. In the Soviet Union, the gigantic farms they established remained few in number and applied the technologies primarily to planting and harvesting wheat. When Khrushchev took charge, those farms provided a template for further developments.

The leaders of the Soviet Union favored these methods because they shared with Americans a vision of modernity defined by technology and a belief in humans' capacity to master natural conditions in pursuit of economic gain. Under Stalin, advocates portrayed nature as an object to "conquer," one to be "transformed and bent to the human will." This Promethean component of Bolshevism stressed grand transformations of nature, a principle it had in common with other modernist belief systems.[35] By contrast, environmental historians stress that societies comprise part of the whole environment, which is not merely a surrounding natural world separate from humans.[36] Dreams of controlling nature reached astounding heights in the largely unrealized postwar Stalinist Plan for the Transformation of Nature. Redirecting environmental forces to serve productive ends, it envisioned enormous irrigation canals, thousand-kilometer belts of trees to shelter fields from wind erosion, and vast land reclamation projects.[37]

Even if visionaries never achieved such ambitious results, later leaders maintained this creed.[38] Khrushchev agitated for many changes, including altering crop structures to include corn. Before collectivization, peasants had selected crops and managed fields according to cultural preferences, climatic conditions, and market forces. After that cataclysm, the government had used its power to modify those choices, although under Stalin it primarily limited itself to extracting staple grains and raw materials for industry. Khrushchev's Virgin Lands campaign was the first challenge to limits imposed by natural conditions. Gargantuan sovkhozes established in Kazakhstan, Siberia, and other sparsely populated regions employed mechanized cultivation methods descended from those brought from America at the end of the 1920s. Insufficient use of measures to combat erosion that were developed at home and modeled on those found in use on North America's Great Plains allowed windstorms to erode the semiarid steppe's irreplaceable topsoil. However, the lands have since remained productive but with harvests dependent on highly variable weather, making them comparable to dryland regions of North America and Australia.[39]

Accordingly, agriculture connects societies to environments in relationships mediated by technology.[40] Beginning in the nineteenth century, technologies and practices transformed this ancient human endeavor by replacing closed systems with open ones dependent on capital investments in machinery, fuel, fertilizer, chemicals, and agronomy.[41] Over centuries, agriculturalists developed closed systems using legumes and organic fertilizer to fix nitrogen. Abundant but useless in the atmosphere, this vital element is rare in compounds in the soil accessible to plants' roots. Once farmers could acquire nitrogen from distant sources, this approach began to give way to open systems that allowed them to concentrate energy. The fertilizer first came in natural forms discovered in the deserts of South America's western coast. After World War I, farmers embraced cheap synthetic alternatives made possible by the Haber-Bosch process, which used energy from petroleum to synthesize nitrogen compounds from thin air.[42]

Khrushchev embraced industrially farmed corn based on open systems, which vied with alternative closed ones. Under Stalin, soil scientists had refined the *travopole*, or grass-field, system of land management. He had mandated it in an attempt to compensate for shortfalls in organic fertilizers that were the legacy of collectivization, and for refusal to invest in synthetic alternatives. Having questioned Stalin's orthodoxy while in Ukraine, Khrushchev considered it incapable of meeting his goals. As early as 1954, he charged, "Many sovkhozes use the land for grasses, which in practical terms means they are planting nothing. There is a lot of grass, but no livestock feed. It is necessary to plant not grasses, but corn, wheat, and other crops that produce more."[43] In principle, he supported crop rotations tailored to local climatic conditions. In practice, his almost incessant advocacy for corn undermined sound policy.

Over time, Khrushchev grew even more devoted to synthetic fertilizers and chemical herbicides. In place of clover and hay, which required little labor and few inputs to ensure a modest harvest, he demanded corn, which produced large yields only with intensive cultivation. The all-out assault on the grassfield system occurred only in the early 1960s, when Khrushchev contrasted the harvests achieved under that system with those of the corn-based industrial farming of Iowa.[44] For the first time, the Soviet Union began to invest considerable resources in synthetic fertilizers, but lagged behind the rising capabilities of rival industrial countries. In 1954, the country produced only 6.4 million tons, a quantity deemed insufficient even to planners' intentions to use modest amounts primarily on cotton and other raw-material crops. A policy proposal indicated an increase in capacity to 17.5 million tons over five years, as well as similar rises in pesticide production.[45]

Khrushchev extended industrial farming to crops beyond wheat. Officials made contacts with farmers, researchers, and manufacturers in the United States. Drawing on what they learned, they adopted technologies for planting, cultivating, and harvesting corn and other crops. These aided in employing synthetic fertilizers, chemical pesticides, hybrid corn seeds, and labor-saving machines little developed in the Soviet Union before the 1950s. Within a few years, they transformed corn from a labor-intensive niche crop grown in a few southern locales into—at least in theory—one vital to modern, mechanized farms.

Even before 1955, experts in the Soviet Union stressed how corn enabled America's bounteousness. Before World War I, American farmers benefited from natural fertility and welcoming climates to grow grain, raise livestock, and sell meat and milk cheaply. Haltingly during the interwar economic collapse and widely after 1940, they adopted new technologies that permitted a dwindling number of farmers to bring in unprecedented harvests of every crop, but of corn most of all. Annual output of the crop lingered around 75 million metric tons in the early years of the century. In 1965, it surpassed 100 million tons on its way to nearly 140 million by 1975. In the meantime, the number of hectares planted *fell* from 32 million in 1954 to 22 million in 1970. This decline was possible only because per-hectare yields rose more than threefold between 1945 and 1980. This bounty gave the United States command of global grain markets, even though most corn left farms only as the beef, pork, poultry, eggs, and milk that Americans ate more often because surpluses drove down consumer prices.[46]

A similar process reshaped farming and consuming in Western Europe. After a postwar agrarian crisis left France a net importer of food in 1950, its farmers invested heavily. Benefiting from subsidies, they adopted new technologies and expanded plantings of barley, a traditional feed crop, by 348 percent, and those of less familiar corn by 815 percent.[47] By 1960, the country had begun to export butter, cheese, and other dairy products. These transformations coincided with social change, as large estates gave way to highly capitalized farms. The accompanying affluence, furthermore, heralded the Thirty Glorious Years of unparalleled prosperity following the war.[48]

Corn became the engine and symbol of industrial farming owing to a self-reinforcing process. American engineers and agronomists found the crop responsive to new technologies and therefore concentrated on it. Experiments led to commercial hybrids that became common in the 1930s.[49] A diversity of landraces, locally adapted cultivars produced by prescientific selection, provided the genetic material breeders needed to isolate genes instilling resistance to pests, drought, and extremes in growing-season length. The resulting hybrids transformed synthetic fertilizers into yields 30 percent higher than those of traditional varieties. Using pesticides and herbicides, farmers engineered uniform fields where corn thrived. Machines tailored to planting, cultivating, and harvesting corn increased labor productivity, making the crop even cheaper to produce. Corn surpassed all other crops by becoming an integral part of a global capitalist economy based on industrial farming.[50] The United States thus arrived at its present condition: dependent on abundant corn grown apparently cheaply with the aid of fossil-fuel energy.

The speed and audacity of Khrushchev's program was the product of circumstances as well as his temperament. He highlighted corn's potential by spelling out a vision of abundance requiring a sudden transformation, rather than incremental growth based on measured improvements proposed by Malenkov. Soon, Khrushchev broadcast abroad his ambitious pledges to "catch up to and surpass America." In the People's Republic of China, Mao Zedong responded by launching the Great Leap Forward to mobilize the masses to compete with capitalist rivals.[51] Combined with calls for aggressive confrontation with class enemies everywhere, Mao's program challenged Khrushchev's claims to lead the communist world and pledges that communism was imminent in the Soviet Union.[52] After 1959, the Chinese occasionally paused their attacks on Khrushchev's foreign policy to savage his domestic reforms, labeling him an anti–Marxist revisionist who was "restoring capitalism."[53] Condemning the Great Leap, he defended himself by critiquing the Chinese for putting false hope in "moral factors." "What are moral factors?" he asked rhetorically. "This is babbling about communism but having to tighten one's belt, that's the 'moral factor': you can call it communist paradise, but there's no chow."[54] The Soviet Union, he supposed, was building communism through abundance secured, in part, by corn and technology.

Ideals and their practical consequences drove Khrushchev to demonstrate that the Soviet Union was progressing toward communism, countering Chinese accusations that it was backsliding. From the late 1920s, the Soviet Union had experienced a production revolution in industry of the type Khrushchev now envisioned for agriculture. Emboldened by American examples, Khrushchev proposed to spur a transformation of farming at a Bolshevik tempo similar to that in industry by planting corn, plowing up the Virgin Lands, applying technology, and reaping the rewards in the form of more and cheaper meat, milk, eggs, butter, and other foods.[55] Given global trends in farming technologies, the model provided by American achievements, and the Soviet Union's past accomplishments, Khrushchev's highly ambitious faith in a sudden, thoroughgoing transformation seems less harebrained than critics have imagined.

Khrushchev adopted policies reinforcing his own preferences for corn and mandating practices observed by the experts that he sent to other industrial countries. Learning of Khrushchev's enthusiasm for corn in early 1955, the editor of the *Des Moines Register* extended an unofficial invitation to visit Iowa. It quickly received a positive response.⁵⁶ Negotiations between the State Department and the Ministry of Foreign Affairs cemented an agreement to exchange a delegation of American farmers and academics traveling as private citizens for an official delegation led by Vladimir Matskevich, the deputy minister of agriculture, and including corn experts Andrei Shevchenko and Boris Sokolov. That summer, this first postwar delegation traveled from Atlantic to Pacific, and visited several Canadian provinces. The officials inspected farms, agricultural colleges, private companies, factories, and other infrastructure necessary for industrial farming. They met farmers, professors, engineers, corporate executives, political leaders, and many others. Above all, they concentrated on the Corn Belt and examined models proving the feasibility of specific research practices, machinery, livestock, and land management.⁵⁷

Meetings with counterparts developed into contacts that persisted despite Cold War tensions. As soon as the winter of 1956, a second delegation tasked with studying hybrid seed-corn production and processing arrived with a mandate to purchase seeds and equipment.⁵⁸ Khrushchev developed a decade-long friendship with Roswell "Bob" Garst, the seed-company executive who served, William Taubman suggests, as the leader's "guru" for all things related to corn.⁵⁹

Garst promoted postwar innovations little known to the visiting officials because of the Cold War–inspired hostility of the preceding decade. Upon first meeting Khrushchev in 1955, Garst asked why, after the Soviet Union had stolen American nuclear-weapons secrets, the country's experts considered information available in any American farm journal a discovery. Bursting into laughter, Khrushchev replied, "You locked up the atomic bomb, so we had to steal it. When you offered us information about agriculture for nothing, we thought that might be what it was worth."⁶⁰ By 1955, improving relations signaled by a summit meeting in Geneva ushered in agreements for reciprocal farm delegations that summer, soon followed by technical, educational, and cultural exchanges.⁶¹

The 1955 delegation found mechanization, efficiency, and financial calculation the guiding principles on American farms. They encountered, in a word, industrial farming, even though contemporaries did not use the term. Corn was the foundation on which everything stood. Reporting in writing and to Khrushchev in person, the experts filtered what they had seen through preconceptions about America. Noting details both mundane and extraordinary, they expressed envy for technological innovations and highlighted differences between the countries' socioeconomic foundations.

For instance, the visitors found that capitalism demanded constant investments in the latest technology because competition promised to ruin any farmer unable to adopt the most productive methods.⁶² They were preconditioned by Marxist precepts to consider this pressure an immutable law of capitalism requiring farmers to maximize profit and ruin their neighbors. They thus highlighted American statistics recording a declining number of farms and the concomitant rise in the surviving ones' average size. "The concentration of production," their report explains, "and elimination of smaller farms by the larger reaffirms [Vladimir] Lenin's axiom that capitalism's fundamental tendency is to destroy small-scale enterprises in both industry and agriculture."[63] The official published narrative asserted that these "intrinsic contradictions" justified the Soviet authorities' emphasis on studying technologies while condemning inequality.[64] It thus avoided discussion of evident prosperity and underscored instances of poverty. In private communications, officials cautiously praised American practices that contrasted with domestic counterparts, delving deeper than they allowed in public statements. In each case, they divulged assumptions about the value neutral-nature of technology.

Technologies allowing private farmers to outperform kolkhozniks revealed injustices ingrained in capitalism. At home and abroad, propaganda touted the Soviet Union as an exemplar of modern and egalitarian agrarian production relations. For instance, Khrushchev had recently contended that only a few elite capitalist farmers employed technology available to every kolkhoz in the Soviet Union.[65] The country's leaders continually extolled equality at home and accentuated inequality under capitalism. Publically, the authorities contrasted the Soviet Union with America, where they discerned shadowy monopolists and malignant market forces, which forced farmers to battle just to remain solvent.

The volume chronicling Khrushchev's 1959 visit to the United States therefore noted the contributions of advanced agriculture to American wealth, even as it highlighted social and economic disorder churning beneath the outwardly prosperous surface of capitalist society.[66] In 1961, Shevchenko, Khrushchev's personal agricultural aide, wrote in praise of the kolkhozes and sovkhozes in the same terms he and his fellow delegates had in 1955, judging them superior to private ownership and capitalism. Americans had succeeded not because of those institutions, which only created inequality, poverty, and personal bankruptcy. "It is not the private farm," Shevchenko explained, "but corn that has helped the United States increase its grain production and, as a result, improve animal husbandry. Yet corn does not serve the capitalist system alone."[67] Far from inferior, the kolkhozes and sovkhozes simply needed to master the crop and modern technologies for growing it. Khrushchev himself contrasted his country's farmers, who labored "for themselves and society," with American counterparts driven from their land into the ranks of landless wage laborers.[68]

The visitors highlighted American farmers' compulsion to specialize, practices facilitating efficiency but driven by savage competition unlike anything found in the Soviet Union. Instead of growing their own food, American farmers bought essential supplies while producing only a few commodities, including corn, risking ruin if the price for a commodity were to collapse.[69] By contrast, kolkhozes strove to supply themselves with staples, seeds, spare parts, and other goods difficult to obtain from the state distribution system. Protected from bankruptcy, they might adapt such specialization and benefit even more. "Having actually witnessed the ruin of small farmers in the United States and Canada, the enormous benefits of our socialist system became clear," the delegation stressed. "Given our large farms and planned economy, we enjoy great advantages over the United States. Our kolkhozes and sovkhozes ... can quickly increase output ... and decrease production costs by more effectively employing machines and organizing production."[70] Although leaders recognized defects at home, they believed the kolkhozes immune to the sort of socioeconomic crises inherent in capitalist farming.

Concluding that Americans specialized in crops suited to local climates, the delegates argued for recreating the practice. Instead of growing some wheat everywhere, American farmers planted the crop only in a few well-suited regions. When selecting any individual crop, they got the most out of temperature, daylight, rainfall, and soil conditions, as those in Iowa did by planting corn from fencerow to fencerow.[71] In the Soviet Union, conditions limited many crops, including corn. Khrushchev put faith in irrigation systems' capacity to overcome the dry climate in the south. He believed that scientific breeding of cold-resistant hybrids promised to mitigate the short growing season in the north. To this end, the delegation sought such seeds when visiting the upper Midwest and Canada.[72] Khrushchev's interest was so well known that in 1961 Secretary of State Dean Rusk mischievously told him about a hybrid maturing in record time. Under orders, the USSR agricultural attaché in Washington searched in vain for information about the purported wonder seed.[73]

A pillar of industrial farming, regional specialization prioritizes output and cost over location, depending on rail networks to distribute the harvest. Under Stalin, planners had favored local self-sufficiency to compensate for dysfunctional transport networks. Khrushchev jettisoned this approach. Hearing of practices in the United States, he grumbled that Stalin's priorities meant that the Soviet Union planned agriculture "completely without a plan."[74] Khrushchev's initiatives testify to new principles of specialization, albeit ones applied inconsistently. Already in 1954, he had explained how the Virgin Lands blazed a trail for subsequent related innovations. The wheat harvests from industrial farms of thinly populated locales promised to free settled regions of European Russia, the North Caucasus, and Ukraine for intensive livestock raising, for which they needed feed. Khrushchev ordered farms there to adopt corn to nourish the cattle, hogs, and poultry needed to make inexpensive meat, milk, and eggs available not just to local consumers, but to urban centers across the country.[75]

The experts who traveled to America had been familiar with corn and industrial farming, but they returned with their convictions confirmed. They did not bring back corn itself or a newfound belief in it. Instead, they were inspired to transform it from a crop requiring vast amounts of manual labor to grow modest yields into one harnessing advances in machine-building, chemistry, and genetics to boost yields and reduce the cost of meat and dairy products. In particular, they saw ways to remedy chronic shortfalls of tractors, harvesters, and other equipment, as well as spare parts for them all. Similarly, they gathered evidence for the importance of double-cross hybrid seeds, a challenge to the disreputable biologist Trofim Lysenko's rejection of the advanced hybrids. Each receives detailed treatment later. Here, it is sufficient to note that the experts were well disposed to these methods because of the principles ingrained in Bolshevik theory and in practice in the 1920s and 1930s. The American agrarian system they observed shared these ideals and applied them to a greater degree, confirming the beliefs of the Soviet visitors and those of political masters including, most importantly, Khrushchev.

In 1959, Khrushchev inspected Garst's farm for himself, but he had first become acquainted with the booster of industrial agriculture in 1955. Then, Garst had whisked the leaders of the Soviet Union's delegation away from their planned route to inspect his farm.[76] Impressed by what they saw, Matskevich and Shevchenko invited him to visit their country, where he soon met Khrushchev. Each loquacious showman found a kindred spirit in the other. Garst shared knowledge of the latest advances for growing corn and feeding it to livestock with Khrushchev, already a devotee of the crop and modern technologies.

Although profiting from the ensuing trade, Garst also believed that engagement reduced Cold War tensions and prevented armed conflict. Between 1956 and 1958, he arranged the sale of more than $1 million worth of hybrid seeds, a temporary moratorium on trade after the Soviet Union's November 1956 invasion of Hungary notwithstanding.[77] Throughout the Khrushchev decade, Garst extolled the progress made in the Soviet Union and its allied countries in planting hybrids and applying synthetic fertilizers.[78]

Garst acted on the belief that these measures offered the means to prevent a predicted global famine. As one contemporary American newspaper explained, he was "a man on a mission, and his mission is scientific agriculture."[79] Echoing Khrushchev's policy of peaceful coexistence, Garst later expressed willingness "to give all that is innovative to the Soviet Union. Let the Soviet Union share with China, India, and other countries, so there are no hungry people in the world and no wars, and so that there is peace on earth and friendship among peoples."[80] These words were published in the Soviet Union only because they reflected official positions, but Garst often expressed the ideas in other contexts. In some ways, he proved correct, as the threat of crisis receded into the future in the 1960s, when technological advances transformed farming in industrial countries and gave rise to the Green Revolution in others.

Garst spread Khrushchev's vision of industrially farmed corn across Eastern Europe. Proselytizing with the leader's blessing, Garst encouraged officials and farmers in Hungary and Romania, countries where corn had long been a staple, to take up the new methods. He thus demonstrated his own belief in the value-neutral nature of technology. After the initial 1955 visit, Garst returned again and again throughout the ensuing decade, always delivering the latest know-how to interlocutors in those countries especially.[81]

Transfers of technology from the Soviet Union to its allies testify to the country's role as a conduit for innovations and ideas. As early as March 1955, Khrushchev told advisers of a vision for the satellite countries. "This is a colossal breakthrough," he enthused. "If we get the Hungarians and Romanians, who are already corngrowers [to go along], then they will be awash in grain and silage, whereas now they have nothing to feed to livestock."[82] Khrushchev soon substantiated the initiative by assigning specialists to establish a research station in Romania and dispatching other resources.[83] His experts surveyed Albanian corn growing practices, proposing to introduce hybrid corn and other innovations as remedies for perceived backwardness. Moreover, the proposal included training economists, establishing test stations, and applying technology to other endeavors.[84]

Experts consulting on modern agronomy and corngrowing technologies also traveled to countries where corn had been little known. Evangelizing in Poland, East Germany, and Czechoslovakia, Khrushchev backed his words by shipping improved seeds to each.[85] In 1956, farmers in Poland doubled plantings to 200,000 hectares. Soon, the Soviet Union delivered tractors and machines specialized for cultivating corn, domestic versions of models first observed in America. By 1959, deliveries exceeded 2,000 tractors and specialized implements.[86] Even though its agricultural sector was advanced relative to that of the Soviet Union, East Germany's authorities invited experts on corn.[87]

Efforts to spread knowledge about corn ranged even further afield. Thousands came to the Soviet Union to inspect selected demonstration farms and the All-Union Agricultural Exhibition. In 1956, some 1,705 visited from the people's democracies of Central and Eastern Europe, while the remaining 545 traveled not only from avowedly capitalist United States, Canada, and France, but also from Mexico, India, Burma, Afghanistan, China, Vietnam, and Egypt.[88] Until relations soured, experts from the Soviet Union worked in the People's Republic of China. Some advised in establishing large mechanized farms resembling those in the Virgin Lands. Others studied potentially beneficial practices they encountered in their travels, while collecting samples of corn and other crops to aid breeding programs.[89]

For years after 1955, Khrushchev and his experts continued to look to the United States, but also investigated New Zealand, France, Finland, and other countries with productive agricultural systems.[90] In 1955, a group headed by Minister of Agriculture Ivan Benediktov was the first postwar delegation to England, where it studied crops, research, chemicals, machine-building, and livestock. Skeptical of the small rented farms under market pressures and of the country's need to import staple grains, the visitors highlighted features similar to those found in the United States. They stressed that machines and efficiency allowed only 5 percent of the working population to cultivate all of England's arable land.[91] They also noted the advent of new wheat varieties resistant to lodging, a key Green Revolution technology.[92] Seeking other avenues for learning, the Soviet Union assigned attachés at embassies in advanced countries to study developments in agricultural science, technology, and output.[93] In the mid-1950s, the Soviet Union made overtures to the Rome-based United Nations Food and Agriculture Organization, or FAO, considering membership as a means to acquire knowledge applicable at home. Although experts recommended joining, authorities maintained standing objections to the FAO's mandatory annual disclosure of data on the harvest.[94]

The Soviet Union's secrecy about proprietary harvest data sheds light on Americans' reasons for clamoring to visit the country. Americans showed intense interest in the Soviet Union and its delegations. Not only Khrushchev's 1959 visit, but also the modest 1955 delegation, attracted large crowds and intense press coverage, as a photo essay in the general-interest magazine *Life* demonstrated.[95] Some followed Garst in sharing information about technology and methods. American visitors to the Soviet Union, by contrast, journeyed not to learn practical methods, but to investigate the unknown. Skeptical of published official statistics, they hoped to measure crop yields and living standards firsthand. One returning member of the 1955 American delegation concluded that the exchanges would be useless unless the authorities provided data required to interpret what the guests were shown.[96] Such concerns did not prevent numerous delegations from traveling under the two-year exchange agreements signed in 1958, 1960, and 1962.

American visitors typically did not consider what they saw useful at home, having brought with them a sense of superior technology and organization. They found few attractive ideas or technologies, with the exception of the occasional crop variety exhibiting desirable traits.[97] Nonetheless, they revived long-standing interest in the Soviet Union's agriculture. Americans questioned the accuracy of available data about harvests, even though the authorities published somewhat more of it after 1953. The government and party in turn harbored suspicions about the visitors' motives, and therefore used any available means to shape the foreigners' impressions. In 1963, Vladimir Semichastnyi, chair of the Committee for State Security, or KGB, reported to the Central Committee that a delegation led by the US secretary of agriculture visibly attempted to "compare what they see with the officially published statistics and economic data." Semichastnyi summarized "documents acquired by operational means" regarding the delegation's mission to learn about "the state of Soviet agriculture and . . . [ask] particular questions to Soviet specialists."[98]

Khrushchev's efforts to apply industrial agriculture in the Soviet Union reached new heights in the unusual Seven-Year Plan adopted in early 1959. This strategic plan called for new capital investment in mechanization, electrification, irrigation, synthetic fertilizers, herbicides, and insecticides designed to boost harvests of all crops, especially corn. In public remarks, Khrushchev lavished attention on these technologies. He recognized the debt to Garst, who had invited farmworkers from the Soviet Union to spend an entire growing season gaining hands-on experience on his farm. Foreign detractors, Khrushchev said, "always chide us communists for only criticizing capitalists, but now they see that we thank the farmer-capitalist [Garst] for the advantageous deal."[99] That training aided kolkhozes and sovkhozes in substituting mechanized production of corn and other crops for the old approaches requiring large amounts of expensive manual labor. They used adaptations of machines and technologies that engineers and experts observed in the United States or mastered by detailed study of models purchased there.[100]

Khrushchev endlessly touted the new technologies and methods. In 1959, he praised the large corn harvest on the kolkhoz in his home village, Kalinovka, which enjoyed his patronage. The Kursk Oblast authorities had ordered the kolkhozniks also to produce hybrid seeds, despite the absence of know-how and favorable climatic conditions. "Why do that?" Khrushchev queried:

> We have passed that stage by. . . . Don't make your backwardness obvious. Seeds should be grown only on seed-production farms. Take America as an example. There, not every farmer raises seed corn. He buys it from a specialized seed company. But here, some want to raise corn for silage and also produce seeds. This is primitive production and must be avoided. We live in the age of specialization.[101]

In choosing a word denoting primitive handicraft production, Khrushchev harkened back to Vladimir Lenin, who used the term as a pejorative to contrast with factories expected to form the foundation for socialist modernity.

In 1961, Khrushchev obliged the Twenty-second Party Congress to adopt the Third Party Program, codifying his pledges to provide the material necessities of communism by 1980. It promised larger harvests and higher productivity on the basis of principles classified as intensification.[102] Corn remained a centerpiece in a program stressing that no single technology or practice could meet every agrarian challenge facing the Soviet Union. Instead, an interconnected set of policies designed to apply existing technologies promised to raise farm output to American benchmarks.[103] Khrushchev maintained this line even after a 1963 crop failure and accompanying food crisis had discredited his leadership. In 1964, he redoubled advocacy for the synthetic fertilizers that had revolutionized farming in the United States and Europe over the previous generation.[104] Whereas wide use had developed over decades there, Khrushchev gave the Soviet Union a mere seven years to match the achievement, encouraging a radical fivefold expansion of fertilizer output to 100 million tons.[105]

Industrial farming became even more entrenched after Khrushchev exited the Kremlin. Capital investments grew from the equivalent of 13 million new rubles during the Fifth Five-Year Plan (1951–1955) to 24 million between 1956 and 1960, the height of Khrushchev's success. Between 1961 and 1965 this trend continued, as investments amounted to 38 million rubles, a figure that reached some 60 million during the five-year plan covering 1966 to 1970. These outlays yielded buildings, electricity, irrigation, and machinery. The number of physical tractors grew threefold, resulting in a fourfold increase in standard 15 horsepower units.[106] Yet these investments returned only modest dividends, even as harvests surpassed historical norms. Maximum yields in years of favorable weather rose from 1.11 tons per hectare between 1955 and 1960—already a substantial improvement over Stalin's time—to 1.85 tons twenty years later. Yields in poor years similarly increased, from 0.84 to 1.42 over the same period. In 1980, farms applied four times as much fertilizer as in 1965. The magnitude of capital investments grew 9.5 percent from 1970 to 1975, and a further 7.3 percent between 1976 and 1980. Yet economists have shown that the rates of return on these investments were low, meaning that output growth failed to keep pace with growth in investments.[107]

In the mid-twentieth century, interconnected transformations of rural societies, farming technologies, and agrarian economies remade countries adhering to capitalist and state socialist models, but also those escaping colonial subjugation. By the late 1960s, this latter trend became famous as the Green Revolution, the product of programs funded by philanthropic foundations with US government backing. Although not a participant in the Green Revolution, the Soviet Union provided a few countries with expertise and long-term, low-interest loans to buy its equipment, although this assistance was in amounts far smaller than grants made by the United States and its allies. A small fraction of

Soviet aid went to development projects oriented toward agriculture and agrarian development, programs testifying to the influence on the Soviet Union of the same transfer mentality that defined the American-backed Green Revolution. In particular, the Soviet Union sponsored development projects to increase its prestige, to cultivate trade ties, and, inasmuch as it advocated an ideal, to privilege principles of centralization and state ownership.[108]

Since the term "the Green Revolution" first emerged in the late 1960s, the standard narrative has privileged technology. In 1943, the Rockefeller Foundation funded the Mexican Agriculture Program (MAP). Seeking higher yields and larger harvests, its researchers investigated corn, but soon moved to breeding the wheat cultivars that earned American Norman Borlaug the 1970 Nobel Peace Prize. They solved a problem paradoxically arising from conditions made too favorable by industrial farming technologies. Existing varieties failed because their tall, thin straws could not support heads of grain grown heavy thanks to synthetic fertilizers, irrigation, mechanical cultivation, chemical pesticides, and scientific agronomy. As the grain reached maturity, the straw collapsed, leaving the harvest on the ground, a disaster called lodging. After 1945, American breeders crossed prolific varieties with dwarf ones from Japan characterized by straw only two-thirds the normal height. The new hybrids aided Borlaug in making further innovations. Facing food shortage and dependence on unaffordable imports, Mexico used the new varieties to become self-sufficient almost overnight.[109] The Rockefeller and Ford Foundations then collaborated to spread these high-yielding varieties (HYVs), which first arrived in India in 1961. By 1966, farmers there planted more than 500,000 hectares of them.[110] The wheat harvest doubled in four years, stabilizing a food supply that had seemed perilous to pessimists pronouncing India chronically malnourished and impossible to feed in the face of a swelling population. Having proven that the formula of improved seeds and capital investments could be transferred from one agrarian society to another, experts soon created dwarf high-yielding varieties of rice. The HYVs triumphed again when so-called miracle rice spread from the Philippines to Southeast Asia and beyond.

This history of the Green Revolution omits important details.[111] These particulars indicate that the Soviet Union belongs in a story that has long relegated it to the margins. In the 1930s, American progressives led by Henry Wallace, a pioneer in hybrid corn and later vice president under Franklin Roosevelt, envisioned exporting the New Deal. Like Appalachia, the Deep South, and other loci of rural poverty, agrarian communities abroad might achieve security and gain access to consumer goods, making the coming era "the century of the common man."[112] This faith confronted the circumstances of the Great Depression, the apogee of a larger interwar global agrarian crisis. Apparently immune to the crisis while seeming to modernize industry and agriculture, Stalin's revolution offered a prototype of noncapitalist development. The MAP originally promised to integrate peasants into society as consumers and citizens by fostering sustainable social change; facilitating higher output, earnings, and consumption; and making radical alternatives less attractive. Quickly abandoning this social mission, the MAP became the package of technological fixes favored by Borlaug and his fellow experts.[113] It addressed low yields, not rural poverty, which the technologies exacerbated because those unable or unwilling to adopt them joined the landless poor streaming to cities. Influenced by the transfer mentality, the program's experts sought models replicable anywhere, disregarding sustainable livelihoods and local collaboration.[114] In reality, technologies constituted only part of a larger package, which included incentives favoring farmers already strong enough to invest.[115]

Cold War mentalities encouraged Americans to believe that desperation drove people toward totalitarian communism. Policymakers believed that they were combating Maoist peasant insurrections and national-liberation movements modeled on the Vietminh. When coined by a United States Agency for International Development (USAID) official, the term "Green Revolution" explicitly contrasted with the red alternatives sweeping Asia, seen as a global challenge to the power of the United States.[116] Even though the Soviet Union neither directed nor financed these, the Cold War rival loomed in American consciousness.

In subsequent standard accounts, however, the Soviet Union has received insufficient notice. Typically, scholars have considered it only as a model for crash industrialization guided by a bureaucratized planning apparatus, the model developed under Stalin and replicated in the Eastern Europe and beyond after 1947. Paralleling American programs to spread agricultural technologies, the Soviet Union adopted the component technologies, including irrigation systems, improved wheat varieties, hybrid seeds, and chemicals. Putting them to work at home, its leaders hoped to formulate a modern, egalitarian agrarian system founded on state ownership of farms, infrastructure, and other means of production.

In the 1950s, the Soviet Union's leaders hoped to spread this version of modern agriculture. At the time, Sovietologists recognized this, but subsequent thinking about Khrushchev's reforms has largely abandoned the idea. Dating from the early 1960s, the only previous work devoted to the corn program considered it an example of development under totalitarian and nonmarket conditions. It argued that Khrushchev's effort was significant because success might enhance his case for the Soviet Union as an alternative model.[117] In reality, the country achieved little in spreading its version of agrarian development because, to outsiders, sovkhozes and kolkhozes seemed to offer few advantages. American policymakers always highlighted those weaknesses, contrasting the apparent advantages of capitalist development, private property, and the beneficence of the US government and American foundations.[118]

After 1953, Khrushchev encouraged a new kind of relationship with a new kind of country then emerging from decades of colonialism. The frequent travels of Khrushchev, Anastas Mikoyan, and other leaders, accompanied by increases in foreign trade, were not a power play designed to encourage imitators driven by ideology. Because the newly independent countries were already enmeshed in a world economy favoring former imperial powers and steered by the United States, their leaders did not enjoy full freedom of action. The Soviet Union's leaders hoped to expand economic engagement with that global order and enhance their country's prestige. They did not seek to replicate their version of socialism in full, but to promote centralization and state ownership, which partners found attractive for pragmatic and historical reasons rather than an affinity for full-scale mobilizational planning.[119] The limited number of agrarian development projects the Soviet Union backed therefore took the form of irrigation systems and other large-scale capital investments, as well as state-owned demonstration farms reminiscent of sovkhozes.

Experiences in India illustrate the limits on the Soviet Union's ambitions regarding agrarian projects, on the comparative size of its actual contributions, and on the results achieved. The country received the largest share of resources the Soviet Union offered to nonaligned states. Largely in the form of low-interest loans, the sum was far less than American aid, which typically took the form of grants.[120] Hostile to anyone proposing alternative paths to socialism, Stalin had shown indifference toward India after it gained independence in 1947. His successors facilitated trade through agreements signed in 1953 and 1958. In 1955, Prime Minister Jawaharlal Nehru and Khrushchev undertook reciprocal visits, each leader greeted warmly by his hosts.[121] Congress Party governments considered—but did not copy—practices in the Soviet Union when creating their distinctive approach to planning, making it one among many influences. In fact, this moment initiated a decade of intellectual engagement that shaped the Soviet Union as least as much as, if not more than, India.[122]

Although steel mills and other industrial projects received pride of place, the Soviet Union funded state-owned agribusiness, irrigation, and other agricultural initiatives. When Nehru visited the Soviet Union in 1955, his itinerary included farms along with factories, metro stations, and other attractions. Khrushchev soon donated 5 million rubles worth of farm equipment to India, including a 1956 grant of some 300 tractors, harvesters, and other machines.[123] This aid went to a farm in the arid northwestern state of Rajasthan. By 1961, the Suratgarh farm had sown 7,000 of its 12,000 hectares of its previously unused land, of which at least 25 percent were corn and other grains cultivated using machines. Fed by a new long-distance canal, the farm's irrigation system was still under construction. Relatively few in number, the machines desperately required spare parts and trained mechanics.[124] Despite such obstacles, the demonstration farm achieved higher yields than neighboring peasant farms, encouraging Nehru and other leaders to envision replicating it.[125] By 1964, the farm was earning a profit, although the director requested that the USSR agricultural attaché aid in securing additional expertise and machinery useful for growing its primary crops: sugar, rice, and corn. Tellingly, the attaché expressed alarm that signs heralding the Soviet Union's aid and Khrushchev's visit to the farm had been removed under a previous director. By reporting the current director's statement that he planned to correct this "abnormal" situation, the attaché indicated the importance of ensuring that the Soviet Union received credit.[126]

Representatives promoted the Soviet Union's ideals and achievements to Indian audiences, yet continually expressed fear that they were losing a contest with the Americans. As early as 1956, officials complained about a lack of knowledge about India and of speakers of required languages.[127] In 1959, the ambassador reported that, although agriculture comprised 70 percent of India's economy, it spent billions of rupees on food imports from the United States. Despite investments in agricultural development and technology, India was becoming subject, he feared, to influence bought with aid.[128]

The Soviet Union sought public recognition via many channels. It earmarked 1 million rubles for an exhibit at New Delhi's 1960 World Agricultural Exhibition, which also attracted exhibits from the United States, the People's Republic of China, Poland, East Germany, and a handful of other countries.[129] Unsurprisingly, responsible officials reported "enormous interest" from the Indian public, government officials, and foreign visitors. Whatever the actual response, the Soviet Union's leaders considered these efforts vital to promoting their version of modern agriculture and agrarian development. Extolling the Soviet Union's space program and other high technology, the pavilion included practical demonstrations of farm machines and displays on new varieties of rice, wheat, cotton, and corn. It also heralded improving measures of quality of life in the Soviet Union's rural areas.

Despite apparent success, exhibit personnel and visiting minister Matskevich expressed fears that their efforts trailed American counterparts. Each lamented the limited number and variety of publications in English, Hindi, Urdu, and Punjabi. Specialists and experts in vital fields of agriculture visited India relatively rarely.[130] By contrast, Matskevich reported having frequently encountered American specialists advising at research stations and universities. "Despite interest in Soviet agricultural development," his interlocutors had "little knowledge,... demonstrating the insufficiency of our propaganda about the specificities and achievements of Soviet agriculture."[131]

These fears persisted through years of continued efforts. In 1962, the agricultural attaché in New Delhi described his own ambitious annual program, as well as obstacles imposed by a piteous lack of resources. While the American government and private foundations were devoting millions of dollars to India, the attaché complained that his only translator had returned to the Soviet Union in May and no replacement had arrived. To date, he had no regular allocations of embassy funds to cover purchases and travel expenses, forcing him to beg superiors for money to purchase an automobile. Gathering new crops, varieties, and research publications, he also conducted outreach to specialists, officials, activists, and workers. Visiting research institutes, exhibitions, and villages, he particularly emphasized contacts with producers' cooperatives. Expressing fears of bad press from Western sources and decrying Indians' lack of knowledge about the Soviet Union, the attaché shed light on his standing mission to raise his country's profile. He proposed sending more experts than the seven dispatched that year. Because much of what Indian researchers knew about Nikolai Vavilov or pioneering soil scientist Kliment Timiriazev came from citations of their work in English-language sources, the attaché proposed—unsuccessfully—to found a journal on agriculture in the Soviet Union aimed at South Asia, Southeast Asia, and Africa.[132]

The Soviet Union's leaders sought to portray their agrarian model in the best light to the widest audience. At international conferences, its experts lionized its development. Academician Aleksandr Tulupnikov thus boasted that under Khrushchev the Soviet Union had added more machines, planted more hectares, and harvested higher yields than ever before.[133] In 1959, the country's pavilion at the Novi Sad agricultural expo in Yugoslavia touted these virtues to visitors in 25,000 brochures printed in a variety of world languages. Employing technology and freed of exploitation, collectivized farming had brought peasants into the modern world. Extolling large-scale farms, high productivity, and advances made under Khrushchev, the pamphlet's writers aimed to capture the minds of sympathetic audiences from capitalist, socialist, and nonaligned countries.[134] Tellingly, experts who reviewed a first draft assailed the materials for insufficient attention to elements vital to the Soviet Union's self-image under Khrushchev: party leadership, partnership between working class and peasantry, material progress, economic emancipation for women, increasing supplies of consumer goods, and recent innovations in planning.[135]

The Soviet Union propagandized to audiences in Africa, Asia, and beyond about its contributions to hydroelectric and irrigation projects. Khrushchev's enthusiasm stemmed from their prospects for spreading modern technology and farming practices. The Soviet Union had begun in the 1920s with a modern dam on the Dnieper River supplying electricity for industry and irrigation water for surrounding regions of Ukraine. Like other capital projects of the era, the dam relied on American expertise and equipment.[136] Such projects applied the same technologies as New Deal counterparts of the 1930s, including both dual-use dams in arid locales of the West and the hydroelectric ones of the Tennessee Valley Authority. These brought development to sparsely populated and impoverished regions. In turn, they inspired ventures integral to postwar international development. Thus, in the 1950s, American engineers devised the Helmand Valley Authority in Afghanistan.[137]

Under Khrushchev, the Soviet Union expanded irrigation and development initiatives to hundreds of thousands of hectares in dry regions across southern Ukraine, the North Caucasus, and the cotton lands of Central Asia.[138] As early as 1955, Khrushchev recounted how Khamrakul Tursunkulov, decorated chairman of a model cotton kolkhoz in Uzbekistan, had told him about a visitor from India. Having inspected the fields, facilities, social services, and accommodations, the Indian began to weep. Khrushchev continued, "Comrade Tursunkulov inquired about what had disturbed the guest. 'These are tears of joy in my eyes,' the man answered, 'for your rich and happy life, but also tears of bitterness for the millions of impoverished and wretched people of the capitalist countries.'"[139] Speaking to a domestic audience, Khrushchev told the story to encourage pride in past achievements and efforts in the future.

Visiting India in 1960, Vladimir Matskevich called for further attempts to publicize such projects in Central Asia, which brought together irrigation, cotton, and development.[140] Khrushchev later raved about the potential of irrigation to feed countries facing food shortages, calling it "a new step toward intensive management of agriculture." Deriding the "bourgeois economists" who predicted overpopulation and malnutrition in India, he envisioned plenty. "If the achievements of science and technology are properly utilized," he explained, "then the potential for food production is simply limitless." Speaking in the year when HYVs arrived in India, Khrushchev suggested that electric power, water pumps, and pipelines heralded a blossoming even of dry land.[141] In projects such as Suratgarh, the Soviet Union shared its expertise in these areas on a small scale.

India was the largest recipient of the Soviet Union's assistance, but other nations welcomed advisers and aid, programs demonstrating the influence of the transfer mentality on both the leaders of the Soviet Union and postcolonial modernizers. In 1958, experts from the State Committee for Foreign Economic Relations recommended expanding aid programs, especially those relating to agriculture. Their confidence growing due to Khrushchev's early successes, they argued the Soviet Union should fund projects to expand plantings, create irrigation systems, develop livestock raising, improve agronomy, and advance related endeavors.[142]

A small component of overall efforts, agriculture was significant because it aided in public relations. In addition to the portfolio of loans and projects negotiated since 1955, leaders proposed new initiatives in Cambodia, Sudan, Ethiopia, Nepal, and Libya designed to develop state economic sectors. On a list of 160 projects, only 11 directly funded agriculture.[143] However, many of the 100 industrial projects also furthered rural modernization. A nitrogen fertilizer plant, food-processing facilities, and irrigation systems numbered among those in the United Arab Republic, a fleeting union of Egypt and Syria. In Afghanistan, the Soviet Union constructed a dam, two hydroelectric stations, and an irrigation canal designed to make 30,000 hectares arable. Other initiatives cleared forest for farmland in Indonesia and Ceylon (today's Sri Lanka), and dispatched agronomists to Burma.[144] In part an irrigation project, the Aswan High Dam received financing from the Soviet Union after the Eisenhower administration backed out due to political disputes with Egypt's president, Gemal Abdel Nasser. Experts stressed to the Soviet Union's leaders that farming was vital "to supporting the sustainability of these countries' economies." They called for expanding hitherto limited aid to facilitate production of staples and industrial crops. Their plans envisioned diversifying monocultures imposed by colonial rulers, setting up credit systems, negotiating trade deals, and employing improved land management, agronomy, and seeds.[145]

Nonetheless, these plans raced ahead of the modest practical achievements. A 1963 agreement with newly independent Algeria required delivery of tractors, plows, planters, and other technology needed for a mechanized farm capable of planting 20 metric tons of improved cottonseeds. The plan quickly stalled due to delays in manufacturing and shipping the equipment, barriers common to industry in the Soviet Union.[146]

The Soviet Union did not offer comprehensive models for industrialization or agrarian modernization, but it did pursue the prestige that large and conspicuous development projects might secure. Khrushchev hoped to create examples of state-owned agribusiness to spread modern farming technologies and support anticapitalist egalitarianism. Contrasting its own ideals with the social problems ascribed to capitalism, the Soviet Union's official narratives of development appealed to some in Africa, Asia, and Latin America. Nonetheless, these appeals were only one part of complex interactions among local and global influences. Potential partners did not choose freely, but instead weighed the potential of sponsorship by the Soviet Union against losses in trade and aid from the United States and its allies, the likely results of overtures to Moscow. A narrative of straightforward exchanges of aid for influence cannot account for conditions in which postcolonial leaders had to seek a delicate balance.[147] In advancing these endeavors, Khrushchev welcomed foreign ideas into the Soviet Union and, coupled with its own additions, sought to export them. Seeking partners, he wanted to expand economic cooperation and enhance his country's prestige. Emphasis on bipolar conflict between static models, by contrast, has encouraged many contemporaries and subsequent scholars discount agrarian development in the Soviet Union's links to postcolonial nations.

Endeavoring to transform farming in the Soviet Union, Khrushchev embraced a belief in the value-neutral nature of technology and a Promethean understanding of nature. These qualities testify to the country's enmeshing in a global web of ideas, technologies, and practices. Beginning in the 1920s, ideology and practice had demonstrated affinity for factory farming, not least of which through a faith in humankind's capacity to master nature and the ability of the new society to tame innovations devised under capitalism. Considering the Soviet Union a superior social and economic formation, Khrushchev believed that applying these technologies promised enormous gains in productivity and living standards.

Taking into account doctrines and previous experiences, he and his advisers observed farming methods abroad, which contributed to the goals he established. Adhering to the transfer mentality, policymakers adopted technologies and then offered them to friendly socialist countries and potential partners in the postcolonial world. Rather than a coherent, comprehensive model, the Soviet Union's alternative promoted state-owned demonstration farms designed to spread elements of its approach, but it achieved little influence. Nonetheless, each effort illustrates the dominance of the transfer mentality and, thus, the kindred nature of ideas about development held by the leaders of the United States and the Soviet Union.

This is not to suggest that the Soviet Union participated in the Green Revolution in any capacity. American government programs and nongovernmental organizations promoted the Green Revolution by shepherding technological change and social transformations in Latin America, Asia, and—to a limited degree—Africa. The goals of those projects—rural modernization, industrialization, and strengthening central governments against unrest among a peasant majority—did not apply in the Soviet Union. They had already begun there long before, reaching a crescendo under the distinctively brutal auspices of Stalinism. Instead, concepts governing agrarian policy in the Soviet Union reproduced related technologies. The Green Revolution began with the mission of integrating peasants into national societies, although this gave way to a model of technological change stripped of its social mission and room for local initiative.

The Soviet Union's distinctive version of industrial farming used the same technologies and interacted with similar social and demographic shifts, while its history of war, forced collectivization, and crash industrialization made its path special. Khrushchev's policies and efforts to put them into practice demonstrate that he and his supporters made a comprehensible attempt to create an industrial farming system. For decades, inspiration from foreign practices had been embedded in principle but limited in practice. Khrushchev attempted to apply these ideals, giving the lie to contentions by contemporary critics and subsequent scholars that his programs were destined failure by climatic or technological limitations.

Seeking to apply American technologies to achieve dividends similar to those achieved in the United States, Khrushchev presumed that the Soviet Union could mitigate the negative effects of the American model, which has proven unsustainable and difficult to replicate. While developing industrial farming in the United States, a range of individual, government, and corporate interests expropriated land, labor, and capital to the detriment of agrarian communities, rural cultures, and the environment. Between the 1920s and the 1970s, farmers expanded harvests seemingly without limit, but this turned out to be a one-time miracle.[148]

As experience was to show, specific policies in the Soviet Union suffered from crude design and haphazard implementation. The depletion of rural communities and agricultural capital wrought by war and turmoil in the century's first half had sapped the potential for agrarian reform. American farming, by contrast, benefited from particular social, historical, cultural, and environmental advantages that Soviet Union did not possess. In light of the American model's successes in the 1950s and 1960s, however, Khrushchev's endeavors to adopt industrially farmed corn appear less harebrained than they seemed to contemporary observers. The kolkhozes and sovkhozes did not harvest the higher yields reaped by counterparts in North America, Western Europe, and beyond, even though their practices shared common roots with those systems. Instead, practices in the Soviet Union developed in response to contingent historical conditions, bureaucratic economic organization, and a rural social and labor crisis, all of which limited these methods' potential.

3 CORN POLITICS

In August and September 1953, Iosif Stalin's successors proposed remedies for his failed agricultural policies. Celebrating corn, Nikita Khrushchev pressed subordinates to require farms under their control to plant it. "Some of you," he commented, "are thinking, 'Do you think you've discovered America? We've been planting corn for many years.'" "And what good comes of it?" he questioned, "We get only small harvests." Increasing yields required innovative methods, he asserted: "It's all in the way you plant it." Adopting experts' advice about techniques derived from American models, Khrushchev declared square-cluster planting an inviolable doctrine. The method requires grouping two or three plants in clusters uniformly spaced on axes both parallel and perpendicular to the direction of the tractor. A tractor-drawn cultivator can thus pass between the clusters in both directions, eliminating weeds without the manual labor or herbicides required when corn is planted in rows (Figure 3.1).

Local and farm authorities did not automatically heed Khrushchev's advice. The leader recounted a recent visit to Ukraine's Izmaïl Oblast, where district bosses assured him that they had trained farmworkers in the methods of a Ukrainian record-breaking corngrower. Meeting a kolkhoz chairman, Khrushchev asked, "'Have you ever heard of Mark Ozërnyi?' He had not. . . . Comrades, the lectures were about Mark Ozërnyi from beginning to end. Our dear chairman must have been asleep and, therefore, Mark Ozërnyi went in one ear and out the other."[1]

During the ensuing power struggle, Khrushchev cemented the authority to mandate this and other practices. By 1955, he had amassed enough to dictate the boundaries of agricultural policy. Nonetheless, the napping kolkhoz chairman and numberless similar impediments curtailed Khrushchev's authority over *practice*, meaning that he was unable to ensure that Communist Party and government administrations transformed instructions on paper into corn in the fields. Coupled with the agrarian social and economic crises inherited

FIGURE 3.1 The fields of the kolkhoz "Third Five-Year Plan" in the Volodarsk District of Ukraine's Stalino Oblast (present day Donetsk Oblast) were planted in square-cluster patterns. Set to this task, farmworkers use tractors to cultivate on both parallel and perpendicular axes in April 1954. Unusually, each tractor driver's assistant, who makes fine corrections to the cultivator's course, is female.

Photo by S. Gendelman, courtesy Russian State Archive of Cinema and Photo Documentation, ID#0-238037

from Stalin, these limits explain Khrushchev's inability to realize his vision of a communist social system characterized by rich diets. This provides grounds for disputing the explanations of previous scholars, who held Khrushchev responsible for flaws ingrained into the system.

Totalitarian, conflict, and revisionist interpretations of politics each successively shaped scholars' views of Khrushchev's power and authority. In the first instance, Sovietologists of the 1950s tended to view the party as a disciplined hierarchy penetrating every corner of the country and controlling society in accordance with Moscow's orders. Disagreeing on whether policy or power was the goal of struggle at the top, scholars accepted that its winner led obedient organizations.[2] Arising after Khrushchev defeated his rivals in 1957, the conflict school challenged scholars to comb the press for evidence of hidden factional infighting considered the cause of shifts in policy.[3] It followed that Khrushchev's power—his security in the position of first secretary of the Communist Party—and authority over decisions waxed and waned according to the results of votes on policy, including agricultural policy.[4] Convinced that Khrushchev had secured power in 1957, revisionist political scientists of the 1970s studied his ability to get things done by building coalitions among various interest groups.[5]

In reality, shifts in policy reflected collective responses to changing circumstances rather than realignments of interest groups for or against the status quo. Khrushchev and his compatriots backed reforms adapting to a rapidly evolving society. When their unease about the pace or direction of change grew, decreasing confidence caused them to halt reforms and even reverse those already in motion.[6]

The consensus about agricultural policy always evolved within boundaries set by Khrushchev. Amassing authority in 1953 and 1954, he became secure enough by early 1955 to make his wager on corn without facing any direct challenge. He maintained that authority until 1964, even as comrades grew dissatisfied and the public grumbled.[7]

Down on the farm, however, Khrushchev's authority never amounted to the degree of control scholars often imagined. "In a collectivized society, it is principally government policy that molds the institutional arrangements, the socioeconomic environment and, to a large extent, the technology with which the farmer must work in cooperation with nature to produce food and fibers," one Sovietologist wrote after Khrushchev's fall.[8] This blamed Khrushchev alone for apparent failures. The Soviet Union lacked an equivalent to the Corn Belt, not to mention fertilizers, hybrids, and mechanization, "but this Khrushchev refused to recognize in pushing his crash program to make corn a major Soviet product."[9] In fact, the leader confronted social conditions, economic realities, institutional arrangements, and local officials who haphazardly implemented even judicious initiatives to adopt those very technologies. Khrushchev's constant struggle to transform policy into practice expressed itself in frequent public declarations about policy and seemingly incessant administrative reorganizations, which many took as signs of Kremlin infighting or Khrushchev's erratic nature.

A few scholars suggested that Khrushchev might not be the only one at fault. One economist later conceded that the crisis arose from "a generation of neglect and impoverishment." Yet reform could come only from above using "the methods of Soviet officialdom, created by the Stalin epoch, ensuring that they [followed] a set pattern, regardless of local circumstances, so as to report what Moscow wanted to hear." Consequently, even good policies became campaigns, and campaigns led to "excesses."[10] Nove identified imbalances in investment, failures of planning, and faulty incentives, each important to later chapters. A more complex, convincing explanation emerges, however, from his contemporary analysis of village-prose fiction, in which local officials were not "merely levers [or] connecting rods" in a mechanism. Instead, they responded to "policies insisted on by their superiors, [which] now had a life and logic of their own... [and] represented a major obstacle [that] distorted some of Khrushchev's measures."[11]

Dividing responsibilities, Stalin's former associates both fostered continuity and initiated change. The new leaders kept the kolkhozes, sovkhozes, compulsory government procurements, and taxes. Aware that abuses and underinvestment caused persistent poverty, they designed tax and purchasing reforms to aid rural families and stimulate the agrarian economy. Announced in August 1953, these initiatives emerged after several years when circumstances in Moscow foreshadowed change. Utilizing power, authority, and a cult of personality, Stalin had defined policy. After 1945, age and failing health had necessitated long vacations on the Black Sea, periods permitting members of his coterie a degree of independence within their respective spheres of responsibility. Chastened by the 1949 Leningrad Affair and resulting bloodletting, Stalin's underlings had checked factional struggle. During the leader's absences, they had developed a style of decision-making presaging the collective leadership announced in March 1953.[12]

After concentrating on ideology and propaganda during World War II, the party reasserted itself in administration. It appeared that Khrushchev, when he became paramount Central Committee secretary in March 1953, had envisioned the party as a power base in the struggle against Georgii Malenkov, the head of the government.[13] Party revitalization had actually begun, however, in October 1952, when Stalin summoned the Nineteenth Party Congress, the first since 1939. Re-establishing that essentially defunct institution, Stalin's initiative later helped Khrushchev strengthen the party's influence over economic affairs.[14] The collective leadership regularly met as the Central Committee Presidium, a body Stalin established at the congress as an alternative to the long-atrophied Politburo. Khrushchev insisted on periodic Central Committee plenums bringing top central and regional officials together in Moscow for managed discussions of policy. Underestimating the resurgent party, Malenkov considered the government stronger on the basis of its preeminence during and after the war. In March 1953, he chose to become head of government and cede his position as top Central Committee secretary to Khrushchev.

The power struggle between Malenkov and Khrushchev was personal and institutional, but it also was a matter of policy. The personal conflict dated to 1949, when Malenkov had retained the mandate to oversee agriculture coveted by Khrushchev on his return to Moscow. The subsequent tussle over Khrushchev's agrotown proposal deepened resentment. Afterward, Khrushchev maintained outwardly friendly relations with Malenkov.[15] Yet the new party boss disliked the polished yet jejune bureaucrat, a feeling that only intensified when Malenkov announced the August 1953 agricultural reforms. In the words of Anastas Mikoyan, Khrushchev "did not forget or forgive" Malenkov for past quarrels or for the popularity accrued by attaching his name to the initiatives.[16] Considering himself best suited to define and communicate agrarian policies, Khrushchev made the Central Committee plenum called in September a forum for outlining his own program.

Claiming authority over agriculture, Khrushchev prepared for the plenum address in ways typical of later speeches on agriculture. He outlined preferred solutions to pressing issues, which aides then transformed into plans of action.[17] In August and September, a group including personal aide Andrei Shevchenko and an editor of *Pravda*, Dmitrii Shepilov, commandeered an office on the top floor of the Central Committee headquarters overlooking central Moscow's Old Square.[18] An advocate for corn even before he toured America in 1955, Shevchenko had served as Khrushchev's adjutant on agriculture in Ukraine.[19] The informal meetings testify to an atmosphere characterized by the sharing of meals and of chauffeured automobiles, incomparable to the suspicion and hierarchy prevailing until a few months before.[20] Valuing firsthand knowledge, Khrushchev dragged officials to model farms near Moscow to observe favored technologies and crops, including square-cluster planting of corn.[21] These farms brought in harvests seeming to justify Khrushchev's faith in the crop, making Moscow an example for neighboring oblasts hitherto unfamiliar with it. Khrushchev expected officials dragooned into these day trips to learn about corn and technology as he did, time and effort that he always felt few ever expended.

At the September plenum, Khrushchev staked a claim to leadership. Without naming Malenkov, Khrushchev explained that the October 1952 declaration that the grain supply was sufficient relied on misleading measurements of grain in the field, the so-called biological yield, which made no allowance for unavoidable losses during the harvest. The proposals that followed lacked the single-minded focus of Khrushchev's later speeches in February 1954, which launched the Virgin Lands campaign, and in January 1955, which initiated the corn crusade. In 1953, Khrushchev touted investments in additional machines and in technologies to prevent waste during harvest and storage. But wheat could not feed the livestock needed to hit targets for meat, milk, eggs, wool, and more. For the first time on such a large stage, Khrushchev commended corn because it produced grain and feed at the same time, later a constant refrain.[22]

In 1954, farms in fifty-three administrative regions sowed some corn for the first time. This included areas, such as Kurgan Oblast (Figure 3.2), far to the north and west of the crop's previous range. This initiative expanded plantings to 4.3 million hectares from the postwar low of 3.5 million reached in 1953.

FIGURE 3.2 A. P. Bereznikov, the chair of the Ural kolkhoz in Kurgan Oblast of the RSFSR, inspects the farm's 1958 corn harvest. In 1954, the crop was new to the locale, which is characterized by relatively low annual rainfall, and achieved mixed results. Within a few years, corn plantings improved, as in the case of this field, where it towers above a man on horseback.

Photo by S. Iudin, courtesy Russian State Archive of Cinema and Photo Documentation, ID#0-272108

The Central Committee received reports of increased feed supplies, but most harvests proved unsatisfactory. Officials often did little more than fulfill planting plans, but few were as outspoken as the Uzbek SSR's head of government, Usman Iusupov, who declared that corn was "food for paupers."[23]

In 1954, Khrushchev concentrated on the Virgin Lands, a return to mobilization campaigns familiar from the Stalin era. The Young Communist League organized volunteers and relocated them to sparsely populated lands in Kazakhstan, Western Siberia, and other locales to establish vast sovkhozes using industrial methods to grow and deliver wheat. The initial results strengthened Khrushchev's authority over agriculture, but also made him overconfident in the reliability of the harvest. Promoting this extensive growth, he diverted investments that might have ensured measured intensive growth in settled regions. Then, extensive growth and rapid change became the stuff of political debate. Khrushchev's apparent success boosted his political fortunes by demonstrating visionary leadership and advancing ambitious policies, qualities offered by no rival.

As the first Virgin Lands wheat poured in, Khrushchev scheduled a plenum for January 1955, the fourth meeting on agriculture in eighteen months. Plans to address the crisis in the livestock sector had been developing at least since July 1954.[24] Dictating his own evaluation of the problem and proposing remedies on November 5, Khrushchev established the boundaries of policy discussion: corn was the solution to chronic feed shortfalls. He left room to determine only how far and fast to pursue this course. As an informal memorandum, the monologue, recorded by a stenographer, guided aides developing the proposals Khrushchev expected to come to his desk. Above all, it communicates unabashed enthusiasm: "We will raise corn because it has proven itself to possess boundless potential as feed." "There is a limit," Khrushchev conceded, "but that limit is distant." He outlined priorities including manufacturing harvesters essential to bringing in the feed crop before the frosts and building silos, structures for storing the fodder needed to nourish cattle through the exceptionally long winters.[25]

Before the January plenum, Khrushchev consolidated his authority by orchestrating the removal of Malenkov as head of government, a decision confirmed by the Presidium. When forcing Malenkov to step aside to the post of deputy premier, the other Presidium members cited his advocacy for investment in food processing and consumer goods, a deviation from the orthodoxy prioritizing heavy industry.[26] Determining the plenum's agenda, Khrushchev demonstrated his authority over agricultural policy. Never advocating consumption at any cost, least of all in 1955, he maintained the orthodoxy even as he raised food and farming to a closer second place. When criticizing Malenkov, Khrushchev did not opportunistically conceal his own preferences. The line he considered correct mandated modest increases in investment in farming, augmented by efficient use of existing resources. This required competent managers who could unlock the latent capacity Khrushchev imagined lying idle, a conviction motivating his frequent attacks on the bureaucracies.[27] Increasing kolkhozes' output and income, the reforms promised self-sustaining reinvestment by the kolkhozes themselves, permitting growth without large diversions of state allocations.

Khrushchev considered corn a primary means for activating this latent capacity. In January 1955, he acclaimed it "the decisive requirement" for increasing output of milk, eggs, and meat. Proposing to quickly solve chronic shortages of feed, Khrushchev reiterated his September 1953 statement regarding the importance of corn to American farmers' high productivity.[28] "It is not accidental that corn has become widespread in a host of countries with advanced livestock sectors," he had explained then. "Yet in the Soviet Union, it occupies an extremely small area even in regions where it grows best," Khrushchev had complained, singling out Ukraine.[29] Soon afterward, Khrushchev had similarly lamented that his country planted corn on only 3.6 percent of cropland, far short of the 36 percent in the United States.[30]

The January 1955 plenum directed oblasts and republics where corn had previously been considered unsuitable to plant the crop. It also required southern locales to plant much more, harvesting some as silage instead of grain. Khrushchev endlessly extolled corn as livestock feed, meaning that even southern farms' grain was meant for hogs rather than humans. He praised techniques for managing moisture to overcome dry conditions and advanced hybrids capable of maturing sufficiently in cool climates. These could push the boundaries of corngrowing north and east. Conceding that climate imposed some limits, Khrushchev expected northern areas to harvest not grain, but corn in earlier stages.[31] Listening to some experts, he considered the milky-wax stage, characterized by mature but not-yet dried cobs, to be equal to grain in nutritional value.[32] That meant that the whole plant was suitable for harvesting, chopping, and preserving as slightly fermented silage, a food for cattle and sheep, albeit one poor in protein. Khrushchev's preferences informed the choice by the delegation to the United States that year to visit not only Iowa, but also Minnesota and Canada, where conditions resembled those of the Soviet Union's northern agricultural regions.[33] In fact, a greater reliance on corn silage mirrored long-term trends in the practices of American and Canadian farmers in these climates during the period.[34]

Attempting to persuade a skeptical audience, Khrushchev dismissed the old northern boundary for harvesting corn as grain, which ran from Western Ukraine to the republic's east and thence southeast into the North Caucasus. He agitated for requiring farms hundreds of kilometers to the north to plant corn that very spring, in imitation of the Stalin kolkhoz in Chuvash ASSR, located along the upper Volga River. In 1954, the farm had harvested silage *and* grain from 35 hectares of the previously unknown crop. "We have tested corn in various regions of the Soviet Union," Khrushchev enthused, "Everywhere, even in northern regions, good yields have resulted wherever it has received proper care."[35] Demanding an increase from the 4.3 million hectares of 1954, he envisioned 28 million by 1960, a proportion of crops approaching 30 percent. In off-the-cuff remarks concluding the plenum, Khrushchev jettisoned earlier caution by declaring that corn could grow almost anywhere. "Now we can not only compete but, strictly speaking, can overtake America because the potential for growing corn now truly extends over the whole territory of the Soviet Union," he boasted.[36] That spring, he even suggested growing corn in Iakut ASSR, a vast, chilly expanse in the northeast.[37] Instigating only timid experiments there, such wild assertions fueled the belief that he expected corn to thrive even beyond the Arctic Circle, evidence for his program's essential hopelessness.[38]

To guide local officials, party procedures spread Khrushchev's message while simultaneously the press publicized it. The party distributed transcripts of plenum sessions to inform meetings designed to convey orders to subordinate committees in each republic, oblast, krai, and district. At least in the theory, this defined their subsequent actions. Important speeches and unsigned editorials featured in *Pravda* and *Selskoe khoziaistvo* [*Agriculture*], the daily published by the USSR Ministry of Agriculture.[39] Publishing this speech, *Pravda* added a headline with a typically martial tone: "The directives of the Central Committee plenum are a plan for battlefield action." Subsequently, newspapers publicized responses under headlines announcing "the battle to fulfill them."[40]

Before the first seed was planted, Khrushchev lamented that bureaucrats had substituted routine for action. Speaking to advisers in early March 1955, he disclosed mounting concerns about practices inherited from Stalin. Radio, film, and print messages about corn had increased in number, but conveyed the wrong message. Singling out the newspapers responsible for setting the tone, Khrushchev decried their failure to convey the policy's essence. As a result, local officials prattled about corn's importance while still considering it a grain, rather than livestock feed. "Why?" Khrushchev asked, "Because they have been taught that way for decades." Long experience under Stalin had conditioned the party and government machinery to respond to any initiative as a temporary campaign. Officials concentrated on the moment's priority with confidence that attention was soon to move on to the next obsession, allowing them to return to business as usual. Fearing this, Khrushchev demanded that aides take to the press to ensure that everyone understood that emphasis on corn would endure for longer than a single growing season.[41] Without naming the offending article, Khrushchev complained that the editor of *Pravda*, Shepilov, had "printed an editorial, [reproducing] text word for word from the plenum [resolutions] in bold type, without explanation," a description fitting that of March 1, 1955.[42] "We must make this matter plain," Khrushchev continued. "What is written in the plenum directives is a point of departure, but we must clarify it so that everyone comprehends what it will look like in action."[43]

A survey of *Selskoe khoziaistvo* from late January until planting in May indicates that the leader's criticism achieved some effect. In February and early March, the newspapers used boilerplate language exhorting readers to participate in "expansive socialist competition for high yields of corn." After Khrushchev's criticism, content began to detail specific practices.[44]

Lecturing his agricultural aides-de-camp, Khrushchev made his case that corn and essential technologies provided the means for raising living standards to meet rising popular expectations. It was at this moment that he assured them that he "believed only in [corn]" and associated innovative methods.[45] Clothing, housing, and food embodied the promise of socialism in Khrushchev's utilitarian conception, which was disparaged by critics at home and abroad.[46] He described popular living standards as an issue that the party could not address simply by pointing to injustices and contradictions inherent in capitalism. Instead, the authorities had to answer workers' questions about material matters. " 'I believe in you,' " Khrushchev's imagined citizen affirmed. " 'I fought for [socialism] during the Civil War, fought against the Germans, [and] defeated fascism. . . . But tell me: Will there be meat? Will there be milk? Will there be quality pairs of pants?' " Failing to provide necessities, socialism would prove its hard-won achievements empty. "Of course this is not [a matter of] ideology. One cannot have the correct ideology but go about without pants," Khrushchev quipped, drawing laughter and applause.[47] The same year, he allowed that pursuing abundance without corn would only cost precious time and resources. "Communism is not something pie-in-the-sky," he explained. "We are not priests who say that earthly paradise is temporary, while heaven is eternal but must be earned by suffering here on earth. . . . We can [ensure high living standards] faster, easier, and cheaper with corn, if only we can learn how to grow it."[48]

Alongside these promises of material riches, Khrushchev articulated a vision of reformed individuals and society, idealism that culminated in the Third Party Program announcing the "full-scale construction of communism." Alternatively dismissed as a simpleton concerned only with butter or as a utopian dreamer, Khrushchev considered butter integral to utopia, and believed that the Soviet Union was soon to have both.

In 1955, the corn crusade yielded mixed results. Khrushchev's authority over policy meant that plantings officially rose fourfold, surpassing 17 million hectares. Its limits were demonstrated by the poor harvests brought in by most districts and kolkhozes. The time separating January directives from planting in April and May was too short to familiarize managers and farmworkers with necessary methods, to deliver high-quality seeds, to secure supplies of fertilizer, and to produce necessary machines. Under compulsion, the sovkhozes and kolkhozes planted corn anyway, turning a potentially good idea into a fiasco.[49]

Problems emerged even before the January plenum ended. Khrushchev warned party bosses against simply giving orders and expecting results. Instead, he explained, they faced the demanding task of "ingraining the importance of corn plantings into the consciousness of each farmworker." Singling out Zinnat Muratov, party chief in the Tatar ASSR, Khrushchev lambasted perfunctory nods to rational planning. Earlier in the plenum, Khrushchev had questioned whether Muratov's plan for 40,000 hectares of corn could meet feed requirements. Having conferred with advisers, Muratov had instead proposed 200,000 hectares, a figure that he expected to satisfy Khrushchev. Instead, the leader denounced Muratov for not calculating the planned size of the republic's herds, concomitant feed requirements, possible corn yields, or the hectarage required to meet the need. Khrushchev fumed that Muratov had plucked a figure from thin air, a timeworn practice incommensurate with scientific management of agriculture. Arbitrary quotas led to "foolishness" and outlandish excuses for failure to produce more meat and milk.[50] Finishing with Muratov, Khrushchev thundered that party leaders "must not be armchair administrators, but must describe [the corn program] to the kolkhozniks so that they understand." Khrushchev bemoaned the ineptitude of local party officials, who endured low status, abysmal pay, and primitive living conditions. He decried those who wasted time on "conversations without content" and "generalized slogans," which neither instilled practical knowledge nor ensured that everyone understood.[51]

Khrushchev's crusade began inauspiciously. That spring, local party committees agitated for corn. As Khrushchev feared, they did little to educate the specialists and farmworkers tasked with the job. Farms lacked the equipment to plant, cultivate, and harvest the corn, as well as the labor power to remedy the shortfall. Consequently, a substantial percentage of the corn they planted yielded little, making it inferior to oats, barley, and hay. Even in Moscow Oblast, officials had to recognize that the year's corngrowing had gone poorly. One reported, "gross violations of the agronomic methods for planting corn," as farms planted low-quality seeds late in poorly tilled and unfertilized soil.[52]

As Khrushchev often complained, entrenched practices made excessive haste endemic. An April 1955 reform rationalized and decentralized authority by granting kolkhozes the right to plan both annual production and long-term development. Gone—in theory—were quotas for dozens of products assigned by republic ministries of agriculture, which had dictated minute details of daily operation one growing season at a time since the 1930s. This new planning procedure issued a kolkhoz purchase orders, leaving the production methods, herd sizes, and crop rotations at the discretion of chairpersons, specialists, and the kolkhozniks, all subject to supervision by the local soviet's executive committee and the party.[53] But only so long as everyone pledged to grow more corn. Aware of its priority, local officials contravened the new procedure because they thought only in terms of single growing seasons. Inheriting this campaigning mentality, Khrushchev nurtured it even as he decried the resulting chaos. Typically, officials whose domains exceeded the plan for planting corn or plowing new lands gained awards and promotions on the basis of these measurements of process rather than of those of output, let alone of efficiency or thrift. Any who acted cautiously risked demotion, even though arrest—always possible under Stalin—had become unlikely.

Oblasts and republics strove to exceed expectations, but failed to reap harvests matching Khrushchev's ambitious goals. Primarily concerned with appearing to plant corn, officials in many administrative regions reported areas ten times those of 1954. Marked by cool, humid climates unsuitable for corn, Leningrad Oblast and a grouping of neighboring northwestern oblasts increased plantings from 18,900 hectares—already a record—to 272,000.[54] Responding to the campaign, officials likely overstated the true quantity by reporting that farms had planted corn in particular fields where another crop actually grew. In Ukraine's Odesa Oblast, one kolkhoz reported that it had replanted fields of wheat and barley killed by a hard winter with corn, a common practice. It then added the corn a second time in a separate count of regular spring plantings.[55] In later years, such ruses came to light more often, but they apparently had occurred from the start.

Other locales boldly reported disaster. Officially, 10 percent of plantings in the RSFSR failed to germinate, with locally higher ratios recorded.[56] Across the Soviet Union, 6.1 million of the 17.9 million hectares were planted in regions where corn had never grown on a production scale. Those regions accounted for 958,000 hectares of the 1.35 million that failed.[57] Perhaps authorities quietly protested in hopes of convincing Moscow that corn was unsuitable to local conditions. Some northern oblasts directly appealed to avoid future assignments. Straddling the upper Volga River north of Moscow, Vologda Oblast reported that farms had harvested feed from a mere 5,700 of 33,000 hectares, a failure rate exceeding 80 percent.[58] Regardless, it continued to receive orders to plant corn. In Murmansk Oblast farther to the north, however, fifty-three tiny test plots failed, causing Moscow to cease its demands.[59]

Conversely, labyrinthine record keeping procedures made underreporting failure rates simple if authorities chose that strategy. Farms planted corn in the worst fields, conserving the best soil and scarce labor for other crops. Such plantings yielded little fodder because the corn became overgrown with weeds. Managers then reported that they had chopped the corn and fed it to livestock, or turned animals loose in the field to eat their fill. In Riazan Oblast, a local inspector found that a party secretary "reported fraudulently" to the oblast committee that his district's kolkhozes had chopped substantial yields of corn and fed it to cattle. Moreover, the secretary had coerced the kolkhozes' chairpersons to report the same.[60] Similarly, farm accountants quietly rewrote plans over the course of the year. Transferring plantings from the grain category to silage or from silage to green fodder, they concealed in the statistics small harvests of corn that had not matured.[61]

Top leaders received analyses judging corn a partial success. Isolated achievements contrasted with the lackluster results of the majority. The conclusion that those corn plantings actually yielding a harvest had surpassed alternative crops by between two and four times encouraged further pursuit of the policy.[62] Officials from Khrushchev on down asserted that corn yielded a good harvest where it received proper attention. They typically refuted claims that corn could not grow with this tautology, and then redoubled their demands. Blaming managers for "the crudest violations of agronomic techniques," they then compared the farm, district, or oblast to a neighboring one harvesting higher yields, which they portrayed—often tenuously—as having the same resources.[63] Thus they heralded Krasnodar Krai for raising milk output by 37 percent in 1955. Silage allowed production to remain high into the winter months when it typically dipped because of poor feed supplies.[64] Addressing the Twentieth Party Congress in February 1956, Khrushchev recapped the first year of the crusade. Corn had provided feed supplies boosting milk production by increasing each cow's output, in some cases several times. Faintly praising a few locales, Khrushchev held leaders of the majority responsible for their poor harvests. "In a considerable number of districts, corn did not produce satisfactory results," he disclosed. "The only reason is the careless attitude of leaders."[65]

By fulsomely praising the successes, the press irked even Khrushchev. In October 1955, he privately expressed anger. "Right now, they are writing in the newspapers that corn turned out well everywhere," he complained. "Yet in places where it did not, [people] think that the Soviet leaders clearly know nothing and are lying to them, and to themselves."[66] Although he did not single out any publication, many bullish accounts had appeared in *Selskoe khoziaistvo*, such as a full page of stories under the headline "We saw corn through to the end!"[67] Wary of popular incredulousness, Khrushchev demanded that the government and the press "tell the truth about where things are good, where they are bad, and why. Then people will understand and adapt their attitudes [toward corn]. It will be bad if we say that it is great everywhere."[68]

In the following days, *Selskoe khoziaistvo* ran critical stories denouncing fraud and waste. One related that an entire district in Orël Oblast had declared the harvest complete, even though the kolkhozes had harvested only 10 percent of the weight of silage they planned. "Judging from the reports, the whole crop has been brought in," it explained. "In reality, substantial areas have yet to be reaped, and are yellowing and drying" in the fields, losing nutrients by the day.[69] Similar articles condemned excessively positive reports reaching Moscow from Stavropol

Krai, Kustanai Oblast in the Kazakh SSR, Kherson Oblast in Ukraine, and Frunze Oblast in the Kyrgyz SSR.[70]

No evidence exists of widespread or vocal protest at any level in 1955. Even though the Presidium was hardly under Khrushchev's thumb, no member spoke out.[71] After surviving a June 1957 attempted palace coup, Khrushchev publicly castigated Viacheslav Molotov in particular for resisting the Virgin Lands campaign.[72] In his memoir, Shepilov recalled similar accusations leveled against him by Khrushchev's backers. Yet no one accused Molotov, Shepilov, or anyone else of objecting to corn. Shepilov later claimed that he had considered each project wrongheaded, even though he had dissented against neither at the time. He wrote, "Like my generation of communists," born around 1905 and educated in politics under Stalin, "I had been schooled in the spirit of utter loyalty to the party and the strictest discipline; to express doubts about the party's directives would have been sacrilege."[73] Despising Khrushchev, Shepilov was justifying his own inaction, but also illustrated the expected regimentation.

That regimentation notwithstanding, officials did not execute the corngrowing policy with enthusiasm or effectiveness. On occasion, they registered mild, unattributable discontent. After the January 1955 plenum, procedure required each administrative region's party committee to meet to propagate the new initiatives. The Central Committee's envoy to the gathering in Riga described how a few local bosses spoke out anonymously. Those on the podium mentioned corn only "timidly."[74] Latvian leaders reported an uncontroversial version of events, noting only that "certain communists . . . expressed doubts about the possibility of fulfilling the corn planting plan."[75] Moscow's envoy judged that even the republic's agricultural officials, including the minister and later historian Aleksandr Nikonov, had substituted mere phrases, "which you cannot feed to livestock," for a specific program of action. When district officials gathered, voices from the audience questioned: "How will we plant?" "How will we harvest?" An unsigned note addressed to republic party secretary Jānis Kalnbērziņs complained that in 1954 Moscow had insisted on corn, which had failed. To repeat the experiment on a larger scale in 1955 demonstrated "foolishness."[76] Years later, Nikonov claimed to have spoken out against demands to plant 200,000 hectares of corn, citing the lack of machines, seeds, know-how, and suitable climate. This protest only met with charges that he was an "'opportunist,' 'oppositionist,' and 'searcher for the easy way out.' Later, 'antiparty element,' among others, was added."[77] Although not signaling the mortal danger they had under Stalin, such labels seriously threatened the career of anyone showing insufficient enthusiasm for the moment's campaign.

The 1955 planning reform purported to protect Latvia from such arbitrary figures. Despite Khrushchev's warning to Muratov against indiscriminate, one-size-fits-all plans, they were common. Repeating the point in March 1955, the First Secretary cautioned, "I am in no way proposing to impose on kolkhozes and sovkhozes a designated percentage of corn." Doing so only to create the appearance of following directives prevented rational analysis of the needs of farms, districts, or oblasts.[78]

Khrushchev had plenty of reason to fear just that effect. A sovkhoz in northerly Komi ASSR received orders in 1955 to plant many hectares of corn, which failed. The next year, new orders forced new failed plantings, a cycle repeated in subsequent years.[79] Unlike sovkhozes, which were operated and financed by the state, kolkhozes nominally possessed the authority to regulate themselves after the 1955 reform. In theory, the district soviet's executive committee oversaw them by formally approving plans, aggregating those for all farms in the district, and passing them along to its oblast or republic superior, which followed a similar procedure. Actual practice little resembled this arrangement, creating conflicts between kolkhozes and district bosses typically won by higher authorities.[80]

Inspectors from Moscow discovered and settled one such conflict in Omsk Oblast. In early 1956, they fanned out to verify the results of corn cultivation in 1955 and preparations for the coming season. In the process, one adjudicated a dispute between the Khrushchev kolkhoz and the Isilkul District soviet and party committee. On April 2, he attended a meeting of the kolkhoz administrative committee, whose members described repeatedly submitting planting plans to district bosses, who on three occasions altered them to include *fewer* hectares of corn. They justified this action by presuming high average yields unlikely in the locale's cool Siberian climate. The kolkhoz resolved to again argue its case before the local authorities while implementing its own plan, which envisioned planting more corn to ensure a harvest sufficient to feed its livestock.[81] The inspector's files do not record the local authorities' response to his report, which favored the kolkhoz. However, local officials usually had to comply with recommendations by Moscow's inspectors. Rather than a typical result of such conflicts, this instance was an exception because the inspector interceded to resolve the quarrel in favor of the kolkhoz, typically the weaker entity. In the unknown number of similar struggles not subject to such intervention, victory went instead to the local bosses.

That 1956 inspection found that administrative regions' aggregate plans called for fewer plantings of corn than in 1955, evidence that local officials initially interpreted Khrushchev's crusade as a campaign lasting only a single season. Leaders in Kurgan Oblast proposed that sovkhozes plant 2,300 fewer hectares. Declaring that the bosses had failed "to properly learn lessons from last year's failures," the inspector "corrected [the] mistake" by revising the plan upward to 70,000 hectares, 27.3 percent more than the 1955 total.[82] In Estonia, republic officials responded to a 1955 harvest reaching less than 33 percent of the target by ordering a cut to 1956 corn plantings, a move reversed by inspectors.[83] Discovering a similar plan in Latvia, inspectors criticized Nikonov and his ministry for having "not yet become an operational aide in managing the republic's MTSs and kolkhozes."[84]

In effect, these practices robbed kolkhoz authorities of the power to plan, substantiating Latvian officials' charges that pressure from above to plant corn countered Moscow's program to foster kolkhozes' initiative.[85] Protest was futile, giving officials little reason to speak out more often or vocally. Kolkhozes and other local authorities did not possess actual authority to design crop rotations in response to government procurement orders and their own needs. Instead, higher authorities imposed quotas of corn on them without regard for specific conditions, despite the fact Khrushchev had warned against doing so.

In 1956, Khrushchev pressed on despite initial setbacks. The cumulative effect of his dissatisfaction with the two seasons led him to reconsider. Kolkhozes and sovkhozes did much of the weeding and harvesting by hand, making the crop labor-intensive and therefore very expensive. In 1957, the total declined from 23.9 million hectares to 18.3 million. In 1958, the area rebounded only to 18.7 million. In a statement of the diminished pressure characteristic of these years, Khrushchev admitted that, although corn remained a priority, it surpassed established crops only if each hectare yielded 25 tons of feed. This was not permission to abandon corn:

> Some might ask, "What's this, you're sounding the retreat? After constantly agitating for corn, now you say this on the matter?" No, comrades, this is not a retreat. I consider corn the queen of the fields. No crop can compare to corn, but because it is a queen, it requires appropriate respect and care.[86]

Allowing for tactical maneuvering, Khrushchev demonstrated that *his* enthusiasm had not flagged. At each stage, Khrushchev used authority secured by demoting his rival, Malenkov, who had represented an alternative interpretation of the consensus on agricultural modernization.[87]

The confrontation that fully entrenched Khrushchev in power resulted in part from differences about corn and consumption. In June 1957, his rivals in the Presidium gathered a majority and made an unanticipated motion to demote him. Protesting, he invoked long-disused rules requiring that the body that had originally confirmed an officeholder, in this case the full Central Committee, ratify any dismissal. Notably, he could count on its members' loyalty because he had appointed many of them to their positions of power.[88] The following day's emergency Central Committee plenum overruled the anti-Khrushchev coterie and expelled Malenkov, Molotov, Lazar Kaganovich, and "Shepilov who joined them" from the Presidium. Driven by common concerns about the pace and direction of efforts to grapple with Stalin's legacy, Khrushchev's rivals had attacked on two fronts related to domestic politics. First, they charged that the collective leadership had degenerated due to his behavior in the Presidium, suggesting that the cult of Stalin had given way to a cult of Khrushchev. Accordingly, his rivals criticized as mere showmanship treks around the country to inspect farms, factories, and other sites.[89]

Second, they attacked Khrushchev's claims that the Soviet Union was competing with the United States to produce consumer goods and food. Scholars have often considered his slogan "to catch up to and surpass America" a vital part of the charges.[90] In May 1957, Khrushchev famously boasted that the Soviet Union was doing so in per-capita output of meat, butter, and milk. Rejecting experts' estimate that this might occur by 1975, he buoyantly named a much nearer date. Redacted published versions of the text repeat a common phrase, "in the next few years," but transcripts of the radio broadcast record that he had specified 1960 or, in the worst case, 1961.[91] In the tradition of Stalin-era exhortations to achieve the impossible, Khrushchev intended this as motivation rather than a statement of fact. He appealed for a population yearning for decent living standards to make demands. Simultaneously stoking elite fears of popular discontent, he presented his reforms as the solution to the problem.[92]

In reality, Khrushchev's rivals attacked not the timeframe, but the slogan itself, which they claimed betrayed an unorthodox emphasis on agriculture and consumption.[93] In 1955, Khrushchev had pummeled Malenkov for proposing to reorient investment from heavy industry to consumer goods industries. Yet no one had challenged Khrushchev's basic claim about competing with America when, as early as February 1955, he had stated that the Soviet Union had "entered into a competition with the richest capitalist country in the world, the United States of America." "We must work hard," he continued, "to overtake that country in output of food per capita."[94] A few weeks later, he had elaborated, "In the competition with America, comrades, there is no doubt victory will be ours. This is because our economy, based on the teachings of Marx and Lenin, develops without the bourgeoisie, the landowners, or the exploitation of man by man."[95] In these uncomplicated terms, Khrushchev envisioned socialism and the communist future as systems of material abundance achieved in the absence of private capital.

Having retained power, Khrushchev faced little opposition to his authority over policy. Long backed by the Central Committee, he now packed its Presidium with supporters. Beginning in 1958, they rubber-stamped major structural reforms in rural governance. Benefiting from the post-1953 price reforms, the strongest amalgamated kolkhozes suffered from interference by the machine-tractor station (MTS). Relieved of the requirement to pay the MTS dearly for services, economically vibrant kolkhozes were to purchase machinery from the government and strike out on their own. Abolishing one of Stalin's primary controls over the countryside and agricultural produce, Khrushchev proved willing to initiate substantive reforms.[96]

The reform did not proceed as planned, an instructive failure of practice to match policy. Khrushchev proposed that strong kolkhozes should proceed first with equipment purchases made using loans subject to repayment over several years. Employing authoritarian methods, party and state officials moved with unwarranted speed to make the reform comprehensive. Then they demanded that all kolkhozes, including those too economically weak to have participated in the first place, repay the loans early at the expense of all other financial concerns. Instead of proof of the haphazard nature of his reforms, this episode evidences the difficulty Khrushchev faced in carrying out even judicious ones.[97]

The corn crusade continued, albeit fitfully. In 1958, a countrywide survey examined how farms organized labor, adopted innovations, harvested the crop, and measured yields of livestock feed. Inspectors in Krasnoiarsk Krai, the region encompassing the vast basin of Siberia's Enisei River, recorded that plantings had decreased from more than 250,000 hectares in 1956 to 195,000 in 1958. Individual farms brought in large harvests, but the majority fell short of even minimal expectations. Attributing increasing meat and milk output to vanguard farms that followed prescribed procedures, the inspectors concluded that, "even given the [climatic] conditions in the krai, observing correct methods for cultivating corn can ensure a high yield." Following the rules, one kolkhoz brought in the largest harvest in the krai, albeit one only 20 percent higher than the 25 tons considered minimally economical. Using prohibited practices, the other farms in Minusinsk District harvested as little as 10 percent of the maximum. Averages for the krai had ranged from 3.3 to 5.4 tons per hectare. Even these figures were suspect because farms imprecisely calculated or even estimated the weight. They harvested at the end of July and in early August, when the corn was still immature and low in nutrients, because later in August they devoted machines and labor to reaping wheat and barley. Inspectors found that farms furthermore chose unfertile land, applied no fertilizer, and diverted the farmworkers assigned to cultivating it to other work. These widespread practices ensured that corn drew on fewer nutrients, competed with more weeds, and yielded little feed. As usual, inspectors blamed local officials who "still do not devote the necessary attention to this valuable feed crop." In their own defense, the culprits cited shortages of machinery, but inspectors concluded that these were diverted to other tasks.[98]

Located in Krasnoiarsk Krai's south, Minusinsk District has a climate milder than that of neighbors to the north, one resembling that of locations on a similar latitude in nearby Altai Krai, where farms achieved visible success in 1958 (Figure 3.3). Citing the cold as proof of the corn program's foolishness, contemporaries and subsequent scholars have privileged outliers accounting for a small fraction of the whole. Predictably, the 1958 inspection found that farms further north in Krasnoiarsk Krai did not harvest the corn they planted. The last frost came on June

FIGURE 3.3 Emelian Emelianenko, director of the Kulundinskii sovkhoz in Altai Krai, RSFSR, inspects the farm's corn in this 1958. The fully mature corn stands well above the height of a man perched on the hood of a passenger car, a "Pobeda" (Victory) produced by the Gorkii Automobile Plant.

Photo by A. Agapov, courtesy Russian State Archive of Cinema and Photo Documentation, ID#1-36294

5 and the first on August 16, leaving only 32 days between with warmth sufficient to sustain growth, far short of the approximately 100 required. However, farms there planted at most 30 or 40 hectares.[99] These amounted to a small percentage of corn plantings in the krai, and an infinitesimal proportion of the country's total, which was composed primarily of hundreds of thousands of hectares in Krasnodar Krai, Stavropol Krai, Ukraine, Moldova, and other southern locales.

In the north, corn was bound to struggle, but the country's average yields remained low because of southern and central regions. In 1959, farms planted 22.4 million hectares of corn, 11 percent of a crop area totaling 196.3 million hectares. Estimates vary, but between 2.81 and 5.7 million hectares of corn were in unsuitable regions, the higher figure amounting to only 12.5 percent of all corn plantings. Corn furthermore comprised less than 6 percent of more than 100 million hectares of crops planted in northerly regions.[100] These northern margins accounted for a modest proportion of overall yields. In fact, plantings in southern locales warm enough but too dry to reliably yield grain and silage without irrigation contributed more to the low aggregate yields. In 1960, harvests of corn grain reached 1.93 tons per hectare. Yields for silage and fodder were only 13.2 tons per hectare, barely 50 percent of the target of 25 tons. Locally, the figure ranged as high 29.4 tons in the Baltic republics. Areas with larger plantings but lower yields included Ukraine (14.7 tons) and central Russia's black-earth region (16.9 tons).[101] The causes were twofold: in areas with an appropriate climate, many farms harvested before the grain had fully matured. In many locales where they might have waited for high-quality silage, they actually reaped watery green fodder devoid of nutrients.

In central and southern regions, inspectors uncovered the same transgressions as in Krasnoiarsk Krai. In Briansk Oblast, local party and government officials had issued directives that were ineffective because, inspectors concluded, they "were not accompanied by organized effort to train tractor operators, kolkhozniks, and sovkhoz workers" in square-cluster planting and related practices.[102] Kolkhozes used the land irrationally, did not know how to plant corn correctly, organized work inefficiently, and calculated yields inaccurately, complaints echoed in reports from neighboring Orël Oblast and Penza Oblast.[103] Instead of investing scarce resources into a crop they mistrusted, farm managers favored crops yielding modest harvests but requiring little labor or machinery. In Novozybkov District of Briansk Oblast, farms quietly eschewed corn in favor of lupine, which they planted easily in poor soil and left unattended until a protein-rich harvest was ready.[104]

Even in lands where the climate suited corn and farmworkers possessed requisite knowledge, inspectors found mixed results. The same practices that doomed corn in cold Krasnoiarsk caused farms in warm Moldova to harvest only 1.6 tons of grain per hectare.[105] Milk output in the latter, however, had risen each year of Khrushchev's crusade, hinting at additional untapped potential. Inspectors concluded the same about Ukraine, as well as about Kyrgyzstan, where low yields resulted from serious substantive failings capped by "gross violations of agronomic procedures."[106]

Given skepticism about official statistics warranted by documented distortions, we must consider yield figures with caution. Nonetheless, they show that even inspections—an alternative to self-congratulatory reports passed through government and party channels—supported the conviction that corn was boosting output of meat and milk. As readers of the 1958 inspection's final report learned, "the achievements in agriculture secured by kolkhozes and sovkhozes might have been much larger had they observed the necessary package of agronomic methods when cultivating corn."[107] Far from plunging ahead heedless of an obvious disaster, Khrushchev was staying a course that seemed to be working, even if subordinates failed to make the most of corn. Moreover, by 1959, farms increasingly began to transform industrial farming principles into actual practices in the form of advanced hybrid seeds, specialized machines, synthetic fertilizers, and more.

Khrushchev's relative caution gave way as the corn crusade entered this new phase. At a December 1958 Central Committee plenum, he surveyed the achievements of the past five seasons and looked forward to the unprecedented Seven-Year Plan slated for approval weeks later at the Twenty-first Party Congress. Instead of accepting weeding and harvesting corn by hand to bring in the low yields, he relaunched the crusade as one of tractors, harvesters, synthetic fertilizers, and pesticides promising more corn at less cost. "It is no exaggeration to say that it has become possible for milk output to increase and completely satisfy demand," he assured, "owing to corn."[108]

Khrushchev again blamed dashed expectations on "bureaucratic" and "negligent" local leaders who made unacceptable recommendations to farms. He ridiculed an official publication from Latvia for advising planting not two plants in a cluster, but five or six. Competition between so many ensured that they crowded each other out, producing less feed than just two. Local bosses shrugged this off, claiming that the climate defeated corn. "But how could it grow when planted in such a way?" Khrushchev indignantly queried.[109]

The published text of the speech omits the subsequent telling exchange with the Latvian party chief. "I looked over at Comrade Kalnberzin," Khrushchev explained, "and he looked straight to the floor. We are friends: our gazes should meet, but [our points of view] have, so to speak, parted ways on the matter of corn." When Kalnbērziņs admitted shame at his republic's failings, Khrushchev continued to badger him. "We cannot feed people shame," the leader thundered,

"Let's have corn instead, because there is shame today and shame tomorrow, but that provides neither meat nor milk."[110]

Returning to the prepared text, Khrushchev denounced officials who, cloistered in their offices, only demanded reports providing superficial knowledge about the situation but not about the actual state of the fields. That was not real leadership.[111] Worse still, those not sharing Khrushchev's vision continued quiet opposition. The head of the Bashkir ASSR party, Zia Nuriev, explained, "There are no longer any skeptics among us. Now we all understand the importance of corn for the livestock sector." Khrushchev interrupted, "They're still there, Comrade Nuriev, . . . it's just that they've started to talk about it less." When Nuriev soon admitted that local yields remained low, Khrushchev countered, "Well, there's your skepticism. . . . Even if there are no skeptics, there's still no corn." He then scolded Nuriev for confusing the names and characteristics of the hybrids planted by farms in his republic.[112]

Condemning any sign of uncertainty, Khrushchev nonetheless sounded a cautious note. "I stand for solving the meat shortage by markedly increasing corn plantings," he declared, "I say markedly, but don't rush to increase them precipitously. Increase them only in locales that have mastered the crop." The answer was not to be found in generalized slogans. Instead, intelligent use of land and capital promised to ensure that farms improved on the past year, when less than 40 percent of the 19.7 million hectares of corn produced grain or silage, a contrast to more than 95 percent in America. In the Soviet Union, farms harvested the majority as nutrient-poor immature corn. This "harmful" practice resulted from "simple stupidity," Khrushchev charged. The almost valueless "so-called green fodder" was actually corn mixed with the weeds that had overgrown the fields.[113]

The Seven-Year Plan prioritized machines and related investments. Boasting of recent annual growth outpacing that of capitalist competitors, Khrushchev pledged that new policies would maintain rates of at least 8 percent. He promised new tractors, harvesters, and other implements, even though state investments were projected to fall below the 100 billion rubles of the previous five years.[114] This was neither irrational nor blind to economics. Khrushchev maintained that kolkhozes had the capacity to compensate for declining state investments with outlays from their income, which was rising thanks to procurement and tax reforms. In 1958, annual state investments had risen to 266 percent of the 1953 figure. In 1959, they decreased, only increasing to offset insufficient contributions by the kolkhozes in 1961.[115]

The plans required increasing per-hectare yields of crops and per-animal productivity in the livestock sector. Not coincidentally, Khrushchev envisioned that effective corngrowing would facilitate both. For many years, government purchase prices had not met production costs for commodities such as eggs and beef, a conundrum Khrushchev hoped to solve with higher output per unit of labor and capital. Decreased production costs offered a better solution than another rise in government purchase prices, which would necessitate unwelcome increases in either subsidies or consumer prices. In 1959, plan documents stipulated building new irrigation systems and applying "scientifically sound land management" to determine which crop rotations and technologies suited each climatic region. In practice, farms utilized irrigation poorly. The new land-management doctrines provided cover for Khrushchev's subsequent offensive to replace low-intensity hay and pastures with capital-intensive row crops, especially corn. To facilitate this, the plan called for ending chronic shortages of synthetic fertilizers by increasing annual output threefold to 31 million tons by 1965. It mandated manufacture of more than 1 million tractors, 400,000 harvesters, electric milking machines, and other labor-saving devices designed to ease burdensome everyday tasks. Corn found its way into this most sanctified of documents as the basis for ambitious targets for harvests of the feed necessary to provide the meat and dairy products needed to meet rising demand.[116]

These programs responded to evident needs. The fertilizer plan, for example, was an advance over previously inadequate investments. The abandoned Sixth Five-Year Plan (1956–1960) had called for a 20 percent increase in output but, by 1958, regions' desperate pleas had made clear its inadequacy. The resulting plans formed part of a larger program to develop a modern chemical industry.[117] Farms poorly used what fertilizer they had, be it synthetic or organic, applying in some cases less than 20 percent of the figures named in their plans.[118]

No one could mistake the link between investments, corn, and output goals. In December 1958, Khrushchev had praised Aleksandr Gitalov, the tractor driver who traveled to Iowa to work on Roswell Garst's farm. Gitalov returned with knowledge about American machines and practices, which Khrushchev ordered him to share with engineers, expecting improved designs able to save time, labor, and money.[119] A few months later, the leader followed up by instructing aides to concentrate on machines that could build on achievements in producing hybrid corn seed using American models, genetic stock, and equipment. "The American example has already proven how to use composite mechanization: that is, planting, cultivating, and harvesting carried out only by machine operators," he explained. Khrushchev demanded more machines, their efficient use, and farm managers competent to guide the work. The farms had to replace, he concluded, "incapable and bureaucratized people."[120]

His orders got results. Thousands of farms jettisoned the old model in which work teams of dozens weeded and harvested manually 1 or 2 hectares of corn per person. Over the following years, groups of two, four, or six machine operators took responsibility for plots of corn measured in the hundreds of hectares. In early 1959, for instance, Belgorod Oblast authorities reported that 1,240 work teams comprising 2,830 members had pledged to cultivate 103,000 hectares, an average of 36.4 hectares per person.[121]

Investments in machines, fertilizers, and other innovations yielded dividends well short of Khrushchev's expectations because he launched the 1959 initiative from an unstable platform. Before, farms had to advance more labor to grow corn than required by wheat or other sown grains. To realize the full potential of corn, they needed machines to plant, cultivate, and harvest efficiently. Khrushchev's orders notwithstanding, factories produced fewer machines than required, a consequence of deficiencies rooted in centralized economic management. When he inquired, he received assurances that output met needs. In fact, it did not. Based on this falsehood, planners had trimmed machine production, retooling factories for other purposes. Khrushchev learned of the shocking shortages only later. By 1957, annual output of corn harvesters had risen to 55,000. By 1960, this figure had sunk to a mere 13,000 even as the cropland devoted to corn surged again beginning in 1959. No matter how often Khrushchev agitated for this pet project, any constituent part went awry as soon as his attention moved on.[122] Manufacture of all types harvesters had more than doubled between 1953 and 1958, but then fell in 1959, held steady in 1960, and then rebounded to the 1958 level only in 1961. Vital to the timely transport and storage of the harvest of all crops, trucks by 1958 rolled off production lines at a rate 50 percent higher than in 1953. In 1959, it fell by 25 percent of the 1958 total, and surpassed the latter level only once between 1960 and 1964.[123]

Even in official analyses, the Soviet Union lagged far behind the United States. American farmers possessed between three and four times as many tractors, trucks, and similar machines, a failure made more glaring by the fact that the Soviet Union's farms planted many more hectares. Even in standard 15 horsepower units, an advantageous measure given the greater power of tractors in the Soviet Union, the deficit remained: 1.79 million to 5.25 million.[124] Regions often pleaded with Moscow for new allotments of necessary machines, as well as more fertilizer and herbicides.[125] Moscow invariably replied that it simply had no surpluses to allocate.

Khrushchev spoke constantly about agriculture to reinforce the importance he placed in it. Contemporary Sovietologists interpreted this as a sign of struggle against a leadership faction favoring heavy and defense industries at the expense of consumers and farms.[126] Although opposed to unbending emphasis on the traditional priorities, Khrushchev never advocated for a reorientation toward consumption, a radical break with the orthodoxy he had defended against Malenkov in 1955. Moreover, no organized opposition contested this issue, or any other. Policy changes reflected shifts in the consensus among leaders, albeit one heavily influenced by Khrushchev himself. His plans for investment in the military and heavy industry moderated in order to remedy long-term underinvestment in consumption.[127]

Yet industrial administrators and officials in government apparently quietly maintained investments in heavy industry. By contrast, Khrushchev astutely argued for aiming to only fulfill the steel plan and to devote the discretionary resources typically spent on overfulfilling that plan to improving agriculture. "If we want to have sustainable growth," he explained, "then agriculture must to some degree outpace the increase in demand: only then will the economy develop."[128] Intent on modernizing industry, the military, and infrastructure, Khrushchev prioritized producing consumer goods and food at a minimum of cost by making use of what he saw as capacity latent in the sovkhozes and kolkhozes. To be unlocked by competent leaders and conscientious farmworkers, this capacity promised to increase food output at a rate faster than the growth of capital investment.[129]

Revelations about shortages of machinery and fraud further soured Khrushchev's view of the structures governing the agrarian sector and related manufacturing concerns. Hoping to discipline them into responsive tools, he further weakened them. The 1955 planning reform had curbed the authority of the USSR Ministry of Agriculture and its republic counterparts. In the early 1960s, Khrushchev sacked the minister, Vladimir Matskevich, and replaced him with a string of successors wholly indebted to himself. Khrushchev relocated the ministry from its Moscow headquarters to a sovkhoz in the surrounding oblast, a move replicated by republic ministries. There, the personnel were to run a model farm while carrying out reduced administrative functions in rudimentary offices. Reluctant to abandon comfortable living in Moscow, personnel made long daily commutes over poor roads. Within a year, 75 percent had sought posts in organizations that remained in Moscow.[130]

Khrushchev drew on American precedents described in many reports and observed by the leader himself during his September 1959 visit to the US Department of Agriculture facility in Beltsville, Maryland, outside Washington, DC, as well as Iowa State University in Ames. He spoke approvingly about decentralized systems in which individual states ran land-grant universities and extension services, whose personnel provided practical advice to farmers. He contrasted this with the endless stream of mindless orders he saw emanating from his own bureaucracy.[131]

Khrushchev also saw America as a model for education. The location of Ames in a farming region allowed students to work on the university's model farm. The USSR Academy of Agricultural Sciences and the Timiriazev Agricultural Academy, the country's most prestigious training institution, were both in

Moscow. As early as 1955, Khrushchev had complained that the former was located on Miasnitskaia Street, scarcely a kilometer from the Kremlin.[132] His reorganizations sought to relocate agricultural education to specialized small towns where students, instructors, and officials would connect with the land. He hoped to thus disabuse trained specialists of disdain for assignments in rural areas, which caused tens of thousands to seek positions in towns, rather than on farms desperate for their know-how. Some new institutions were constructed outside cities, but Khrushchev's plan to move the Timiriazev Academy never came to fruition because of resistance and the stupendous cost.[133]

Emboldened, Khrushchev challenged the party and government officialdom he blamed for the poor results, a mistrust exacerbated by a spate of fraud in reporting deliveries of meat and milk. He designed innovations to empower local officials and require them to connect to the farms. Consolidating the rural districts created in the 1920s into larger production administrations specializing in agricultural management, Khrushchev expected competent, hands-on leadership responsive to local conditions. Another in a string of innovations bewildering to observers, the advent of these administrative units reinforced the party over the government, regional specialization over self-sufficiency, and local decision-making over Moscow's authority.[134]

Khrushchev's design notwithstanding, the administrations perpetuated entrenched failings. Bureaucrats had long interfered in kolkhoz activities, privileging the appearance of fulfilling the planting plan at the expense of measurements of output and productivity. The new units proved too large to administer because of poor roads and unreliable telephone networks.[135] In 1963, the first year of the new administrations, inspectors flagged widespread "top-down commands" and "administrative abuses" in disregard of the 1955 planning reform. Local authorities imposed plans for corn, wheat, milk, and other products on the kolkhozes at the beginning of each planting season with little regard for the farms' needs and capacities. Bosses subsequently subjected farms to constant plan updates and demands for extra produce. They threatened to cut off any kolkhoz that challenged their dictates from its only source of credit.[136] In other cases, the production administrations ordered kolkhoz managers to fabricate paperwork concomitant with their demands, while telling the kolkhoz specialists to actually do "what you think is best." Production-administration officials typically complained that they were simply relaying orders forced on them by superiors. In any case, these "fraudulent and formulaic" actions inflicted real economic harm.[137]

Deception extended to the matter of corn. Production administrations forced kolkhozes to plant more than they had planned, overextending labor, seeds, and machinery. One farm resultantly planted ten times its initial plan, but was incapable of bringing in these 3,000 hectares in a timely manner. Not surprisingly, the harvest was poor.[138] An academic expert wrote to describe his horror at discovering that a production administration near the city of Kirov had forced farms to plant some 9,000 hectares in late June simply to fulfill the plan. The corn would yield nothing when planted so late, least of all in such a northerly locale. When questioned, the administration's head baldly admitted, "'Well, in a month we'll plow it all back up and prepare it for sowing the winter grains.'" In the eyes of such "fraudsters," or *shablonshchiki*, "results mattered for remarkably little compared to the appearance of a fulfilled plan." The academic's letter concludes:

> For them, filling out the report is most important. . . . If a farm manager does not complete the planting plan for some reason, or changes what is dictated, then [superiors] will quickly punish him. If the plan is fulfilled but the harvest does not turn out or the farm takes a loss, he will receive . . . at most a scolding at the next routine meeting.[139]

Having gathered such complaints for a year, the Central Committee and Council of Ministers passed a March 1964 resolution admonishing local authorities. Restating policies enacted nearly a decade prior, the declaration curbed the production administrations' power to review plans, mandated that they deliver procurement orders to the kolkhozes covering multiple years, and banned midyear adjustments.[140]

Khrushchev recognized the obstacles that William Taubman terms "the infernal unreformability of Russia."[141] Meeting with Cuban leader Fidel Castro, Khrushchev pointed to the pool at his vacation home on the Black Sea to explain why real change proved elusive. Having conceived of reapportioning party personnel into specialized agricultural and industrial committees during a swim there, Khrushchev wrote a memo to the Presidium, giving members time to consider the proposal and suggest modifications. "A week later," he recalled, "each copy returned without a single change; everyone agreed." He implied not that his comrades considered the proposal flawless, but rather that they refused to risk challenging him. Even good ideas could not achieve much. "There is such inertia in Russia that it is almost impossible to overcome it. You'd think that I, as First Secretary, could change anything," Khrushchev supposed: "Like hell I can! No matter what reforms I propose or carry out, at bottom everything stays the same." He then elaborated using one of his beloved metaphors: "Russia is a tub full of dough," he explained. "You put your hand all the way to the bottom, and its seems you are the master of the situation. You pull it out, and there is a barely perceptible indentation. Then, before your eyes, it closes up, leaving only the dough."[142]

Khrushchev faced no open challenge to his authority over agrarian policy, let alone to his power. Initiatives fell short of his ambitions in part because of disorderly implementation. When he attempted to overhaul the administration and its practices, he achieved little more than making new enemies, evidence that he had less influence over practice than contemporaries imagined and historians have consequently presumed.

Amid the reorganizations, corn testified to Khrushchev's continued authority over policy. He mentioned it less often than in 1955, but it remained a priority. Officially, corn plantings exceeded 28 million hectares in 1960 on the way to some 37 million in 1962. Supplies of machines, seeds, and fertilizers increased, as did knowledge about the crop, but farms could not keep pace with the expansion. The results thus lagged behind expectations. The more corn farms planted, the more they stretched already thin resources. Prior improvements in milk yields and other measurements even gave way to decline.[143] As in 1957 and 1958, the crusade moderated a little in 1963 and 1964, when plantings fell by an aggregate of 20 percent.

Still advocating for corn, Khrushchev conceded that farms should expend increasing supplies of synthetic fertilizer and chemical herbicides on whichever crop was most profitable in local conditions. Farms had previously devoted these resources almost exclusively to cotton, flax, and other raw materials for industry, for which they received higher prices. Now, they might apply them to corn, wheat, sugar beets, or any other crop. Hybrid seeds required fertilizer to realize their potential, yet even in 1962 only about 25 percent of supplies went to corn and only 25 percent of corn received vital nitrogen fertilizer.[144] "'Why is Khrushchev, who agitated for corn so much, now sounding the retreat?'" the leader asked rhetorically. "We must not be afraid to reevaluate crop rotations and, if necessary, to limit corn cultivation in dry zones, planting high-yielding wheat, barley, pulses, and sorghum." Instead of considering corn or any other initiative the sole solution, Khrushchev envisioned a complete package of measures designed to "intensify" production, getting more from each hectare, hour worked, and unit of capital.

Even so, he could not help but make categorical statements about corn. "This is not relevant on irrigated land," he asserted, "On irrigated land, corn produces higher yields than any other crop."[145] In fact, research found that in warm, dry regions of the south, irrigated wheat produced 28 percent more than without irrigation, while irrigated corn yielded 220 percent more.[146] Although the talk of intensification emerged only in 1961, this did not signal a departure. It simply applied new terminology to party resolutions mandating the industrial farming principles defining the decade's policies.

Because Khrushchev faced no challenge to his power and authority over policy, he could concentrate on corn and industrial agriculture. While redesigning bureaucratic structures, he frequently exchanged one adviser for another. Fully responsible for both foreign and domestic policy after June 1957, he designated Central Committee secretaries to oversee various policy areas. He frequently intervened in agriculture, however, quickly tiring of each successive appointee and expelling them from the heights of power just as quickly as he had promoted them. Khrushchev quickly sent Nikolai Beliaev, previously a hero of the Virgin Lands campaign, to head the party in Kazakhstan. An associate from the Ukrainian days and a key backer in 1957, Aleksei Kirichenko was dispatched to head the Rostov Oblast committee and, soon after, a factory in Penza.[147] Khrushchev repeated this pattern with the ministers of agriculture who succeeded Matskevich, constantly seeking someone who could provide competent hands-on management. Looking beyond Moscow, Khrushchev filled central administrative posts with farm managers and party secretaries from farming regions. The acclaimed director of a sovkhoz in Lipetsk Oblast, Ivan Volovchenko was tapped to head the diminished ministry in March 1963.[148] Each man gained or lost power in proportion to the backing of Khrushchev, the only individual whose opinion mattered.[149]

Khrushchev's overwhelming authority over his comrades' job security drove them to force him into retirement before he could strip them of power and ship them off to run a hydroelectric plant in Kazakhstan, the fate of Georgii Malenkov in 1957. Only in 1964 did an organized faction, by then composed of almost everyone around Khrushchev, coalesce to challenge his power. Always attentive to his moods, colleagues endured the threat posed by his dissatisfaction. Remembering years heading the Kyiv party committee and then the Ukrainian Central Committee, Petro Shelest recalled escorting the leader on visits to the republic. As soon as Khrushchev departed, the phone rang: Leonid Brezhnev and Nikolai Podgornyi, Shelest's predecessor and patron, were calling from Moscow to inquire about any remarks Khrushchev made regarding Kremlin intrigue.[150] Even the next most powerful men in the country feared Khrushchev and therefore endeavored to avoid his ire by anticipating his moods and preferences. Given the rapid fall of aides who failed to meet Khrushchev's expectations in agricultural policy, everyone had ample motivation to primarily consider appearances. It paid to show him the model farms and productive cornfields while concealing shortcomings, no matter how numerous.

Even as party and government constituencies lined up against Khrushchev, he maintained the authority to determine the boundaries of policy. At the February 1964 Central Committee plenum on agriculture, Khrushchev's protégé, Volovchenko, delivered the keynote address on machinery, irrigation, land management, and other industrial-farming methods. Khrushchev's periodic questions and interruptions demonstrate that his own views overlapped: far from feeling his power threatened and his authority constrained, he still shaped formal policy.[151] As was his habit, he dictated memoranda to the Presidium conveying observations made during his travels, as well as proposals to rectify the chronic fertilizer shortage.[152] At the time, outsiders interpreted missives published as attempts to influence an opposition entrenched at the summit of power and to overcome declining political fortunes. They were nothing of the sort. Unpublished for many years, the final memoranda date from June and July 1964. They contained proposals to replace the expiring Seven-Year Plan with a new one covering eight years, which finally provoked defiance.[153] Unpopular initiatives to split party committees and impose term limits to ensure personnel turnover combined with slow economic growth and foreign policy failures to ruin Khrushchev's credibility. Although his successors scrapped the party reforms and production administrations, he had determined what was possible until the very end, ingraining into practice industrial farming principles.

Beginning with the September 1953 Central Committee plenum, Khrushchev quickly amassed authority over agrarian policy, which he soon used to introduce the Virgin Lands project and then the corn crusade. Khrushchev determined the direction of agricultural policy, leaving open only the question how fast to travel his chosen route. At each subsequent stage of the corn crusade, Khrushchev's authority provided the decisive political force maintaining pressure to plant the crop.

Assigning blame for the evident flaws of post-Stalin agrarian reform, scholars have incorrectly heaped almost all of it on Khrushchev. Hardly blameless, he plunged ahead while only begrudgingly heeding bureaucratic, climatic, and economic obstacles. Nonetheless, he has not received deserved credit for reorienting a system in crisis toward improving rural citizens' lives and livelihoods. Impatient and given to simplistic thinking, he constantly reassessed the situation and his reforms, responding by enacting new ones or overturning past ones. Pushing boundaries, he had to remain within certain ones inherited from Stalin: a dominant role for the state in purchasing farm produce, the primacy of collective and state-owned farming over private property, and aspirations to provide a diet bounteous in meat and milk to ensure social stability and enhance the prestige of the Soviet Union. Traditional interpretations have held Khrushchev to an unrealistic standard by supposing that his policies ensured that a more-or-less compliant and efficient bureaucracy implemented intended practices. Implicit in these judgments are expectations of actions—disbanding the kolkhozes or at least nurturing private property—contravening Khrushchev's fundamental principles and, therefore, unthinkable to him.[154]

The reality was much messier—and more interesting—than supposed by labels applied to his policies, including "harebrained scheming." Bureaucratic, climatic, and economic obstacles exacerbated faults in the policies themselves. Party discipline required everyone down to the local committee and soviet to outwardly comply with the campaign of the moment, often by simply aping slogans: "Plant corn!" "Plow up virgin soil!" and "Catch up to and surpass America!" From the farm and the district up to oblast the republic, even officials who harbored doubts, which they made evident by their foot-dragging, declined to vocally protest. Sharing little of Khrushchev's boundless faith, they demonstrated low regard for corn by reserving fertilizers and fertile land for other crops. A quiet struggle ensued between Khrushchev and an officialdom long conditioned by practices inherited from Stalin. Rather than adopting the methods recommended by Moscow or developing alternatives tailored to local conditions, officials superficially fulfilled the minimum requirements of the corn-planting campaign. This expedient responded to the incentives to take care, above all, to maintain the appearance of compliance. Dismissing corn as a passing fancy, officials assumed Khrushchev would move on to a new campaign within a season or two.

The resulting poor harvests harmed later efforts. Guilty by association with the initial failures, the initiative after 1958 to put machines, chemicals, and other technologies to work faced an uphill battle. Despite Khrushchev's orders to produce machines and fertilizers, shortages of them remained unremedied well into the 1960s because of failings in the manufacturing sector. By 1964, skepticism about corn had survived a ten-year campaign and multiple administrative restructurings. Time had transformed the crop into a rallying point for discontent, emboldening Khrushchev's onetime protégés—now opponents—to dismiss the crusade. Silently expressing doubts by evading and dissimulating, those administering the economy hamstrung the crusade and, by extension, the larger agrarian program. They thus doomed Khrushchev's vision of socialism-becoming-communism, defined as it was by a comfortable living standard complete with a diet rich in pork, bacon, beef, eggs, milk, yogurt, cheese, butter, and other valued provisions.

Outstanding farms grew satisfactory—sometimes even impressive—harvests of corn, which for a time boosted output of meat and milk. They served as examples to the great majority that reaped little benefit. Khrushchev and his supporters developed a tautology in response: corn grew so long as good managers employed proper methods. The crop could not fail, it could only be failed. Although the shibboleth shifted blame from upper to lower ranks, the charge contained a certain truth. Denouncing so-called skeptics and haranguing officials, Khrushchev tried to convince them that corn was not a passing fancy, but had come to reign as queen of the fields. Lacking the power or inclination to replicate Stalin's purges of subordinates, Khrushchev faced an officialdom that spoke and acted expediently even as they demonstrated indifference to the details determining success or failure. Merely fulfilling the planting plan, officials gained from going along with the moment's campaign. This relative freedom to respond to commands in such superficial ways, moreover, further testifies to the limits of Moscow's power in the undergoverned countryside.

4 BETTER LIVING THROUGH CORN

During ten years in power, Nikita Khrushchev made seemingly endless speeches about agriculture and habitually sang the praises of corn as a means of allowing the Soviet Union to close the gap with the United States in agricultural output. Such enthusiasm for the crop's potential encouraged outlandish metaphors. "I am an advocate for corn," Khrushchev told a group of Polish farmers, "I love it not because it is corn, but because it offers farms the most potential." "If you took corn away from" farmers in the United States, he continued, "they would go hungry." He observed, "The United States is riding a racehorse, which is corn, and we must catch them on the same racehorse."[1] "It is impossible to select a better horse than corn," he told the Central Committee, employing the same metaphor.[2] On another occasion, he declared that corn was "a tank for use by soldiers, by which I mean kolkhozniks. It is a tank with the capability to overcome barriers ... in the path to creating plenty for the people." "Therefore," he continued, "I call on you to befriend corn by learning how to cultivate it."[3] Following his cues, the press routinely termed it the "miracle crop" and "queen of the fields."

These and many similar metaphors appear puzzling, even absurd, but Khrushchev and the official press intended them to address specific audiences. To urban consumers disconnected from the land, these statements promised inexpensive and abundant meat, milk, eggs, wool, and other goods. Never setting foot in a cornfield, consumers were subjected to messaging designed to reinforce Khrushchev's pledges that plenty was at hand. Furthermore, press campaigns encouraged farmworkers to view corn as a contributor to material comfort for all, part of a larger process of altering their incentives to work. Thus Khrushchev's policies altered the relative balance of the three mechanisms by which any system of production influences people to act: coercion, moral incentives, and material incentives. Moreover, imagery making corn a symbol of abundance helped to define Khrushchev's historical legacy.

Seeking increased influence over everyday activities on farms, Khrushchev steadily supplanted coercion with material incentives while augmenting existing moral ones. In the 1930s and 1940s, propaganda supplemented the government's coercive control over the harvest, which had testified to kolkhozniks' second-class status. Campaigns touting extraordinary feats of labor offered material rewards to the few. During the war, patriotic feeling had motivated some, while many desired to support loved ones at the front or to defend Orthodox Russia, an ideal recognized and co-opted by Iosif Stalin.[4] Each inspired farmworkers to grow the crops needed to feed soldiers and workers in war industries.

In the 1950s and 1960s, Khrushchev's policies placed kolkhozniks under a style of governance new to the countryside, an effort to remedy the persistent weakness of party and government authority there. Emblematic of these evolving strategies, corn propaganda appealed to ideals, supplementing new wage and pay policies that distributed material rewards more widely. This new fusion of moral and material incentives was designed to prompt kolkhozniks to produce as much as possible, a strategy long familiar to workers in other sectors.

In the Khrushchev era, propaganda campaigns resembled Stalin-era predecessors, even though they embodied new conceptions and responded to evolving circumstances. As historians have come to understand, the Soviet Union did not simply coerce subjects who wanted to resist but were rendered incapable by the power of official ideology. Instead, citizens were motivated by the appeal of official ideals due to their own commitment to those ideals.[5] Religious dissenters and other exceptions aside, most members of society negotiated with the authorities constantly as they navigated daily life. From the Stalin era, authorities inherited a conception of the kolkhozniks as socialist-realist subjects responsive to appeals to patriotism and socialism. In the 1950s, social scientists and policymakers reimagined them as akin to the classical liberal subject, who was perceived to rationally pursue personal interests.[6]

Designed to shape citizens' beliefs and actions, the mass media campaign surrounding corn testifies to processes of governmentality. Originally developed by Michel Foucault after observing liberal Western Europe, the concept of governmentality can explain practices used by the Communist Party to mold the society of the Soviet Union, a commonality highlighting links between these industrial societies. The party sought to enlist individuals in the project by appealing to their own interests, opinions, and priorities, a process of mobilizing that made them not passive objects, but active agents of governance.[7]

Mass media, moreover, typified the strategies the party used to influence conduct. Journalists navigated many complexities while carrying out their assignment to help govern society by shaping audiences' everyday decisions. Rather than a crude bludgeon, the mass media was an elegant—if not entirely effective—system for encouraging individuals in a collectivist society to self-mobilize.[8] The press, however, did not necessarily influence readers in the ways its leaders and their political masters imagined. As interviews given by expatriates during the late 1940s indicated, many read the official press skeptically and supplemented it with rumors. Nonetheless, media narratives and education shaped the attitudes even of those skeptical of authority.[9]

The press encouraged all citizens to view corn as source of the material plenty that Khrushchev pledged to soon provide to all, but it also pressed farmworkers to act by contributing to the endeavor to grow it. By showing that such labor had value, campaigns attempted to counter the body of evidence accumulated over decades testifying to the inferior position of kolkhozniks, who primarily worked to the benefit of the state and society, not themselves. Appeals in the mass media to principles used an established, even formulaic, language of words, symbols, images, and practices designed to influence the world. Both the world and the language, however, changed over time. Language operates in a reciprocal relationship with existence, which allows it to both shape the world and in the process be shaped by the world in a dynamic interaction.[10]

The mass-circulation newspapers *Pravda* and *Izvestiia* often featured corn, but it appeared most often in publications targeting farmworkers. The USSR Ministry of Agriculture published *Selskoe khoziaistvo* [*Agriculture*] for farmworkers, specialists, and local officials in agricultural districts. In 1961, it changed format and became *Selskaia zhizn* [*Rural Life*], but its audience and function remained consistent. The technical monthly *Kukuruza* [*Corn*] offered local officials, farm managers, specialists, and other personnel information on best practices and the latest advances. First published in April 1956, it sought "to aid kolkhozes, MTSs, and sovkhozes to complete the undertaking during the Sixth Five-Year Plan to expand the planted areas and raise yields of corn, the most important source for the increased production of grain and the creation of a sufficient feed supply for livestock."[11] The period also witnessed an extraordinary outpouring of books, pamphlets, academic papers, and other publications. A 1960 bibliography on the subject listed more than 4,000 titles published since the Imperial period, the vast majority of which had appeared after 1953. Subdivided by topic, it included sections ranging from basic cultivation methods to recommendations for individual locales.[12] A later tally recorded an additional 1,075 books on the subject printed between 1960 and 1964 alone.[13] The annual national bibliography of journal and newspaper articles similarly catalogs a flood of press beginning in 1955.[14] These catalogs primarily documented Russian-language publications, excluding translations and original materials in the country's numerous other languages.

Furthermore, authorities enlisted other forms of media. The initial 1955 burst of agitation for corn saw the authorities print more than 100 posters in runs ranging from a few thousand for local examples to tens of thousands for those emanating from Moscow and Kyiv. The print runs of seven of these exceeded 100,000, with an aggregate of 1.125 million posters.[15] In addition, Soviet Radio often broadcast these messages. A survey of scripts in the organization's files for 1955 finds that almost instantly after the January plenum the semiweekly broadcasts on agricultural themes began to include information about corn.[16] In the 1950s, television was in its infancy in the Soviet Union, but short propaganda and documentary films were quite common, even in rural areas.

The mass media broadcast Khrushchev's praise for corn and repeated his message in waves varying in intensity. Amplified by the official press, the leader's statements encouraged farmworkers—kolkhozniks, sovkhoz workers, specialists, and managers—to embrace their work because, although arduous, it promised to strengthen the country and provide abundance.

To those steeped in this mindset, improving the Soviet Union's farms furthered the cause of world peace, a transmutation that appears odd from a post–Cold War perspective. While Khrushchev first preached his corn crusade in April and May 1955, authorities were gathering party organizations, workplaces, and other groups to sign the World Peace Council's Declaration against Preparations for Nuclear War. Conceived by the authorities to address foreign and domestic audiences, the council denounced the Cold War and portrayed the Soviet Union as an advocate of nonaggression and disarmament in the face of capitalist powers' militarism. Fresh memories of World War II ensured that authorities had little difficulty prompting millions of citizens to sign the document.

Although seemingly pro forma, the signing accompanied pledges from individuals and groups to work more productively, fortifying the country against aggression from abroad. At some of the more than 2,500 meetings in Stavropol Krai, for instance, kolkhozniks promised to grow record harvests of corn. "War is hateful," exclaimed a brigade leader at one meeting. "I witnessed the enormous destruction and saw the numberless victims," he continued, perhaps referring to his wartime military service or the enemy's occupation of the area in 1942. "The strength of the fighters for peace is incomparably greater than that of those for war. We will defend peace by working. Battling to implement the January Central Committee plenum's directives, we pledge to raise a corn crop of 32 tons per hectare."[17] Across the country, laborers forging steel, hewing coal, digging sugar beets, and churning out many other commodities echoed the sentiment. A new priority, corn featured in reports that party committees in Balashov, Briansk, and other oblasts dispatched to Moscow documenting work on both campaigns. As a report from Vladimir Oblast concluded, "Every person who plants a hectare on time strengthens the cause of peace," a particularly important step in preparing to plant "a crop new to us: corn." The signatures and accompanying acts of individuals were, party officials concluded, "a drop in the sea, but the sea is made up of these many millions of drops."[18]

Kolkhozniks had been targets of moral incentives in many forms since collectivization, but only a few individuals earned fame and accompanying material rewards. In the mid-1930s, outstanding industrial workers garnered press attention by emulating coal miner Alexei Stakhanov, busting daily production norms to earn the title of Stakhanovite and preferential access to consumer goods. These industrial workers embodied complex processes in which they, local authorities, and top leaders each pursued particular ends. Stakhanovites also emerged on the kolkhozes, often in the person of women machine operators. Rather than simply seeking to limit the negative effects of kolkhoz membership as so many kolkhozniks did during and after collectivization, a few actively became exemplars of the new agrarian order. In so doing, these female counterparts to the typically male industrial workers achieved their own ends even as they recited officially sanctioned lines. In print, in film, and at public meetings, famed farmworkers such as Pasha Angelina embodied policies that seemingly had transformed the countryside. Active subjects, these women won moral and material rewards even as they risked ostracism from their communities for cooperating with the hated authorities.[19]

Under Khrushchev, the material rewards linked with moral incentives became accessible to many more because farmworkers' incomes improved. These years saw the Stakhanovite give way to the vanguard worker, whose example of highly productive individual work was to embolden neighbors to strive for outstanding results, including astonishing yields of corn. Much like the women of the 1930s, vanguard workers performed in the drama by speaking publicly and received medals, awards, or honorary seats in soviets of various levels. That honor, although only a symbolic vote in a rubber-stamp assembly, carried with it prestige and access to higher authorities.

Models encouraged the rest to improve in hopes of receiving similar accolades. As one local leader stated the idea, the goal was "to provide incentives to vanguard workers and spread word of their achievements." In turn, this "moral influence affected not only the best, but the entire work group, and has great importance in educating peers."[20] Nonetheless, these expectations ignored the considerable official support necessary to turn a competent laborer into a vanguard worker. Moreover, industrial workers during this period did not respond to such campaigns with much enthusiasm because of their relatively established labor processes and sense that these initiatives were archaic.[21] On the kolkhoz, by contrast, labor processes were changing with the expansion of industrial farming, a sign that such skepticism may have been weaker down on the farm.

Press portrayals of the vanguard corngrowers changed over time. Although too many appeared to describe even a significant proportion in detail, a few stand out because they exemplified changes in methods for cultivating corn and in expectations about labor. One of the first, Mark Ozërnyi of Ukraine's Dnipropetrovsk Oblast had already gained fame and rewards in the late 1940s. In September 1953, Khrushchev paid tribute to him as a model worker who had earned the coveted medal of a Hero of Socialist Labor and the Stalin Prize. Khrushchev had met Ozërnyi during his tenure in Ukraine, when the kolkhoznik had grown record-breaking harvests of grain.[22] From 1953 until Ozërnyi's death in 1958, newspaper articles, pamphlets, and books encouraged readers to emulate his techniques.[23] Before mechanization of planting, cultivating, and harvesting spread after 1958, the approved methods required expensive manual labor to complete these vital tasks, raising the cost but producing high yields if conscientiously applied. Almost all information about Ozërnyi was conveyed through text, rather than text and images as would be the norm in subsequent years, when new conventions in the mass press took hold. Thus, with the exception of posters, the press primarily used formal portraits of Ozërnyi. The photos depicted an extravagantly mustachioed man somewhat past middle age in a medal-bedecked suit jacket.

In part, these campaigns told farmworkers what to do—plant corn—and that the quality of their labor mattered. From the very beginning of the corn crusade, the press constructed a dialogue between Khrushchev and farmworkers who seemingly welcomed his plans to grow corn and achieve abundance. Established practices required the press to demonstrate that rank-and-file workers in every corner of the country responded with enthusiasm to the appeal by the January 1955 Central Committee plenum to plant corn. On February 3, the text of Khrushchev's plenum speech appeared in *Pravda*, followed in subsequent issues by appropriate replies. On February 5, articles with datelines in Altai Krai and the Kazakh SSR trumpeted "great enthusiasm" for efforts to wrest high yields of corn from kolkhoz and sovkhoz fields. Appearing under a banner reading "The Soviet People Warmly Welcome the Directives of the Plenum and Are Expanding the Battle for Their Realization," such stories evinced official expectations that almost every farm was to increase corn plantings. Thus the Altai Krai kolkhoz Path to Communism pledged to plant 1,000 hectares, a first step toward a planned fourfold increase in its supply of fodder.[24] Related stories reported pledges from Latvia to produce more milk and from Stalingrad to build more tractors.

Repeated ad nauseam at every phase of the growing season, stilted and formulaic statements reinforced the essential message: plant more corn to produce more food. A February 6 front-page article explained how farmworkers in

Voronezh Oblast discussed the plenum resolutions at meetings and, in response, took on higher obligations in a competition to grow more corn. It reported the specific case of the Molotov kolkhoz. Having only recently begun growing corn, the farm had been saved from a poor harvest in 1954 by the boundless productivity of corn, which made up for low yields of wheat and rye and allowed the farm to produce more milk, pork, and other produce.[25] The newspaper placed this, along with similar pledges, beneath a banner that read, "Laborers of Town and Countryside Announce Their Preparedness to Realize the Resolutions of the Plenum of the Central Committee of the Communist Party of the Soviet Union."

This campaign additionally showed readers the most modern methods in use on kolkhozes and sovkhozes, an image that even an industrial worker far removed from the fields could view with pride. Even *Pravda* and other general-audience newspapers emphasized modern technologies in stories heralding manufacture of more tractors and machines of more varied design. On March 7, 1955, a picture of a state-of-the-art corn planter appeared, followed on April 25 by another of a similar implement in use in a field in Ukraine. Dry descriptions of a machine's technical specifications seem unlikely to have captured the imagination of a factory worker or office clerk. Instead, photos of equipment in use quickly conveyed the message eloquently and without words. The photo sought to convince audiences that the farms were planting corn that very year in the most efficient way possible, notwithstanding the fact that they actually planted much of it by hand.

Augmenting newspapers, journals, pamphlets, and propaganda posters, exhibitions created physical spaces where farmworkers learned about the latest methods. On August 1, 1954, Moscow celebrated the reopening of the All-Union Agricultural Exhibition for the first time since 1941. It occupied a complex in the city's northern reaches built in high Stalinist style that, as *New York Times* correspondent Harrison Salisbury judged, resembled the white-columned buildings of the 1893 World's Columbian Exposition in Chicago. Noting that the Moscow exhibition drew visitors from distant regions of the country and the city's curious public, Salisbury concluded that it most resembled an American state fair, "but without the hot dogs and popcorn."[26] During that month, daily attendance reached 50,000, but demand for tickets exceeded supply by as much as 20,000. In the first two months, more than a million visited. By the end of the year, the total reached nearly 8 million.[27]

The exhibition, however, also constituted a public-relations exercise, giving average citizens a glimpse of promised future abundance. From the moment of the 1952 decree mandating the exhibition's reopening, its designers sought to display agricultural achievements and spread knowledge.[28] Yet critics wrote to complain that the exhibition's pavilions were "homogenous," and the tours arranged for the best farmworkers from regions across the country taught the participants little.

Penning a letter to authorities about the 1954 exhibition, one employee concluded that allowing regions to display only their very best encouraged "braggadocio" and even "deception." Instead, he proposed that some less-successful farms be included as a teaching tool. Moreover, he complained that the exhibition's timing, during the late summer and early fall harvest season, was poorly suited to the demands of the agricultural year. "Who is the fundamental consumer of the exhibition?" he asked.[29] Surely it spoke to farmworkers, the writer assumed. The exhibition showed only the best farms and operated during the harvest, however, because it also spoke to urban viewers—and foreigners—of the promise of agriculture in the Soviet Union, a mission of equal, if not greater, importance.

Rather than providing a mere diversion for the curious, the annual exhibition popularized agricultural achievements and called on symbols of wealth, abundance, and progress in the competition with other societies. Urbanites were numerically its most common audience. The attendance figures for 1954 are instructive: only 306,000 attendees arrived as part of five-day tours offered only to agricultural personnel. Foreign visitors, moreover, numbered just under 10,000. In the most generous reading of the statistics, these two groups amounted to only 20 percent of the total attendance.[30] The minister of agriculture, Vladimir Matskevich, boasted that the exhibits spoke also to a general audience by "demonstrating the achievements of socialist agriculture ... especially [those] since the September [1953] Central Committee plenum, which laid the foundation for rapid agricultural growth."[31] By 1956, the annual reformulation of the exhibits had caught up, emphasizing "the significance of expanded plantings of corn in the task of increasing grain production, ensuring the feed supply, and raising livestock productivity."[32]

The exhibition propagandized corn intensively, encouraging those who would take on the task of planting it. It highlighted exemplary figures, from the already famous Ozërnyi to Mariia Gorlova, a young woman from the Molotov kolkhoz in Voronezh Oblast whose work team grew a record harvest of grain.[33] Adding new buildings each year, the agricultural exhibition's displays grew to include a range of specialized pavilions. A May 1955 letter dispatched to the Central Committee complained that the exhibition boasted specialized pavilions devoted to breeding dogs and raising rabbits, but not to corn.[34] Although Khrushchev endorsed the proposal to open such a pavilion, this came to pass only in 1958, when the pavilion began to exhibit best practices for growing corn and featured displays about farms that brought in large harvests. Although designed for specialists, the pavilion also touted corn to passing exhibition attendees. Accompanied by a picture of exemplary cobs of corn, stories about the pavilion advertise it as a place "revealing the proven record of advanced kolkhozes, sovkhozes, and districts in growing high yields of corn," one where the visitor was to learn of heroic efforts to add up to 16 million tons of grain to the harvest.³⁵ Overall, the exhibition's pavilions spoke to multiple categories of guests, including specialists in their respective fields, but also popular audiences attracted to underlying conceptions of science, technology, and progress. Termed a "city of wonders" by one journalist, its themes were optimism toward a future that promised communist abundance and material satisfaction.³⁶

Beginning in 1955, the Soviet Union launched the corn crusade proper, sent its delegation to America, and strove to apply industrial farming technologies. As a result, vanguard farmworkers achieved fame for planting and harvesting many hectares of corn using machines, growing more livestock feed at less cost. The most prominent were Aleksandr Gitalov of Kirovohrad Oblast in Ukraine and Nikolai Manukovskii of Russia's Voronezh Oblast. Featured in stories and photos in the press, they spoke before Central Committee plenums and performed other ceremonial duties.³⁷ Returning from a growing season spent on Roswell Garst's farm in Iowa, Gitalov achieved particular renown by spreading the word about the farming methods the American corn impresario applied on his farm: corn monoculture made possible by chemicals, synthetic fertilizers, and machines. For his part, Manukovskii had tested mechanized corngrowing on his home kolkhoz in conditions more typical of those on farms in the Soviet Union (Figure 4.1). Press coverage announcing the beginning of a 1959 competition staged between the two indicated that they pledged not only to achieve the highest possible yield without using any inefficient manual labor. They also vowed "to convey our experience to young machine operators, carefully teaching them the newest methods."³⁸

The press portrayed Gitalov and Manukovskii as typical citizens, but also individuals whose outstanding dedication to work made them models for all. Describing Gitalov at work in Iowa, the press characterized him as a practical farmworker who learned by doing. "I received the assignment to study American methods of farm management," he recounted, "and the best way to achieve this was to sit myself behind the wheel of a tractor."³⁹ Manukovskii became the subject of documentary films designed to spread practical knowledge. Part of a larger film about his oblast, a segment about him titled "Machine Operator of the [Lenin] Kolkhoz Explains the Brigade's Work on Cultivating Corn" concentrated on just these aspects of Manukovskii's job. In this 1958 film, the tractor driver, cleancut and dressed in a suit, appears to be an amateur at being on camera. Speaking plainly and with a southern accent, he describes with evident pride the effectiveness of growing corn with machines but without manual labor.⁴⁰ Given the ubiquity of documentary and propaganda films, and the dozens of similar films preserved in the archives, we can conclude that this type of film reached a broad audience.⁴¹

FIGURE 4.1 Decorated kolkhoznik and corn grower Nikolai Manukovskii operating a multi-row corn planter on his kolkhoz in Novo-Usman District of Voronezh Oblast, RSFSR. Manukovskii achieved renown for pioneering the use of machines for each phase of the planting, cultivating, and harvesting of corn.

Photo by A. Zenin, courtesy Russian State Archive of Cinema and Photo Documentation, ID#0-276076

The press portrayed women vanguard kolkhozniks in similar ways, albeit distinguished by gender-specific visual language. Historian Victoria Bonnell has identified three types in photos and stylized representations of women: After 1917, artists created the *baba*, evoking a pejorative term denoting backwardness and ignorance. Around 1930, the thin, stern *kolkhoznitsa* was designed to appeal to urbanites during the era of collectivization. By the end of the 1930s, a more overtly feminine peasant woman, or *krestianka*, became common. Artists depicted a fuller figured woman at work on the kolkhoz as well as in the home, where her family enjoyed the fruits of her labor. Combined with the reality that women comprised the majority of kolkhozniks, this final form reinforced the identification of women with the peasantry as a social category. By the late 1930s, even portrayals of woman workers also acquired many characteristics of the *krestianka*. After the war, the images evolved further, keeping up with the high-Stalinist times by displaying every more outlandish symbols of prosperity and traditional markers of femininity.[42]

In the Khrushchev era, the press frequently represented that earlier generation, which included Evgeniia Doliniuk. Born in 1914, she was the champion corngrower of Ukraine's Ternopil Oblast. From Khrushchev's speeches to the pages of *Pravda*, official statements often portrayed her achievements as models for others. Already famous in 1955, she was the subject of a poster with a print run of 250,000, the largest of the year.[43] In a 1959 feature, *Selskoe khoziaistvo* trumpeted the printing of a poster about Doliniuk by featuring a candid picture of her, decked out in her two Hero of Socialist Labor medals, proudly displaying large stalks of corn and signs heralding her high yields. She is dressed simply, in clothing of a single color, with a light-colored headscarf tied behind her head.[44] An accompanying poem captured Doliniuk's celebrity:

> Not for nothing in her native land,
> Does Auntie Zhenia and her work team,
> Boast a reputation for corn, both feed and grain.
> Let it be said, "Let the beauty [corn],
> Grow everywhere, all around,
> As in Auntie Zhenia Doliniuk's field!"[45]

Each of these elements is reminiscent of the imagery denoting the *krestianka*, hardworking and enjoying the prosperity that accompanied success.

Much like Gitalov and Manukovskii, Doliniuk not only often appeared in the press, but also participated in Central Committee plenums and other political events.[46] In fact, the three were pictured together, as in a January 1961 photograph accompanying the coverage in *Izvestiia* of that month's plenum. Like the men, she praised the advances made toward growing corn using advanced technology borrowed from America. "Comrades, Mr. Garst explained that it is necessary to have hybrid seeds, fertilizers, and—what's more—chemicals for destroying pests and weeds," she explained. "We have the first three conditions on every kolkhoz." But farms needed more of each, especially the chemicals promising to enlarge harvests, decrease labor requirements, and reduce the cost of production.[47]

As a record-breaking corngrower, Doliniuk became the focal point of a people's academy, where she taught corngrowing methods to farmworkers from her home oblast and, soon, to those drawn from afar by her fame. According to officials, short-term hands-on training proved more effective than newspaper articles and informational pamphlets alone. Schools of Vanguard Knowledge, such as Doliniuk's, appeared in 1957 and spread quickly. By 1959, leaders of youth corngrowing brigades organized by the Young Communist League attended training on her kolkhoz consisting of one-day sessions in each of the four phases of the agricultural calendar: wintertime preparations for spring, planting, cultivating, and harvesting. The publicity attracted hopeful youth from other oblasts of Ukraine, as well as from the RSFSR, Kazakh SSR, and Kyrgyz SSR.[48] And they got results. Party officials reported that as many as 100 of the trainees had doubled the yield they had grown the previous year.⁴⁹ Responding to that success, Ternopil Oblast leaders expanded the program in 1960 and 1961. By early 1962, other oblasts were adopting the formula, creating eighty-nine such schools. During their operation, they trained some 21,000 leaders, activists, and heads of work teams.⁵⁰

By the early 1960s, journalists ventured beyond the field to reveal more about vanguard farmworkers' daily lives. This exemplified a new focus on the individual characteristic of the Thaw, when "the little person" became a subject for the official press.⁵¹ In contrast to stories about Doliniuk, Manukovskii, Gitalov, and Ozërnyi, those portraying Liubov Li detailed both her work and daily life in pursuit of didactic ends. Li stood out from the many other vanguard workers of the time for several reasons. She lived in the Uzbek SSR, a non-Slavic republic, and was of a non-Slavic nationality herself. Her family name, a Russianization of the Korean one often rendered in English as Rhee or Lee, indicates that she was a member of the diaspora deported by Iosif Stalin before World War II from the border with Korea to Central Asia. After 1960, Li achieved prominence via newspaper and magazine profiles hailing her achievements in the fields, no fewer than eight in *Selskaia zhizn* alone.⁵²

The photos accompanying the story show Li with head covered by a shawl or cap against the broiling Central Asian sun, devoted to the task of tending her fields (Figure 4.2). Yet these stories also painted an apparently candid portrait of her work and life. Like the young, vigorous women described by Bonnell as the hallmark of the late 1930s and the postwar period, Li—a generation younger than Doliniuk—was always depicted dressed in the overtly feminine manner expected of the *krestianka*. Moreover, the accompanying stories adorn that portrayal by highlighting her role as mother to two young sons. In the first six months of 1963 alone, *Kukuruza* published two feature articles, adding to her prominence. Although far from the only woman featured at the time, Li was distinguished by the prominence of her familial duties in these profiles. Emphasizing humanity and personality, one begins with Li returning home from the fields to be greeted by her sons, who present her with the daily stack of letters filled with inquiries and good will from around the Soviet Union.⁵³

As a Hero of Socialist Labor, a deputy in the USSR Supreme Soviet, and a model citizen, Li served as an example for all. The mass media's portrait of her approached the ideal, for instance in her ostentatious rejection of religion. "Beyond the ocean, in corn's old homeland," one of the articles informs the reader, "people still believe in god and miracles. There was a time when, for instance, Peruvian maidens made offerings of bread baked from cornmeal to the sun." Central Asia witnessed nothing of the sort: "Liuba Li does not believe in any

FIGURE 4.2 Famed kolkhoznik and corn grower Liubov Li tends to the corn plantings on her kolkhoz, "Politotdel," in Verkhnecherchik District, Tashkent Oblast, Uzbekistan. Her informal pose emphasizes the everyday nature of the scene.

Photo by K Razykov, courtesy Russian State Archive of Cinema and Photo Documentation, ID#1-100462

god. She prefers her inspirational labor to him."[54] The publicity spreading a vanguard corngrower's fame intersected with related efforts to shape the way the audience thought and, thereby, how its members acted. In this case, the story made a modest contribution to the virulent campaign against religion Khrushchev pursued in the early 1960s as part of the push to "construct communism." Reversing Stalin's policy of accommodation with and management of religious hierarchies, Khrushchev closed monasteries and launched propaganda offensives against everyday practices, an effort to achieve ideological purity and ensure that the "builders of communism" received a properly atheist upbringing.[55]

Corn was a symbol of abundance and, because it was often portrayed with feminine trappings, of fertility. A drawing from the February 20, 1961, edition of *Izvestiia* is only one of many examples. Typical of portrayals of women, it pictures a well-dressed farmworker hoisting an ear of corn of outsized proportions over her head, as if it were an offering to the gods whose existence communist doctrine rejected. The crop's prominence in mass media encouraged farmworkers to view their harvest as a contribution to plenty for all. For this reason, Khrushchev exhorted them to devote themselves to this work. In turn, the press called on them to emulate outstanding vanguard workers, to build a communist society promising a better life, and to prepare themselves to contribute to that higher stage of socioeconomic development. In these the latter two cases, the message intersected with another aimed at an audience not of farmworkers, but of consumers. For them, corn symbolized today's striving toward tomorrow's abundance. Khrushchev promoted a vision of "constructing communism" in which the crop was the source of the material abundance that was a critical underpinning of the ideal social structure.

A source of material wealth and a signal for the potential for progress unlocked by industrial farming, corn—in an apparent paradox—also appeared in visual and print forms as something miraculous, *chudesnitsa*, a sign not only of scientific and technological advances, but also of nature's goodwill. Cultural historians studying advertising have found that prerevolutionary Russian examples, like contemporaries around the world at the turn of the twentieth century, connected the modern and the magical.[56] As the Soviet Union's form of advertising, propaganda juxtaposed corn and modernity, for instance, by turning it into a rocket.

Images associated with the corn campaign combined with related symbols of abundance and progress to evoke pride in the achievements of the Soviet Union and optimism for its future successes. The illustration dominating the front page of *Komsomolskaia pravda* on January 1, 1960, presented corn as a part of this larger message. A popular public holiday substituting for impermissible Christmas, New Year's Eve celebrations offered citizens opportunities to bid farewell to the old year and welcome the new one. In this spirit, the newspaper depicted the achievements of 1959: the nuclear-powered icebreaker *Lenin*, launched with much fanfare, embodied technological progress. A hydroelectric dam, factories, and tractors conveyed a sense of economic development, while new housing blocks signaled rising living standards. A rocket speeding toward the cosmos called attention to the space program, which basked in the light of the 1957 launch of *Sputnik* and subsequent flights. Even a glance at the rocket reveals that it combined symbols of technological modernity with those conjuring up Khrushchev's agricultural revolution: it was constructed of a standard nosecone emblazoned with the red star, hammer, and sickle. In place of rocket engines, however, appear milk cans, while its body consists of layers of grapes, cotton, wheat, and corn. Each crop signified that material abundance was soaring: wine, clothing, bread, milk, and meat at that moment seemed to be becoming ever more available.

Impressive annual growth rates in industry and agriculture were surpassing those of capitalist competitors. The efforts to catch up to the United States looked like a miracle, not a calamity. Informing the public about such progress, the messages encouraged its members to look forward to unheard-of abundance. Reinforcing the point, a March 1962 issue of *Selskaia zhizn* made corn part of the space age. Accompanied by the caption "Animal husbandry and its 'sputniks,'" or its satellites, the sketch depicts a globe of fat and happy cattle, hogs, and chickens orbited by corn, pulses, and sugar beets, all feed crops for which Khrushchev lobbied.

Corn was often depicted as "queen of the fields," a trope signifying its function as a symbol of abundance. An image appearing in the official forum for satire, *Krokodil*, evoked traditional Russian decorative painting. It reinforced the regal nature of corn by portraying the anthropomorphized figure with a crown. The caption labeling it "queen-corn" and its companion on the tractor "tsar-pea," by this time an important complement to corn because it provided livestock with the protein that corn lacked. The crops and tractor are symbols of abundance and progress. This "Queenly Procession," as the title labels it, appears a bit silly, but it is wholly in keeping with the themes of Khrushchev-era propaganda.[57] In another instance, the "queen" was depicted with "the court of her highness" comprising bowing cattle, hogs, sheep, turkeys, geese, ducks, and chickens.[58]

The leitmotif of corn as a miracle crop was most visible in the often-used tag "*chudesnitsa.*" As early as 1957, it served as the title of an animated film telling the story of the hero, corn, which overcame the doubts of the other crops that it could grow in cooler northern climates. A reviewer writing in *Selskoe khoziaistvo* concluded that the film's creators had "shown the viewer the exceptional value of corn for the national economy. They have created a remarkable film about the 'queen of our fields.'"[59] The term also appeared several times each year in *Pravda* and still more often in *Selskaia zhizn* between 1961 and 1964.

From Khrushchev's speeches to mass-audience visual imagery, the era was pervaded by the idea that corn was something special, exceptional, and even astounding. As early as March 1955, Khrushchev had claimed that communism was a practical stage of material wealth, rather than something "pie-in-the-sky," but he also described corn as if it possessed extraordinary productive powers.[60]

Echoing his enthusiasm, the press introduced audiences to the potential of the then-unfamiliar crop. The following month, a story in *Komsomolskaia pravda* raved: "Corn is the key to increased grain production and to plentiful meat and dairy products. One little kernel of this miraculous plant, sown by caring hands, gives two or even three full-weight cobs; that is 1,000 to 1,500 grains and 4 or 5 kilograms of green mass." Translated into food, "this means from 1.5 to 2 liters of milk, from 60 to 80 kilograms of butter, 2 or 3 cans of delicious canned corn, or approximately 100 grams of pork. And this from just one little kernel!"[61]

When early results fell short of expectations, Khrushchev countered that corn could not work miracles without proper cultivation and care. Failures reflected poorly not on the crop, but on the local bosses and farmworkers who planted it and then expected a bounty without having to weed, irrigate, harvest, and feed corn to livestock. The leader complained that they "plant corn and then wait for everything to happen on its own. No, corn is definitely not a fairytale crop. It gives high yields only to those who take the correct approach, work on it, and use the necessary technologies."[62]

The press emphasized the extraordinary qualities of the crop by referring to it by a variety of similar monikers. "Hero crop" evoked the knight-errant, or *bogatyr*, at the center of premodern Russian epic tales.[63] Artists transformed this evocative language into a visual representation. In March 1962, *Selskaia zhzin* combined the hero crop leitmotif with Khrushchev's horse metaphor: a knight emblazoned with a red star and armed with a lance, astride a horse made of a corn cob. The banner overhead reads, "To the front!" or "*Na udarnyi front*," another of the many instances in which growing corn was compared to combat. It furthermore evoked the term "shock worker" [*udarnik*] employed in earlier decades. The image accompanies a lengthy quote from Khrushchev further reinforcing the unusual and wondrous nature of corn: "Corn is a blessing to humankind. Skillful cultivation of this valuable crop provides great wealth to the country and the people." The accompanying story noted that a hectare of corn yielding 50 tons of silage, a high but not record-breaking harvest, might become 1.56 tons of pork or 10.4 tons of milk. Designed to catch the eye, the drawing attracted the reader to a story designed to shape readers' attitudes to corn by reinforcing its links to abundance. Moreover, the imagery was repeated elsewhere. *Krokodil* had emblazoned a similar motif on the cover of its January 10, 1959, edition. That cavalier was outfitted in the uniform of a Soviet Army soldier, but the lance and horse elements were the same. Armed with textbooks on agronomy, the soldier declared that he had "saddled up a fiery steed," while the cover repeated Khrushchev's statement, "Corn is the racehorse we need."

In the spirit of promoting corn as something extraordinary, the authorities opened new stores and cafes featuring foods made from corn, designed to popularize the unfamiliar crop. These included cafés named Chudesnitsa on Moscow's Garden Ring near the USSR Ministry of Agriculture and on Leningrad's Nevskii Prospekt.[64] Stores dedicated to selling foods made from corn appeared, including one on Moscow's Leninskii Prospekt. Its effect was magnified far beyond that corner of the city by extensive media coverage. A 1963 profile in *Kukuruza* heralded it, stating, "When you cross the threshold of the new store, you automatically get the sense that foods made from corn are richly represented here." Applying the tag "queen of the fields," it goes further, suggesting that this richness was "regal" or "queen-like," and enthuses, "It is almost unbelievable that so many delicious things can be made from corn." The consumer could purchase cornmeal, cooking oil, porridge, cornflakes, popcorn, canned corn, cakes, candies, and more, at least fifty different foods in all. Packaged to attract attention, each carried names evoking richness and confidence: Miracle, Rocket, Golden Cob, and Amber. In keeping with the spirit of the era's journalism, the article focused on everyday citizens. According to the writer, the store received high marks from customers for the quality of the service and of foods for sale.[65] The customers, all women, embodied the ideal consumer. Shopping and other homemaking chores were not only the responsibility of women according to acculturated gender roles, but also were so depicted in the press.[66]

State enterprises produced the foods featured in this shop in quantities too small to substantially alter the average citizen's diet, but large enough to reinforce the symbolic relationship between corn and plenty. In fact, this link emerged in the late 1930s, when canned corn and cornflakes adorned the pages of the *Book about Healthy and Delicious Food*. The foods' presence in literature establishing the normative Stalinist cuisine made them part of an ideal, even if they rarely graced real store shelves.[67] Even after years of effort, in 1963 factories produced only 15,500 tons of cornflakes and froze 1,000 tons of corn on the cob, in addition to distributing canned corn, popcorn, and other culinary rarities. Spread among a population numbering 225 million, this quantity of breakfast cereal amounted to a paltry 70 grams per person for the year, hardly enough for a single meal.[68] Much like the farming technologies designed to ease corn production, these foods were also the product of importing foreign technology from the United States and Great Britain. These origins notwithstanding, the foods were subject to problems seemingly ever-present in the economy of the Soviet Union: shortages of raw materials, inadequate distribution, and poor quality control.[69]

Khrushchev's speeches and accompanying advertising campaigns, including in the pages of *Kukuruza* and *Selskoe khoziaistvo* encouraged readers to view the previously unfamiliar foods as nutritious and valuable. "Corn in the form of flakes is also very nutritious," he announced in June 1962. "The Americans and the English are masters of this. What is breakfast to an Englishman? Often it is cornflakes and milk."[70] Accounts by those who lived through the period suggest that this campaign achieved some success. Writing in American exile about memories of the Thaw, essayist Pëtr Vail and critic Aleksandr Genis noted that, whereas the hallmarks of the Stalin era were solid and monumental—the Metro, the war, high culture—those of the Khrushchev decade were eclectic and domestic: the mass-produced, standardized five-story apartment block and popcorn.[71] Yet even the substantive elements of Khrushchev's legacy reproduced the theme of corn. So integrated were the ideas of corn and abundance during this period that they appeared even in the era's Metro architecture and other corners of Moscow, as well as in other cities, a reminder for those who had never seen a cornfield (Figure 4.3).

Rather than appealing to citizens to consume corn itself, Khrushchev and the mass media alike represented it as the source of beef, pork, milk, butter, cheese, and eggs as well as wool for clothing. In a trend common to societies experiencing rural modernization and urbanization, people living in cities did not produce their own food as rural residents did. Industrial workers, clerical personnel, officials, students, and other consumers depended on buying it either from

FIGURE 4.3 A mosaic in the track-level vestibule of Moscow's Kievskaia Metro Station, which opened in 1953, depicts the abundance to be found in Ukraine, including corn, cabbages, pumpkins, watermelons, and sunflowers of fantastic size.
Photo by Aaron Hale-Dorrell

peasant vendors in the market or the still limited but increasingly sophisticated state system.[72] As part of the process of producing food, corn was hidden from a population increasingly composed of urbanites comparatively removed from rural society and agricultural production. Khrushchev's endless speeches on corn had many purposes, one of which was encouraging those out of contact with agriculture to appreciate the crop as part of a larger effort to offer an abundant, rich, and varied diet. Newspapers and other media took up this theme, and set about promoting corn to the average citizen-consumer.

The press portrayed corn as a welcome part of the food supply to make it familiar. It Sovietized it, rather than giving it characteristics particular to cultures familiar with the crop, such as Moldovan or Georgian, or to dominant Russian culture. Sovietized folk culture adopted corn as its own. In 1960, a Moscow festival of Ukrainian culture included the performance of a dance, "In the Corn Field," by a dance troupe (Figure 4.4). Women dressed as peasants and men garbed as tractor drivers flourished corn tassels as props.

Publications attempted to integrate corn into daily life and culture for readers. In 1963, the issues *Kukuruza* contained 443 stories. Although more than 67 percent were on technical issues or policy, some 8 percent, an average of 3 per issue, were classified under the rubrics "Corn on the Table," "Satire and Humor," and

FIGURE 4.4 At a November 1960 performance at Moscow's Tchaikovskii Concert Hall, a dance troupe from Ukraine performs "In the Corn Field" as part of a festival of Ukrainian art and culture.

Photo by M. A. Trakhman, courtesy Russian State Archive of Cinema and Photo, Documentation, ID#0-367480

"Read This, It Is Useful to Know," the last instructing readers in the history and science of corn. These articles, images, poems, and songs provided ways to accustom the audience to corn's presence in fields and on tables. The January edition, for instance, augmented the story lauding the store on Moscow's Leninskii Prospekt by including a brief explanation of the presence of the crop in American Henry Wadsworth Longfellow's poem "The Song of Hiawatha." Other articles in the 1963 volume extolled corn's nutritional value and offered recipes on how to use cornmeal and other foods processed from corn, both frequent features in the wider press. Others offered readers jokes and games or recounted how ancient Mesoamericans domesticated corn. In 1962, for instance, songs appeared in the May, June, September, and October issues.

By comparison, as recently as 1961, only 20 stories fit in similar categories, reinforcing the conclusion that the monthly sought to reach a broader audience than those interested in narrowly technical issues. Reaching a substantial audience, it grew its print run from 44,600 in 1960 to 65,800 in 1964. Subscribers included mostly libraries attached to kolkhozes, technical colleges, and research institutes, meaning *Kukuruza* likely reached a numerous audience. Yet this is also indicative of the development of a more lively visual style characteristic of the Thaw and visible in, among other places, the portrayal of Liubov Li. In *Kukuruza*, this meant covers adorned with color and designed using geometric forms reflecting the influence of modernist graphic design revived to challenge the conservatism of the Stalin era. In contrast to static, stiff portraits characteristic of earlier years, after 1961, *Kukuruza* brought playful imagery and illustrations into the text, adorned the headlines, and added content not directly pertaining to science, technology, or production.

Corn also received Lenin's blessing, furthering the effort to make it Soviet. The prominence of his maxims, writings, and life story re-emerged under Khrushchev, as he and the party packaged struggles against what they called Stalin's cult of personality as a return to a pure Bolshevik past. Beginning in 1958, a new edition of Lenin's voluminous collected writings appeared.[73] The cult of Lenin reached new highs in iconographic representations of his life and veneration of his writings as if they were sacred, both of which reached into every corner of public life.[74] Such reverence even touched corn. The press reproduced a letter the revolutionary leader wrote to the first head of the State Planning Agency in October 1921. Lenin praised corn's potential to aid efforts to feed the then famine-ravaged country, desperate to recover from destruction suffered during World War I and the Civil War. He ordered the government to secure supplies of seeds and to educate the peasantry about the crop's value. As *Kukuruza* told readers, the documents "demonstrate the enormous importance Vladimir Ilich vested in corn as a practical resolution to economic challenges."[75]

Naturally, corn also featured in the Khrushchev cult that showered praise on him and his policies with increasing frequency after 1957. In a typical case, a 1960 conference of corn growers in Belgorod Oblast passed a resolution declaring their dedication to Khrushchev and his corn policies. Invoking the many achievements attained in 1959, they pledged even greater efforts the next year. Communism was the course Lenin had charted, the text declares, and Khrushchev was bringing the country closer to that destination. "We are proud that victory in peaceful competition with the United States will be ours," it explains, "Much depends on us, the corngrowers, and we will not spare our efforts."[76] Communist plenty and the mission to vie with the United States for superiority represented moral incentives for workers, ideals that encouraged—in theory—kolkhozniks and workers of all sorts efforts to work harder, produce more, and bring each goal one step closer.

Corn propaganda spoke to two audiences. Newspapers, pamphlets, posters, and films comprised the mass media effort to market corn to urban audiences who had little contact with the agrarian economy. They transformed corn from a vague presence in some border regions into a symbol of prosperity and abundance, raising expectations that full store shelves and kitchen tables groaning under the weight of produce were just beyond the horizon. When in the early 1960s, these promises seemed broken, resulting bitterness provoked the many jokes citizens told about corn and its champion. When compelling Khrushchev to step down, his former comrades cited advancing age and poor health, reasons fueling another of the many jokes about the so-called "corn-man." "What illnesses forced Khrushchev into retirement?" it asked. "A hernia: he lifted up agriculture, but strained himself and let it slip. Indigestion: he planted too much corn and too many peas. Heart disease: he chased after the United States but couldn't catch up. Logorrhea and foot-in-mouth disease: cause unknown."[77] So linked to Khrushchev was corn that when his former comrades forced him from power in 1964, they dispensed with the imagery of corn-based abundance almost overnight.

In addition, the propaganda machine promoted corn to those responsible for the actual labor of growing it. It expended untold effort to convince farmworkers, specialists, students, and other citizens who worked on the project that it was worthy of their efforts, making it equal in prestige and importance to the gargantuan industrial projects of the 1930s or to the Virgin Lands campaign. By these means, Khrushchev, the party, and the government sought to alter the system of incentives at work in the countryside. For years, the state had ruled the countryside through administrative fiat and possessed, in theory, virtually absolute power via the formal legal apparatus and local implementation of the law.[78] In point of practice, the period witnessed the persistence of undergovernance in rural areas, meaning that state and party power sufficed to control output, but not to transform mentalities, everyday practices, and social norms surrounding work.

In the past, scholars presumed that once collectivization had been carried out, Moscow gave orders and subordinates on the kolkhozes and sovkhozes acted to bring them to fruition. The authorities thought to reshape people, economies, society, and basic interactions with the natural world. By the 1950s, the leaders had not realized their ideals of high modernism and productive modern farming. Although local officials had the power to dictate, they had less control over rural residents than scholars long supposed.

The Khrushchev era introduced numerous changes in the way the party and state intervened in the countryside. Local officials maintained capacity to direct the kolkhozes, but even by the early 1960s, the authority of Moscow still faced considerable limits on efforts to realize policy. The new propaganda reproduced elements of the old by appealing to moral incentives: be the best and contribute to socialism. Its scope and ambition were wide because it attempted to make up for the relative limits of actual control over the countryside. Since the 1930s, the harsh realities of work and daily life on the kolkhozes had demonstrated to peasants that they were second-class citizens. Under Khrushchev, this began to change. The propaganda began to correlate with a changing reality. It no longer simply told kolkhozniks that their grain was a contribution to building socialism or to winning the war. Everyday reality changed, as wages rose and opportunities to consume followed. This is not to suggest that everything changed, or that change came quickly. Well into the 1960s, kolkhozniks continued to lack the internal passports and state pensions enjoyed by others. These delays notwithstanding, the old model had failed and was giving way to a new agreement with the kolkhozniks, a substantial—if incomplete—reform initiated by Khrushchev.

5 GROWING CORN, RAISING CITIZENS

The Komsomol, or Young Communist League, sponsored corngrowing competitions in which members, typically between fourteen and twenty-seven years old, cultivated corn and earned prizes. Exhorting competitors, the organization's leaders and the press made corn a "front" on which "an immense army" of youth was to "struggle for high yields." Conferring symbolic awards, authorities employed moral incentives while offering material rewards in the form of consumer goods. By some measures, these efforts to influence attitudes and actions achieved considerable success. During this decade, the competitions engaged hundreds of thousands of youth who tended millions of hectares of corn and harvested tens of millions of metric tons of grain and livestock feed. In some oblasts and republics, Komsomol-led detachments grew as much as one-half of the corn.[1]

The Komsomol had long appealed to youth members to become militant participants in building the communist future.[2] During the Russian Civil War, the Bolsheviks had mobilized members of this new youth wing to the frontlines during moments of crisis, but the nascent organization's levies had rarely achieved their goals.[3] These efforts later provided precedent for mobilizations during the First Five-Year Plan (1928–1932), when Komsomol members had contributed to industrialization. For instance, under the slogan "On the march for metal," youth had joined in constructing the titanic Magnitogorsk steelworks.[4] Descended from this tradition, Khrushchev-era projects were to inspire new generations to emulate predecessors. Beginning in 1954, the Komsomol mustered volunteers for the Virgin Lands campaign.[5] By 1955, it plunged into efforts to grow corn in virtually every corner of the Soviet Union, proclaiming that its members were "On the march for corn."

The Komsomol designed corngrowing competitions to appeal to youth in new arenas. The organization thus made its own contributions to Nikita Khrushchev's fight to augment the Communist Party's authority in economic management. Naked coercion abated, replaced by a combination of material incentives and appeals to enthusiasm, socialism, and patriotism as motives for diligent work even on farms. Such moral incentives took physical form as banners, awards, trophies, titles, and other objects with worth, but no value in the strict sense. The organization thus responded to pervasive fears about the post-1953 generation's commitment to socialism and labor. The official ideals of the period took perhaps their most potent form in the commandments enshrined in the 1961 Moral Codex of the Builders of Communism, which exhorted youth in particular to act honestly, work conscientiously, and honor principles of collectivism and patriotism.[6] At the same time, the competitions incorporated tightly controlled activism, elements associated with Khrushchev's dictum that the Soviet Union was becoming a "state of all the people" while it was "constructing communism."[7]

Mobilizing youth for work that promised to fashion active citizens, Komsomol leaders preached Khrushchev's corn crusade using militant language reminiscent of that characterizing past campaigns. "Fighters on the corn front," authorities declared, should put themselves "in the vanguard of the competition to achieve high yields of 'the queen of the fields.'" They then took measures to ensure that local committees and members did so. In 1955, activists organized 100,000 work teams composed of approximately ten people each, meaning that the competition enlisted at least 1 million to perform manual field labor.[8] That feat was made possible by expanding Komsomol membership, which between 1949 and 1958 doubled to some 3 million, growth driven partly by expanding numbers of students and urban youth.[9] Yet the Komsomol also began enrolling more rural youth. In 1951, the Komsomol reached only 878 kolkhozes and 100,000 rural members. By 1959, the organization boasted five times as many primary groups on kolkhozes.[10] It expanded to incorporate generations too young to have shared in the wartime experiences that defined those a few years older. As in many countries, postwar prosperity and stability shaped the generations who came of age beginning in the 1950s.[11] Participants in the 1955 competitions, born between 1928 and 1940, were too young to have been combatants. By 1964, the competitors were even more likely to be members of the postwar generation.

Before 1955, the Komsomol had held episodic contests for farmworkers, but the corngrowing competitions benefited from more resources and support, which the organization sustained over many years. Off the farm, it established prizes for youth who distinguished themselves in machine trades or coalmining. Another competition offered prizes to MTS employees who harvested and stockpiled the most fodder.

The corngrowing contests, however, engaged more participants and lasted longer. Beginning haltingly in 1954, they soon grew into national campaigns, which continued until 1964. Comparatively successful, these corngrowing contests contrast with apparently similar efforts in industry, where socialist competitions failed to strengthen labor discipline. Reversing Stalin-era practices, Khrushchev acknowledged a need to encourage workers in all sectors to consider their own interests aligned with those of the government and party. Competitions combined material and moral incentives to encourage productivity and discipline, but workers disregarded them because they believed bosses falsified the results to reward the undeserving.[12] Because they involved youth, the Komsomol's corngrowing competitions had a distinctive character. Moreover, rapid changes in agrarian production indicate that the corngrowing competitions were different from campaigns in industry, where shop floor practices remained comparatively stable.

In 1954, local corngrowing competitions in a few administrative regions predated Khrushchev's launch of the full-scale crusade. The organizers set out to overcome the skepticism of farm managers and local bosses by appealing to younger generations, who tend to challenge the status quo by supporting innovation. Responding to Khrushchev's September 1953 praise for corn, authorities held contests in the republics of Belarus and Latvia, as well as the RSFSR's Omsk, Briansk, and Arkhangelsk oblasts, all located beyond the crop's traditional range. In Belarus, 1,400 work teams set out to master the crop.[13] In neighboring Latvia, the Komsomol organized 800 work teams to grow 10,000 hectares of corn.[14]

In the latter case, sources point to popular reticence to participate, which the Komsomol strived to overcome. Requiring particular caution, reports compiled by the Latvian Committee for State Security (KGB) testify primarily to officials' nervousness about resistance to corn. They detailed informants' intelligence about "negative attitudes" toward corn and the kolkhozes themselves, only formed at the end of the 1940s. Some kolkhozniks complained that they did not actually own the farms because they were excluded from decision-making. Whereas the kolkhozniks approved a plan for 30 hectares, district officials soon arbitrarily raised the total to 46 hectares and then pressed for 110. Other complaints spoke of how "the party forced them to do it," a sign of some sort of "madness." Many pointed to land and soil that they considered unsuited to the crop. The party "has begun to chase after America, where they truly grow corn, but there they have a different climate." As a result, the farms did the absolute minimum to fulfill the planting plan. They selected long-disused plots of land and expended only the bare minimum of effort on preparing the soil before planting corn.[15]

In enthusiastic tones typical of reports to Moscow, the republic's Komsomol described efforts to overcome such obstruction, as well as technical mistakes arising from the unfamiliarity of the crop and methods for growing it. "Many kolkhoz chairmen, and even some district leaders, did not believe in the success of corn," the organization's leaders explained, listing the same tactics described by the KGB. "With the help of party organizations," they boasted, "these difficulties were overcome." Aided by mass-produced pamphlets and the support of the Komsomol, youth work teams grew corn. Those raising the highest yields won cameras, radios, wristwatches, and other consumer goods scarce in the countryside.[16]

Komsomol bosses committed their organization to the corn crusade from the moment Khrushchev launched it. At the January 1955 Central Committee plenum, the youth organization's first secretary, Aleksandr Shelepin, pledged to lead youth using tested strategies. Addressing the Komsomol Central Committee a few weeks later, Shelepin stressed obligations inherent in that promise. Repeating Khrushchev's formula that the advantage of corn was its quality of simultaneously producing grain and silage, the Komsomol boss lamented, "Until recently corn has been confined, undervalued, and planted only in southern regions, and even there in insignificant quantities." Extolling the previous year's small-scale competitions, he exhorted his subordinates to spread them to every farm by organizing work teams and training members dedicated to the task. Shelepin proposed publishing an open letter urging Komsomol committees and members to lead this corn crusade.[17] On February 24, 1955, the executive committee of the Komsomol Central Committee approved such a text, designed "to mobilize youth for the struggle to raise corn."[18]

The following day, the letter dominated the front page of *Komsomolskaia pravda*, which called for youth to join the Komsomol "On the march to raise corn!" Addressing members and nonmembers alike, even those not engaged in farming, the letter envisioned mass participation by students in technical and high schools. Even Pioneers, aged between nine and fourteen years, might grow some corn. The language and style set out the boilerplate phraseology that *Komsomolskaia pravda* continued to use in subsequent years. Providing detailed guidance for the spring planting, the newspaper called for "Komsomol zeal" from all who would engage in "the battle for high yields of corn." It ended with a clarion call: "Join the competition! Let work team compete with work team, brigade with brigade! Spare no strength in growing high yields of corn everywhere this year!"[19]

From there, the message spread far and wide. The Komsomol leaders' letter appeared in the Moscow-based mass-audience newspapers *Pravda*, *Izvestiia*, and *Selskoe khoziaistvo*. Beginning the following day, the call to action spread to regional and local newspapers.[20] The Komsomol Central Committee ordered distribution of 300,000 copies of a flyer relaying the message.[21] Pamphlets, posters, and other materials proliferated, raising awareness by making the message, as a Komsomol functionary stated, "simple and clear, accessible to every kolkhoznik and youth."[22] Officials considered these publications essential, blaming their absence or inadequacy for the unsatisfactory state of corn production in a given region. Soviet Radio added to the wave of corn propaganda, broadcasting on the morning of February 28, 1955, about the Komsomol Central Committee's "communiqué calling on Komsomols and all rural youth to cultivate corn, spreading it to every corner of our Motherland."[23] Although the receptiveness of youth to radio messages is unquantifiable, icons of the competition such as Ukrainian champion Anna Ilchenko later recalled their influence.[24]

The Komsomol appealed to everyone, not only to activists, a fact setting corn apart from the contemporary Virgin Lands initiative. Relocating to remote regions, participants in that campaign disrupted their lives by leaving behind school, friends, family, and established ways of life for months or even years. As many as 300,000 Komsomols went for a growing season, and some 100,000 settled permanently in Northern Kazakhstan, giving it a distinctively youthful atmosphere.[25] By contrast, corngrowing offered opportunities to contribute while requiring less sacrifice because, as Khrushchev repeatedly asserted, the crop could grow nearly anywhere. The Komsomol energized youth already working on farms, as well as students and Pioneers. Leaders of the party, government, and Komsomol alike stressed that corn was the equal of the *tselina*, as the Virgin Lands are known in Russian. As early as January 1955, Ukrainian party chief Aleksei Kirichenko reasoned that, encouraged by Khrushchev, his republic had made "corn our own sort of *tselina*."[26] A 1959 Komsomol Central Committee meeting heard declarations that "Party, Komsomol, soviet, and economic organizations everywhere seek out new reserves and potential, figuratively terming them their '*tselina*.' For some, this is draining marshes, for others it is planting more cropland, and for still others it is designating corn their '*tselina*.'"[27] For good reason, leaders and activists claimed that they had made corn "a Komsomol crop" in much the same way they claimed the Virgin Lands as their own.

In 1955 and 1956, the Komsomol competitions brought to life planning reforms limiting ministries' control over kolkhozes, re-emphasizing the role of the party and its youth wing in organizing production. Komsomol leaders in Moscow required each local committee to hold a contest and report results at the end of the growing season. Central authorities dictated the form and function, but left details about procedures and prizes to the regions. The Komsomol Central Committee's Department of Rural Youth undertook only to send out inspectors, reprimand those found to have mismanaged some aspect of the campaign, and summarize the competition results. Echoing Khrushchev, Shelepin exhorted local organizers to get their hands dirty, both literally and figuratively. In 1955, he warned subordinates that superficial leadership of the coming planting season would earn only reprimands. Without organization, education, and technical proficiency, the efforts of youth would remain an empty gesture.

"The plans should be specific," he cautioned. "Otherwise, everything will remain on paper and become only idle blather. The development of livestock raising requires not plans and resolutions, but action. Cattle and hogs are unable to read resolutions: they require feed."[28]

Komsomol committees educated and organized rural youth. In Moscow Oblast, officials trained local committees and activists in necessary techniques. Speakers recapitulated Khrushchev's words: "Comrades!" the oblast Komsomol secretary declared, "Expanding plantings of corn is the most important means of raising grain yields.... The value of corn lies in that it simultaneously provides both grain and green forage good for making silage." To sway skeptics, he described kolkhozes that in 1954 had grown high yields. The secretary continued, "This is a matter of the honor of the Moscow Komsomol organization, of every Komsomol organization, and every Komsomol member."[29] "We must actively organize things," another speaker stated, by designating "the membership of the work team, plot of land, fertilizer, cultivation, care, and harvesting."[30] More than mere sloganeering, these instructions defined the relatively unfamiliar task. To the surprise of some, these expectations did not disappear after the first year. Thus the Lithuanian Komsomol considered a 1956 competition with neighboring Belarus a success because it had increased yields of feed and, thereby, output of milk and meat. All of this was the "result of significant work every day."[31]

Carefully managed messages in newspapers simulated popular initiatives to reinforce these instructions. The May 5, 1955, issue of *Komsomolskaia pravda* conveyed authorities' expectations, ensuring that no one underestimated the urgency of the crusade. "We will help raise corn!" was purportedly an open letter written by students at a technical school in Voronezh to peers at similar institutions. Its tone imparted to the corn crusade a sense of adventure reminiscent of appeals to participate in the Virgin Lands. Dated April 29, a draft text in the Komsomol Central Committee's files diverges only slightly from the published version, which declared corn a "miraculous" source of grain, feed, meat, and milk.[32] Even if students had written an initial version, Moscow officials fashioned it into part of a strategy to incentivize participation by editing and approving it.

In the mass-circulation newspaper *Komsomolskaia pravda*, authorities criticized practices considered ineffective, even harmful, while describing approved methods for organizing the competitions and working the cornfields. In 1955, such efforts proved particularly necessary because corn suddenly became a priority for many who lacked experience. One kolkhoz Komsomol secretary described her attempt to grow corn the year before, shedding light on approved methods. She outlined how to cultivate corn, organize work teams, and hold a competition, calling attention to common problems. Pointing to needed improvements, the author described how at a conference she attended, "The activists sharply criticized the oblast Komsomol committee because they had neglected to guide youth corngrowers." Looking to the new contest on the horizon for that year, she wrote, "Hopefully, the past mistakes that left the [1954] competition only on paper will not be repeated."[33]

The competitions typically began with challenges, even though oblast and republic officials closely managed these apparently spontaneous initiatives. Reports such as that from the Azerbaijan stressed that district and kolkhoz committees had shown initiative, leaving out mention of prompting from above. "The Komsomol members and youth of Belokan District," the document reads, "challenged others in the republic to join the socialist competition to grow 10 metric tons of grain and between 80 and 90 tons of green fodder per hectare." Committees had to meet expectations about organizing, which they demonstrated by relaying to Moscow the number of participants, hectares cultivated, and tons of corn harvested. Ensuring "substantive success in this patriotic task," the Azerbaijan committee's "active engagement" had organized 6,000 participants to cultivate 37,625 hectares, more than half of the republic's total of 70,000.[34]

Purportedly popular initiative also fueled programs for Pioneers and schoolchildren. In response to a challenge initiated by a group in Leningrad Oblast, Pioneer troops tended small plots of corn from Stavropol Krai in the south to Kirov Oblast in the north.[35] In April 1955, the Arzamas Oblast Komsomol committee reported on a partnership with the local department of education, which ensured that "schoolchildren in nearly every district willingly help the kolkhozes grow corn."[36] Elsewhere, groups of children led by peers and guided by responsible adults promised to grow plots of corn up to a few hectares in size. An award-winning detachment from the school in Grinev, Brest Oblast, comprised Pioneers from the seventh grade, approximately fourteen years old, who cultivated a plot of a single hectare.[37]

Komsomol leaders chastised local officials unwilling to demonstrate hands-on leadership. In Gomel Oblast, one district's competition "proceeded lifelessly" because the local committee lacked initiative and failed to promote enthusiasm in youth. In a common incentive strategy, district or oblast committees awarded trophies to the work teams that completed planting, cultivating, and harvesting with the highest marks for efficiency and effectiveness. Teams receiving this praise served as examples to the rest, much like vanguard workers. In Gomel, however, district authorities acted only when the oblast committee demanded progress reports.[38] By contrast, a Kaluga Oblast secretary extolled one district committee for "exhibiting practical leadership." He wrote that the committee "regularly tallied the results of the competition between work teams and sponsored articles in the district newspaper by leaders of work teams, with the goal of broadcasting exemplary practices." This committee's head "played a major role by often visiting the work teams and offering them practical assistance."[39]

Such reports outline an ideal of leadership. In keeping with Khrushchev's example, leaders needed knowledge about production methods in order to educate others. In a report about the successes and failures of the Komsomol in Mari ASSR, one inspector reserved her fiercest criticism for the secretary. "[He] understands the state of affairs in the oblast poorly," she charged, because "in the past five months, he has not taken a work trip to assess any district." Falling short of the ideal of hands-on management, his work lacked "efficiency and a businesslike manner" as well as "creativity and imagination." Because he failed "to delve deeply into the issues, he often could not clearly express his own judgment."[40] Detailing their efforts, the leaders of the Latvian Komsomol contended that such exceptional effort was necessary because corn was new to the republic's kolkhozes. They therefore had to "regularly oversee the Komsomol organizations' contributions to corn cultivation. To that end, the bureau heard reports by district and kolkhoz Komsomol secretaries during its meetings."[41] Monitoring subordinates furthermore entailed observing work firsthand, ensuring that the necessary efforts did not remain merely "on paper."

In Ivanovo Oblast, the committee knew only the total number of work teams organized, but not the number of hectares they grew. This proved that its leaders "weakly oversaw the district committees and primary organizations" by superficially counting measures of process. In the oblast's Sokol District, only half of farms had Komsomol work teams. Even where youth had organized to grow corn, they had exceedingly modest responsibilities barely matching the number of hectares expected of children. On the Chapaev kolkhoz, the work team cultivated a field of 3 hectares, whereas the local school group had pledged to cultivate the same number.[42] Coming to light only when outside inspectors visited, these situations cast the oblast committee leaders in a poor light. They had either knowingly allowed them or failed to discover them when carrying out their oversight.

Requiring more than outward appearances, the Komsomol Central Committee charged with "formalism" those local organizations failing to provide evidence of substantive results. In some cases, oblast authorities dumped responsibility onto others. The Saratov Oblast committee, for instance, relayed to Moscow the results of their competition, but bemoaned districts that either did not hold a proper contest or did not report any results at all. The committee thus shifted blame onto subordinates who "had approached the implementation formalistically."[43]

In April 1955, the Department for Rural Youth praised a few oblasts and republics that had organized widespread competitions, but condemned the remainder. "A host of oblast committees," the department explained, "unsatisfactorily implement [the February 1955 directives] by formalistically creating work teams." The department singled out one oblast for organizing only 300 work teams on its more than 1,000 kolkhozes and sovkhozes. Worse still, local officials allowed work teams to pledge themselves to plant a superficial amount, rather than as much corn as they were able.[44] Similarly, an inspector condemned a district committee in Moscow Oblast, which had organized work teams only "in formalistic fashion."[45] Six weeks after the Komsomol launched its effort to contribute to the corn crusade, the committee had formed only eighteen such teams, given them inadequate support, and assigned them insufficient numbers of kolkhozniks.[46] In Sokol District of Ivanovo Oblast, the committee had failed to back up its efforts to form work teams with action, ensuring that they "existed formally, on paper alone." The oblast committee had neither organized the corngrowing contests nor investigated progress "on the ground," permitting a shortage of initiative and oversight from all directions.[47]

Once republic and oblast committees had established a competition, the Komsomol Central Committee also expected them to publicize it, encouraging youth to participate. They employed what the committee termed *glasnost*, which in this context meant employing the press, lectures, meetings, and other means of disseminate knowledge about corn and the contests. The Central Committee's 1955 letter required each committee to sponsor meetings about the competition. In response, the Belgorod Oblast committee obliged its subordinates to meet and discuss the campaign in an effort "to encourage extensive *glasnost* in the competition."[48] The Central Committee consistently invoked the term in subsequent years, as in a 1958 directive on the competition. An inspector sent from Moscow to Ukraine's Dnipropetrovsk Oblast condemned the local committee's "serious shortcomings," as a result of which "many work teams in Sinelkov, Dnepropetrovsk, and many other districts do not know the competition rules."[49] In Penza Oblast, "a few bureau members tally the results without inviting participants or Komsomol activists. They tally the results formalistically, without a deep analysis of the work of each work team. There is no *glasnost*."[50]

Assigned a value by leaders comparable to that of the Virgin Lands initiative, the competitions embodied renewed efforts to increase the Komsomol's prestige and to encourage youth to join projects that authorities hoped would define a generation. Although official reports indicate expectations and outcomes, they provide little insight into participants' motives. Leaders' speeches, newspaper accounts, and related sources portray those motivations acceptable to the official line: the prospect of winning praise and awards for the collectives to which youth belonged. Material rewards, youthful enthusiasm, and social belonging may explain why others joined "the struggle for high yields of corn."

Replicating official terms, young participants emphasized that corn was their contribution to the project of bringing prosperity to the Soviet Union. To succeed, they had to test the status quo embodied in the authority of the kolkhozniks, kolkhoz bosses, and local authorities who scoffed that a crop as strange as corn could not grow. One youth work-team leader in Belarus recalled, "When we began to plant corn, [local leaders] didn't pay attention to us, or they laughed at us. But we wanted to prove that corn could grow, regardless of it all, even in imperfect climatic conditions."[51] A kolkhoz Komsomol secretary described a similar response to the organization's call to arms. "I remember the meeting where kolkhozniks deliberated over preparations for the spring planting," he recalled, "It was difficult to prove to them that the crop could grow on our farm, in the conditions and soils of our region." The secretary continued, "Even the kolkhoz's management doubted the chances for success. To prove that corn could grow on our kolkhoz all the same, the Komsomol members decided to aid in the task."[52]

In 1955 and 1956, each oblast and republic Komsomol committee set rules and distributed awards largely autonomously. Subsequently, each participated in the annual All-Union Socialist Competition of Komsomols and Youth in Raising High Yields of Corn that was first cosponsored by the Komsomol Central Committee and the USSR Ministry of Agriculture in 1957. The rules established first and second prizes for reaching certain per-hectare yields of grain or silage, as well as appropriate material rewards for each. For oblast and district Komsomol officials, winning brought a bonus equaling one month's pay. The committee received either an automobile for first a prize or a motorcycle for a second, both for official use. The Komsomol organization of a kolkhoz or sovkhoz might win a radio, a set of musical instruments, or an assortment of sporting equipment. Outstanding work-team leaders might receive a wristwatch, a cash payment, or an all-expenses-paid trip to Moscow to visit the All-Union Agricultural Exhibition.[53] Contributed jointly by the Komsomol and the ministry, a typical annual budget for prizes was 1.3 million rubles.[54] Although small in comparison to the overall government budget, the goods represented a substantial material benefit to any winning individual or group. Other incentives brought recognition in the form of invitations for 1,000 outstanding representatives to attend a 1958 Conference of Youth Corngrowers in Moscow. In addition to the honor of selection, the event awarded them an expenses-paid visit to the capital, a trip likely difficult for the average rural resident to undertake independently. Outstanding Pioneers, like their older counterparts in the Komsomol, could also win prizes such as a trip to the country's prestigious Pioneer summer camps.

The prizes augmented the moral incentives encouraging participants to consider themselves part of a larger project and to embody socialist virtues. The Komsomol and its committees widely distributed banners, pins, and certificates similar to those the party gave adult laborers. In promoting competition, the committees constantly appealed to patriotism, socialism, and pride in the organization's history of achievements. Even if they achieved lesser effect, these nonmonetary awards spread further than the material prizes they supplemented. Many earned an award and received an accompanying lapel pin emblazoned with the words "For raising high yields of corn."[55] For work teams, brigades, farms, and districts, traveling trophies signaled to passersby a successful collective effort. Such banners, as a district in Ukraine reported in 1955, "contributed to raising the activity level of Komsomols and youth."[56] Others won higher awards, up to and including designation as a Hero of Socialist Labor.

The story of one such medal winner, Anna Ilchenko, testifies to the forces that motivated participants, even if the recognition she received made her story unusual. Her story was typical in that the work team was predominantly composed of women who cultivated corn with intensive manual labor.[57] In April 1958, success in growing corn brought Ilchenko from Ukraine to Moscow for the Komsomol's Thirteenth Congress, which gave her the rare opportunity to speak before the highest ceremonial gathering of youth leaders. "From this podium," she declared, "I want to thank the party for its care for us, the kolkhoz youth. In the name of all the girls in my work team, in the name of the kolkhozniks of Cherkassy [Oblast], allow me to thank Nikita Sergeevich Khrushchev for teaching us to love and to raise a crop so valuable as corn."

Ilchenko's brief speech illustrated the Komsomol's mission with events from her life on a kolkhoz. She described the labor and daily life of the nine young women who were members of her work team. Having finished high school, they faced a choice: remain at home and work on the kolkhoz, or leave their village in search of education and work in the cities, as many of their peers did. Ilchenko recalled that they were told, " 'Don't stay on the kolkhoz or you'll never get married!' " Having nonetheless dedicated themselves to the farm, she and her comrades had proven the skeptics wrong. They earned a technical education and found suitable husbands in a rural community short on eligible bachelors. The young women, now valued farmworkers, married "not drunks, but tractor-drivers, agronomists, and combine operators," men with skills and prospects. She described a Ukrainian tradition allowing young women to give pumpkins to suitors they chose to reject. "We plant them between the rows in our cornfield for that purpose. The first year, we gave out several to the local loafers. Then we decided that that was a waste of pumpkins, so instead we gave them to the kolkhoz cows so as to produce more milk."[58]

Most important, Ilchenko and her friends had mastered growing corn. In fact, by 1957 she had become a Hero of Socialist Labor.[59] Such achievements opened to Ilchenko and others like her the path to party membership, awards, and prizes.

Because of them, she found herself on the Congress podium, serving as an exemplary leader of her local Komsomol group and winner of the annual corngrowing contest.

Ilchenko's choreographed speech also popularized the competitions by reaching audiences across the Soviet Union. She described her team's 1958 goal to grow 13 tons of grain per hectare, several times the average yield. After Ilchenko quit the podium, the Komsomol first secretary, Vladimir Semichastnyi, read out a note from a challenger, Anna Muntian, leader of a work team in Moldova, who promised to raise 13.5 tons per hectare.[60] The contest between the two groups of women to determine who could grow more corn illustrated on a small scale the much larger competition among districts, oblasts, and republics.

As Ilchenko's story of marriage suggests, the Komsomol exhibited concern for youth as farmworkers, but also for their way of life. She embodied an ideal in which young women pursued education, skills, and a suitable husband. The Komsomol oversaw moral education as well as labor, commending efforts to guide participants' personal development and admonishing any who failed to instill communist values in youth as they supervised their work. Committees praised outstanding activists who carried out the corn crusade while fulfilling other vital missions. Furthering the mission to educate members, Komsomol activists in Moldova augmented conscientious work in the fields by studying political events and engaging in "cultured leisure," a term denoting drawing, reading classics and political tracts, playing team sports, and other approved recreation.[61] An organization in Lithuania praised by local officials combined the two seamlessly. Emphasizing corn and culture, its members had created a theater group, dance troupe, string ensemble, a book club, and one-page newspaper.[62]

The Komsomol was to educate youth in citizenship and police drunkenness, theft, rudeness, and similar disorders. In one instance, kolkhoz committees in Minsk Oblast established "posts" designed to keep watch over communal property and behavior. Activists sought out individuals who worked poorly, conducted themselves in a lazy or rude manner, or who were drunkards and thieves, all behaviors evidently prevalent. The surveillance made such malefactors, local Komsomol officials explained, "afraid not only of Komsomol activists, but of rank-and-file youth." Those youth leaders "battled" poor labor discipline, as well as "Komsomol members' amoral conduct, hooliganism, and profanity."[63]

The Komsomol's corngrowing competitions achieved more successes than related methods. In another type of campaign, an individual, group, farm, district, or oblast identified an obligation, a target for production beyond that set established in the plan. These common pledges seem most often to have fallen by the wayside once actual work got underway. Receiving fulsome praise in the press, vanguard workers and farms always fulfilled their obligations. Rank and file workers, however, received little publicity and seem to have had little incentive to follow through. Individuals and groups made similar pledges "in honor" of an upcoming plenum or congress with similar frequency, as those of the "Zavet Ilicha" kolkhoz in Rivne Oblast did in September 1959 (Figure 5.1). In another instance, months before the Twenty-first Party Congress, *Pravda* featured editorials calling for a "nationwide competition." Banner headlines accompanied high-profile pledges to grow cotton, hew coal, harvest corn, and more.[64] Responding to the moral incentives inherent in these campaigns, workers apparently took on these burdens formulaically, quickly forgetting them. The Saratov Oblast Komsomol committee "accepted elevated obligations and carried out major organizational work on the mobilization of youth for their fulfillment," resulting in 400 work brigades, 1,800 teams operating combines, and 2,094 corngrowing work teams.[65]

Yet these numbers provide little sense that these teams' actual production was of equal concern. Other oblast committees sent similar numbers to Moscow, but bemoaned district committees' failure to follow through. One decried "serious shortcomings... in mobilizing youth to fulfill their socialist obligations." It cited how Komsomol members on several kolkhozes "had outlined specific tasks and named substantial obligations, but those did not serve as a plan for actual work." As a result, only two district organizations had met their obligations.[66] Heightened enthusiasm and organized agitation did not, in the end, translate into increased production unless backed with incentives in the form of prizes and awards.

Despite the prevalence of fraud and efforts to conceal it in agriculture in the late 1950s and early 1960s, these phenomena were apparently rare in the Komsomol corngrowing contests. Evidence of systematic falsification is common in the archival records of the Central Committee of the Communist Party. Little evidence of dishonesty exists in the equivalent Komsomol files.

Isolated incidents did occur. In 1960, the North Ossetia ASSR committee nominated for a national corngrowing prize a youth, Boris, who had grown the highest yield in the republic. Dispatched to Moscow in support of the nomination, an evaluation of him followed the standardized form and, later, might have eased admission to an institution of higher education. The document featured the results Boris and the nine other members of his work team achieved: in tending 90 hectares of corn, they had surpassed their obligation of 7 metric tons per hectare by growing 17 tons.[67] A few months later, in January 1961, the oblast

FIGURE 5.1 On September 16, 1959, the kolkhozniks on the kolkhoz "Zavet Ilicha" in Mlinovsk District of Rivne Oblast, Ukraine, harvest the ripened corn cobs from a field. Written in Ukrainian, a nearby banner reads: "Our gift to the Twenty-Second Congress of the Communist Party of the Soviet Union." A kolkhoz official weighs the baskets of cobs, which the women have harvested by hand.

Photo by A. Platonov, courtesy Russian State Archive of Cinema and Photo Documentation, ID#1-54062a & 1-54062b committee retracted its nomination in an explanatory note, a document required when making an admission of failure or wrongdoing. Despite attestations signed by the proper officials, further investigation had discovered a yield not of 17 tons, but of a still-impressive 13.2 tons. The note does not reveal how the incorrect information had been approved, but the circumstances suggest that the separate kolkhoz and Komsomol officials responsible for verifying the results had colluded to sign documents turning with the stroke of a pen an impressive yield into a record-breaking one. The oblast Komsomol committee recognized its "complete responsibility for the mistake," reporting that its standing executive committee had "held a serious discussion about it." The infraction's magnitude forced oblast Komsomol officials to assure superiors in Moscow that they "had taken measures so that such an incident would not be permitted again."[68] This type of fraud likely occurred, but authorities' extensive inspections apparently uncovered such episodes infrequently. Beginning in 1959, when cheating on agricultural output statistics reached its height, the Komsomol conducted rigorous inspections that might well have exposed fraud, but did not.

By the early 1960s, the competitions expanded beyond the Komsomol's area of responsibility. The RSFSR began to hold them for mechanized corngrowing using methods popularized by Nikolai Manukovskii and Aleksandr Gitalov. Engaging adults too old for Komsomol membership, the introduction of these contests indicated that authorities considered the Komsomol competitions effective. The contests offered rewards primarily to those who used machines to plant large fields of corn, a job typically done by men, rather than the manual labor performed by women.[69] In May 1961, the party secretary for Kabardino-Balkariia ASSR praised the competition for incentivizing workers "to fulfill their duties in cultivating high yields of corn" and lauded a commission established to monitor the contest. The competition among work teams, brigades, kolkhozes, and districts measured the quality and speed of plowing, planting, cultivating, and harvesting, much like the parallel contests run by the Komsomol. For this, each group's leaders earned prizes and other material rewards, as well as moral ones: a pin declaring the wearer a "Superior Corngrower of the Kabardino-Balkaria ASSR" or a place on the republic's public honor roll of outstanding workers. The republic capital held an annual Corngrowers' Day to celebrate those "who achieved the highest targets in the socialist competition," an incentive deemed "the most important."[70]

The corngrowing competitions evolved to conform to the ethos of participatory governance characteristic of the early 1960s, changes permitting Komsomol authorities to address concerns arising from earlier contests. Moving to put them "on a social basis," they established district, oblast, and interoblast Councils of Youth Corngrowers to administer them. The councils brought together several levels of Komsomol leaders to adjudicate the results and organize what they called social inspectors to carry out reciprocal inspection campaigns among the competitors. These approaches replaced Komsomol officials who might occasionally inspect the fields with a whole group of peer competitors from a neighboring farm, district, or oblast. More numerous, the groups also had little incentive to turn a blind eye to so-called shortcomings, as officials often did.

This change testifies to the emphasis on activism arising in the late 1950s. Seeking to improve governance and economic efficiency, Khrushchev promoted measures purporting to replace administrators with volunteers as part of a process in which the Soviet Union, as it constructed communism, was becoming a so-called state of all the people. Having overcome class conflict and created socialism, the country looked forward to the imminent arrival at that higher stage. Under communism, administrative functions were to devolve to nonprofessional groups, leaving the Communist Party as the unifying force. Having been strengthened by Iosif Stalin, state bureaucracies were to begin to wither away as Vladimir Lenin had promised, giving each citizen a stake in economic and legal procedures. The Komsomol itself was nominally a social organization, but was in fact fully part of the party-government complex.

Ostensibly independent institutions, the corngrowers' councils reserved leadership positions to party and Komsomol figures who responded to the demands of party and state, producing changes in form but only small differences in practice. Standard terminology called this "reconstruction of agricultural leadership by developing democratic forms of management."[71] Lasting until 1964, the Councils of Youth Corngrowers embodied this principle by ensuring that agricultural officials, Komsomol leaders, members of the organization, and rank-and-file farmworkers determined the prizewinners.

In the late 1950s, social inspectors verified the contest results and the quality of competitors' work, representing a first step toward popular participation. Predating the councils, these groups themselves drew on earlier precedents. Official descriptions and press accounts emphasize their apparent public character, which derived from their mixed membership comprising Komsomol activists, agricultural specialists, and farmworkers.

In growing numbers, groups called light cavalry conducted inspection raids, practices and terminology with roots in the Stalin era.[72] In May 1958, the Stavropol Komsomol newspaper *Molodoi leninets* published a message from the krai Komsomol committee whose staccato sentences printed in capital letters mirrored the format of official telegrams. It *"called the attention"* of city and district *"commands"*—the Russian *shtab* denotes a military staff—*"to agricultural laborers' pressing tasks."* It required detachments to gauge progress in sheep shearing, feed supplies, the grain harvest, and *"the careful cultivation of corn."* Thus the telegram ordered local detachments of light cavalry *"to establish Komsomol inspection posts in crucial places, using various forms of public oversight to uncover shortcomings, rapidly remedy them, and report on them to the krai light cavalry command for publication in* Molodoi Leninets.*"*[73] The krai and district Komsomol committees, moreover, described the raids in similar terms.[74] Much in the same way that the competitions and their prizes encouraged participation, these strategies relied on moral incentives: the sense that this activism was a contribution to a higher purpose.

This apparent activism became more widespread as the Councils of Youth Corngrowers proliferated beginning in 1961. Like their forerunners, the councils remained under the direction of Komsomol leaders while including activists, farmworkers, and specialists. They attempted to improve the competitions' effectiveness in incentivizing participation and productivity. Their reciprocal inspections differed from the earlier raids by putting competitors in charge of governing themselves. Work teams inspected their neighbors' plots, while district councils traveled to nearby districts and oblast councils to surrounding oblasts. A June 1961 Komsomol Central Committee plenum highlighted the councils and described how they should function. One local secretary described how they "rendered practical aid" to work teams, "directed socialist competition," and cooperated with the Komsomol committees. In one exemplary district, members of the council themselves both worked and carried out inspection campaigns. To complete a time-sensitive job, they organized a district-wide *voskresnik*, a voluntary workday on Sunday, typically a day off. Thus the secretary offered guidelines for how the councils should function. Komsomol-led oversight posts and social inspectors had put 2,000 specialists, vanguard workers, and other activists in the fields to carry out such mutual inspections.[75] Similarly, another Komsomol chief described how the councils, encompassing some 6,000 officials and activists, had conducted raids revealing "many examples of Komsomol committees' unsatisfactory aid to corngrowers."[76] Only this kind of "mass political work" ensured that "fighters on the corn front" received the support they needed.[77]

The oblast and interoblast councils reported to the Komsomol Central Committee's Department of Rural Youth about the winning teams, districts, and oblasts.[78] The interoblast councils gathered, condensed, and passed on much of the same information that the Central Committee had gathered from regional committees between 1955 and 1960. In theory, the councils adapted and responded to individuals and their concerns with greater ease than the established bureaucracy. Members were more likely to work in the fields and represent a variety of professions. For instance, the council for the RSFSR's Central

Black-Earth Zone agreed that the constituent oblasts' delegations, while headed by the oblast Komsomol secretary, should also encompass district officials and at least one rank-and-file farmworker.[79] Komsomol officials nonetheless filled all leadership positions, meaning that the councils remained firmly under the youth organization's control.

The councils relied on reciprocal inspections among groups, a practice suggesting that prior administration by the Komsomol had not fully achieved its goals. Implicitly drawing that contrast, the Krasnoiarsk Krai agricultural department's representative to the interoblast council praised this measure by noting that they had "revealed the state of things, aided in thoroughly explaining the work, and proven themselves impartial judges."[80] In theory, mutual inspections meant that authorities could not report a satisfactory situation in their area in disregard of actual conditions. The practice put more observers in the fields than the Komsomol could dispatch on its own, ones who had less incentive to accept the status quo.

By eliminating the possibility that neighbors might reciprocate in turn, the circular assignment of inspectors assured that competitors did not ignore neighbors' failures. Instead, one oblast's council inspected a second, but that second oblast's council inspected a third: to illustrate, in 1961 Voronezh Oblast's inspectors traveled to Belgorod Oblast, the inspectors of which traveled to Kursk Oblast, and so on through Briansk, Orël, Lipetsk, Tambov, and back to Voronezh.[81] The Komsomol Central Committee and its Department of Rural Youth praised the councils in similar terms. Considering Khrushchev's 1962 reforms that reshuffled local Communist Party bureaucracies, the department recommended maintaining the councils because they had made substantial contributions.[82]

Despite the councils' efforts, failures to use the recommended methods remained entrenched. The Central Committee charged, for instance, that subordinates in Turkmenistan "had set adrift the work of the Councils of Youth Corngrowers, formulaically organizing socialist competition" and failing to offer education. Furthermore, half of the fifty specialists in the republic with special training in corngrowing did not work on growing the crop.[83]

Demonstrating the Komsomol's attempts improve the efficiency of economic administration, these efforts achieved only limited success. In January 1961, the Komsomol Central Committee highlighted model committees' efforts to "render active aid to party and economic organizations in restructuring agricultural management on the basis of democratic forms of administration."[84] The councils were contributing substantively to the corngrowing competitions and encouraging activism, but concurrently revealed the limits of those initiatives. They held competitions and judged the results, previously the purview of oblast Komsomol committees. In theory closer to the practical work in the fields due to the broader membership, the councils did not ensure substantive improvements in output of corn. The councils remained under the control of the same Komsomol secretaries who, as before, were personally responsible for ensuring that outcomes appeared positive. Those members actually drawn from workers and specialists had little executive authority, undermining the public activism purported in the councils' mission.

After Stalin died in 1953, discussions of socialist legality had quickly emerged. By the 1960s, talk of social organizations and a state of all the people gave rise to related phenomena. Comrades' courts and people's patrols involved everyday citizens in policing transgressions against the criminal code as well as against social norms, expanding citizens' role in regulating their own communities. The blunt coercion of the Stalin era fell into disuse, supplanted by social pressure and other, softer forms of control.[85] Yet these controls left individuals many loopholes to exploit.[86]

As the corngrowers' councils suggest, ostensibly independent institutions employed the same party, Komsomol, and government officials, meaning that they continued to reflect the demands of the party more than the people. In this, the councils and similar organizations responded to social forces only when they aligned with official priorities, a reality that indicates that even these did little to limit the possibility of evading party dictums if self-interest demanded it.

In practical terms, the Komsomol corngrowing competitions further illustrate how farms transformed corngrowing by applying technologies and other elements of industrial farming with increasing frequency after 1960. The technologies Khrushchev expected farms to apply slashed the number of farmworkers required in the fields, even as the number of hectares cultivated rose strikingly. In Ukraine in 1954, 4,500 work teams cultivated 100,000 hectares, an average of about 22 hectares per team. In 1958, these numbers were 20,000 teams and 400,000 hectares, dropping the average slightly to 20 hectares per team. In the transformative year of 1959, some 24,335 teams cultivated a total of 2.03 million hectares, an average of 83.5 per team. In 1960, the number of teams began to decline, falling back to 20,000, even as they cultivated 3.2 million hectares, an average of 160 hectares.[87] In 1962, the Komsomol organized 90,000 teams and planted 18 million hectares, a figure nearing one-half of the total, 37.2 million.[88]

The transformation seems even more remarkable in light of an estimate of the number of workers in each team. In 1955, a work team comprised ten, even twenty or more manual laborers, who perhaps got help planting or harvesting the crop, but cultivated it by removing weeds row-by-row using hand tools. By 1960, the work teams known as "links" were gone, replaced by new teams, *agregaty*, or collections of tractor drivers and assistants, who operated the machines. Such a collection was composed of one driver and an assistant, or at most a handful of drivers and assistants. The result, a few workers with machines, chemicals, hybrid seeds, and other capital cultivated between 100 and 200 hectares and replaced the labor of at least 100 manual laborers.

In the Khrushchev era, leaders again viewed productive manual labor as fundamental to the proper upbringing of citizens of the socialist present and the prospective communist future, a conception seen rarely since the early 1930s. Propaganda campaigns and school programs inducing students to join the corngrowing competitions sought to inculcate in the younger generation a work ethic and, thereby, commitment to official ideals. Authorities feared that youth who were growing up in a postwar society becoming more urbanized and educated had become increasingly distant from ideology and disdainful of hard work. Stalin's reforms to the education system introduced seemingly conventional classroom learning and preparation for higher education. The postwar years tested the balance between official expectations and the experiences of youth, who mixed loyalty to the Soviet Union with questionable devotion to official dogmas about socialism.[89]

In some cases, corngrowing formed part of proper *vospitanie*, a term donating the education of a moral citizen. More broadly, the system of higher education sought to nurture in youth a comfort in close peer groups, enduring structures bringing youth together and keeping them part of the collective.[90] A report on a group of schoolchildren in Penza in 1959 described how they had worked for four years in the fields in their spare time. By making certain they contributed to the local kolkhoz, the school had "conducted major work in the labor education of its pupils."[91] Literature tailored for teachers explained how to encourage their students to participate in growing corn. *Kukuruza*, *Narodnoe obrazovanie* [Popular Education], and other periodicals reinforced this message.[92] Corngrowing by schoolchildren and teenagers illustrated new educational policies and practices supported by Khrushchev. Providing youth with hands-on work experience, these programs aimed to teach practical skills and inculcate appreciation of manual labor.[93]

As in other areas, the example of the United States influenced leaders or, at the very least, induced them to express their beliefs. When the 1955 delegation of agricultural experts returned from the United States, its members stressed the importance Americans ascribed to teaching teenagers to work. Members reported to Khrushchev and other leaders that rural communities nurtured an appreciation for manual labor. Secondary-school students spent summer vacations driving tractors or detasseling corn. The first secretary's belief in the importance of such labor was made apparent by an unguarded exchange with the hybrid-corn expert Boris Sokolov, who explained that Americans expected youth to work on the farms. "They really do habituate children to work," he exclaimed, describing how the sons of the researchers and university professors he met worked on nearby farms. "This is not," he clarified, "because they don't have money to feed themselves, but because [Americans] consider that [young people] should have work experience." To this, Khrushchev retorted, "And *here*, if a professor has a son and he finishes secondary school, he doesn't want to go to work!"[94]

By the late 1950s, Khrushchev's education reforms required students completing secondary education and seeking to enter higher education to perform manual labor first. The story of a student in Stavropol sheds light on this initiative and the comparatively subtle forms of coercion faced by youth. Aiding kolkhozes short on labor, authorities assigned students and Komsomol members to weed or harvest corn by hand, a remedy to the aversion to work diagnosed by Khrushchev. The students' incentives for doing the work were threefold: they received pay, while refusal to work risked disapproval by peers, and, more consequentially, punishment by educational intuitions or Komsomol committees. In the words of one who participated, the work was "voluntary-compulsory."[95]

In the autumn of 1958, students from a Stavropol medical college traveled to a nearby sovkhoz to lend a hand during the corn harvest. This otherwise unremarkable event became the subject of a story in the krai youth newspaper, *Molodoi leninets*. Its title, "*Izhdivenets*," set the tone by naming its subject a dependent, with the implication that the individual was undeserving.[96] The author leveled this charge at Iosif, a first-year student in dentistry. Alone among his comrades, Iosif did a poor job of harvesting corn, the newspaper claimed, because he resented having to pick cobs from stalks by hand. When challenged to work conscientiously, Iosif haughtily retorted, "If you don't like my work, do it yourself, and I'll go home," at which point he left the field and returned to the city. For this, he received a mild reprimand from the college's Komsomol committee. The newspaper charged that this incident naturally resulted from a lifetime of coddling and unearned advantage, the perceived ill to be combated by Khrushchev's educational reforms then coming into full force. They required students applying for higher education to have proof of experience in production. In this vein, the author of "*Izhdivenets*" implied that Iosif had acquired the supposed experience recorded in his labor book not by earning it, but by using family connections to work for a month in a health resort and for a few irregular hours in a machine shop.[97] A few days before, *Stavropolskaia pravda* had similarly named Iosif "barchuk," a young lord, someone who haughtily disdained work. "Let him first work a bit in production," it counseled, after which, "having perhaps learned to appreciate labor, he will be mature enough for the college."[98]

The two articles signal the rising significance of work in education, but do not tell Iosif's side of the story. He did so in a letter to the editorial board of *Molodoi leninets* and appeals to the krai party committee. His entreaties provide additional details: first, after the events of November 1, he received a reprimand from the Komsomol committee. The newspaper articles followed and only after that, on November 19, was he expelled from the youth organization and from the medical school. Moreover, he received this punishment not for rudeness or shirking work, but for poor academic performance, a charge he considered baseless given his prior good standing.

Second, Iosif's version of the events of that day in the cornfield added crucial details the newspaper omitted. He directed his rude remarks not at a team leader or a fellow student, but imprudently at Sergei Maniakin, the head of the krai party committee's agricultural section, who happened to inspect the farm that day. When Maniakin questioned Iosif about unharvested corn in rows near to his own, Iosif responded that those rows were not his responsibility, speaking words similar to those later taken out of context in the newspaper. Not rejecting all work, he stated only that he considered those rows his fellow students' duty. Within a few minutes, he acknowledged that his words were "tactless" and "hot-tempered," and asked Maniakin to forgive him. However, the official "did not accept my plea for forgiveness," Iosif later wrote, "and began to threaten me with expulsion." The youth conceded that he left the farm, but for a reason he considered justifiable. He had neither been forced to leave nor quit in anger, but instead had secured the blessing of his superiors to travel the 30 kilometers to the city of Stavropol, seek out Maniakin in the office of the party committee, and beg his pardon once again. When Iosif arrived, Maniakin was absent. Iosif instead told his story to a party official who encouraged him to learn a lesson from the incident and return to work on the farm, which he did. Only later, when the newspaper articles appeared, did Iosif learn of the danger to his otherwise bright future. In his letter, he hinted that the stories and punishment happened at the instigation of "an influential person," that is, Maniakin, who fabricated a case against him.[99]

Iosif's pleas to the newspaper's editors and Komsomol authorities apparently achieved little. Dated January 20, 1959, the letter caused the head of the student section at *Molodoi leninets* to write to the krai party committee inquiring about the incident. In his memorandum, he expressed sympathy for Iosif. "Judging by the impression [Iosif] made during a face-to-face meeting," he wrote, "and according to his story, the behavior of ... Comrade Maniakin was not entirely objective." Unfortunately for Iosif, nothing seems to have improved. In March, the krai party committee reaffirmed the validity of his expulsion from the Komsomol and the college.[100]

First, the incident illustrates the capricious and personal nature of authority. Iosif was a relatively well-connected young man: his father was a respected senior faculty member at another college and his aunt served as a deputy in the soviet of the nearby town, where she administered the health resort where Iosif had worked. His privileges and connections, however, could not protect him against a figure with the authority and connections to newspaper editors required to carry out this vendetta.

Second, the story illustrates the official unease about links between work and education. During later years, students in higher education were commonly assigned to harvest-time work. Many tried to avoid it, and few found the work itself a positive experience, especially in light of conditions on the kolkhozes. At most, a few bonded with peers during these forays into the country: "We were young," one recalled in a later interview, "we were with the girls, there were dances, and we celebrated birthdays and organized picnics."[101] Publicizing the incident involving Iosif, the local press had devised a morality tale for others, demonstrating the consequences of refusing to join enthusiastically in collective endeavors. In this version of the story, Iosif was unprepared for the work of harvesting corn because he had not been properly raised to appreciate manual labor.

Authorities also created student production brigades to tend crops for an entire growing season, a program designed to accustom youth to regular work, teach about agriculture, and inculcate appropriate values. Under supervision, groups of fifty or more high-school students banded together to cultivate corn and other crops. Living, learning, working, and having fun together, students took responsibility for the work. In return, they gained wages, knowledge, and appreciation for an honest day's work. Like the contemporary education reforms, this program responded to rising concerns that formal classroom education encouraged students to eschew manual labor. First formed in Stavropol Krai's Novo-Aleksandrov district in 1955 and spreading rapidly thereafter, the brigades soon earned praise from national leaders. On a visit to the krai, Khrushchev noted the "outstanding" achievements of students in the brigades who, "combining study with moderate work, achieve high yields of corn, wheat, and sunflowers."[102]

By 1958, the practice of forming brigades had spread and earned frequent praise, ensuring that other oblasts and republics replicated it. That year at an interregional conference dedicated to the program, an official of the krai educational department drew attention to the perceived defects of secondary schools. Citing Khrushchev, he denounced their "detachment from real life," which they passed on to students by preparing them only for higher education and white-collar careers, thereby shaping attitudes toward work. Teachers and parents threatened students who behaved or performed poorly, saying, "'If you don't do well in school, you'll have to go work on the kolkhoz.'" The new requirement for manual labor coincided with a renewed emphasis on "constructing communism." The education administrator lamented that students listened passively to lectures about the importance of labor, which achieved little. "Communist upbringing cannot be divorced from labor," he stated, "or detached from real life, from the workers' real struggle to construct the new society."[103] The Komsomol Central Committee's representative at the conference reiterated that the present generation needed "to prepare . . . for a life of useful labor [and gain an education] that inculcates . . . the high moral qualities required in a communist society."[104]

Students in the final years of a rural secondary school volunteered for a summer with a brigade, an experience similar to that of urban contemporaries. Whereas local party elites might dispatch their children to exclusive camps on the Black Sea, students in prestigious English-language schools in Moscow and Saratov stayed in the surrounding countryside. Living with their school group in tents, they tended the camp and bonded through collective maintenance of the site. Later, they remembered it as a chance to have fun and grow up a bit through exposure to life away from home and parents.[105] The production brigades put more emphasis on work, but even this was comparatively light: six hours each day for the same wage a kolkhoznik earned in eight or more hours. The remainder was devoted to rest and recreation.[106] School personnel and farm specialists oversaw the brigades as they cultivated corn, especially special plots growing the hybrid seeds earmarked for planting the next year. Beginning with a single group in 1955, the program swelled to 122 groups in 1956 in Stavropol Krai alone, and then 239 in 1957. In 1958, 25,000 participants in more than 300 groups cultivated 34,000 hectares of corn, along with other crops. In 1959, the same number of students cultivated approximately 50,000 hectares, or approximately 5 percent of the krai's corn crop.[107]

The summer camps sought to shape students' actions, but also to transform their attitudes. The chair of a kolkhoz in Novo-Aleksandrov District likened the program to volunteering for Virgin Lands campaign. Parents were pleased with students' promises "to never let dear friends down, and to work as never before." On a practical level, the chairman noted, "They learn to follow the schedule in work and in leisure, conducting themselves as they should." The transformation was obvious: "We once had *beloruchki*: there were those who didn't want to work while in school, but that time is past. Now there are no more *beloruchki* here."[108] The chairman's idiomatic expression, which literally described someone with white hands not calloused by labor, implied that young people had earlier expressed haughty disdain for dirtying them.

The imagery of transformation through labor expressed a belief that work forged individuals suitable for creating and then flourishing in the new society. As a krai party official explained, "The most valuable part of the brigades is that they produce new people."[109] The mission to remake citizens recurred during transformative moments in the Soviet Union's history. The party proclaimed that socialism would reshape basic human nature and thus the way people acted, making them suitable to collectivist society. Prominent during the early postrevolutionary years, the undertaking re-emerged under Khrushchev as the requirement to educate those who were to build and then thrive in the coming communist society.[110] Lest the brigades prioritize work alone, the participants—at least in the ideal—also received time and means for intellectual, political, and artistic development.

Speakers at a conference called to praise the program unsurprisingly spoke of the results in glowing terms, but the students also confronted real challenges. "In comparison with last year, the brigades work in the best conditions," recognized a 1958 report on the brigades of one district. The local party committee had provided comfortable shelter, timely transportation, and hot meals, all evidently absent before. Yet such improvements did not eliminate "instances of callous attitude toward the student brigades," when leaders of farms and other organizations refused to provide such necessities. Agronomists did not instruct or monitor students. Farm chairpersons commandeered them for work outside their assignments.[111]

Some observers expressed misgivings about the participants' motivations. "Little has yet been done to organize labor and achieve educational goals," one noted. "The students consider pay the most important objective of participation. They understand the educational goals poorly, as well as the purpose of the brigade: to complete tasks the party and government assign." Declarations that the students worked hard notwithstanding, others noted that some exploited opportunities to ditch work in favor of recreation. In one school's brigade, lax discipline allowed many students to "run off to the orchard or the pond during work periods," resulting in "plots that are poorly cultivated."[112]

Lofty rhetoric about reshaping teenagers into vanguards of the communist future aside, some youth viewed the brigades as a chance to spend the summer away from home and earn a little money. As some urban youth later reported, they considered work secondary to other pursuits at summer camp, which ranged from swimming and singing to breaking the rules. Even when they worked with enthusiasm, they did not always work well. Life at the camp "meant doing some fun work on the nearby kolkhozes," one recalled. "I remember that they sent us city kids to weed carrots. We good-naturedly weeded and weeded." But when they finished, "not a single carrot was left. We pulled them all up!"[113]

Seeking to develop new people, these brigades foreshadowed the campaign for communist labor brigades that followed. Resolutions of a Central Committee plenum and the Twenty-first Party Congress generated propaganda heralding the construction of communism. As a speaker at the Stavropol Krai youth festival put it, the student brigades were only a first step: "Those who work in student brigades demonstrated heightened interest in studying the parts of the curriculum on agriculture." Moreover, "their attendance rates increased significantly and, by using their work experience, they set an example of study and conduct for the younger students."[114] These traits made them successful participants in the communist labor brigades, which combined obligations for output with rules for conduct and for instruction in politics. As one official stated:

> Who are the young people who have given their all to working to make communism a reality? They are those educated in our schools. . . . They worked in the summer on kolkhozes and now, when they have received their own assignments, they understand that they must contribute their knowledge and youthful vigor. They must not work any old way: instead, they must work as communists, as Lenin taught us, as the Communist Party teaches us.[115]

Komsomol organizers sought this result, broadcasting their efforts. At the Thirteenth Komsomol Congress in April 1958, a young woman from Stavropol described her experience: "Working in the brigade provides great moral satisfaction. The guys are correct to say that [it] encourages self-reflection, teaching us life skills, hard work, and constructive pastimes," including team sports, the arts, and group trips.[116]

Over the years that followed, the links between labor and proper upbringing remained important. In 1961, the Krasnodar Krai party committee reported that these principles guided the local Komsomol to improve its economic activities. "Growing corn is now our most important economic endeavor," it indicated. Although this work lagged in rural areas, it remained the organization's mission to "instill in all youth the value of labor." In this spirit, the local party called for continued efforts to realize the principle that "those who do not work, do not eat," in part by "creating an atmosphere of condemnation of and intolerance for spongers."[117]

Mobilizing youth to participate in the corn crusade, the Komsomol and related organizations sought to have youth plant the crop, but also to shape the attitudes and conduct of a generation. Augmented by material incentives in the form of cash prizes and goods, the competitions offered moral incentives by appealing to higher ideals, and also a banner or a lapel pin to embody them. Making corn an alternative contribution, nearer to home than the distant Virgin Lands of Kazakhstan, the Komsomol depicted agriculture as the younger generation's defining mission. The organization and its press used martial imagery to portray the project as a battle for high yields and, by extension, the material abundance Khrushchev promised.

These efforts evolved over the decade. From local beginnings in 1954, the competitions became more centralized and emphasized the role of the Komsomol, an arm of the Communist Party, in directing economic activity. In the late 1950s and early 1960s, the Komsomol responded to the axiom that the Soviet Union was becoming a state of all the people by favoring social activism by volunteer councils. Although promising to popularize oversight over the everyday management of the competitions, these councils were autonomous in name only because they operated under the same Komsomol officials who had previously been in charge.

The Soviet Union's leaders portrayed labor as a virtue and a component of bringing up the new people who would construct and then be at home in the future communist society. Educational programs emphasized that work on growing corn and other tasks inculcated good morals and communist values. Student production brigades assembled for the summer offered moral and material incentives to youth participants, another effort to mold the conduct of the rising generation. Participation in the groups obliged students to remake themselves through labor, preparing them to remake society—all while producing food and other useful agricultural products. Although voluntary and paid, work performed by students also carried a latent but palpable element of coercion, as the case of dental student Iosif and his school group's corn harvesting suggests. Moreover, high ideals and enthusiasm could ensure success in the cases of those who won competitions and received awards, but these few stood out among the great many average and below-average corngrowers.

These mechanisms of governmentality worked, much as the press and other devices did, by trying shape young citizens' actions and how they understood the world.[118] Paralleling Khrushchev's educational reforms of the late 1950s, the corngrowing competitions allowed leaders to soothe their fears that young people in an increasingly urban, educated, and technological society had little appreciation for manual labor, which they avoided at almost any cost. In emphasizing moral and material incentives, Khrushchev-era policies de-emphasized the coercion that had hung over all rural laborers under Stalin.

6 FROM KOLKHOZNIK TO WAGE EARNER

In October 1955, Nikita Khrushchev seized on reports about accounting practices presented by the Soviet Union's delegation of agricultural experts to the United States. Caring only for gross output, Iosif Stalin had forbidden kolkhozes to calculate production costs. In a capitalist economy, Khrushchev noted, such calculations were "a matter of life and death." "Look at what we've come to," he complained, "that this basic truth has become some sort of secret to discover." "It explains the irresponsibility that has plagued our system from top to bottom," he fumed. "Everyone acts like a bureaucrat and no one cares about results."[1] On another occasion, he bluntly denounced a sovkhoz for inefficiently converting corn into meat. "How does this happen?" Khrushchev demanded. Because the farm's bosses acted "wastefully," he concluded. "Forgive me for my rudeness, but if this [sovkhoz] were a commercial enterprise subject to the forces of capitalist competition, a farmer who spent 8 kilograms of grain to produce 1 kilogram of pork would be left without trousers [bankrupted]. And here? The director of this sovkhoz, well, his trousers are just fine . . . because he does not have to answer for this disgrace."[2]

In Stalin's time, systematic coercion worked in parallel with moral incentives couched as appeals to mobilize in service of socialism or the Motherland, practices that evolved under Khrushchev to offer substantive material incentives. Coercion decreased but did not disappear, while mobilizational tactics defined official propaganda and Khrushchev's many speeches on agriculture. His efforts from 1957 onward to provide kolkhozniks a steady wage were designed to encourage the intensive labor necessary to realize the potential of corn and industrial farming practices. Within a few years, labor and pay procedures on the kolkhozes began to converge with those found on sovkhozes.[3]

Subsequently adding systematic financial calculation of kolkhozes' production costs, the reforms demonstrated concern their financial health. This in turn facilitated management using the principles of *khoziaistvennyi raschët*, the system of intra-enterprise accounting commonly known as *khozraschët*. As an endeavor to govern the countryside using financial mechanisms and material incentives, Khrushchev's approach constituted the application of an alternative theory of how collectivized farming worked.

The reforms attempted to change the policy component of an entrenched political economy, a relationship among production, law, custom, and government authorities. For their part, kolkhozniks avoided work and harbored suspicions of kolkhoz and government. Khrushchev attempted to implement the corn crusade and other policies in a context shaped by entrenched authoritarian practices alienating peasants, who maintained a well-founded mistrust of the existing order built up over years of exploitation under Stalin.[4] Implemented by oblast, district, and kolkhoz authorities, Khrushchev's incentives project influenced the actions of the foot soldiers of the corn crusade who planted, cultivated, and harvested.

In practice, authorities had to constantly respond to kolkhozniks who, although lacking formal input into policymaking, determined the success or failure of the harvest. No initiative could succeed if it violated their moral economy, in which case they would disregard the rules and force superiors to respond. Labor historian E. P. Thompson developed the concept to describe a group's understanding of the principles of justice. He based it on an ethical code of the eighteenth-century English poor that sanctioned collective action against merchants who, awaiting higher prices, refrained from selling bread in times of dearth. Yet Thompson cautions that individuals act in response to complex motivations based on "custom, culture, and reason," rather than to a mechanistic force predetermining a unified collective response.[5]

Persistent community values help explain how, as kolkhozniks pursued individual ends, they justified circumventing formal rules governing the collective farm, of which they were only the nominal owners. Even under Stalin, they had expanded personal plots beyond legal norms, refused to fulfill labor-day minimums, observed religious holidays, and penned complaints about local bosses.[6] The concept of resistance, however, does not adequately characterize such nonviolent actions because it constitutes a coherent, single-minded collective historical subject—peasants who resisted—from sources lacking one. Instead, individuals drew on deep-seated norms sanctioning a range of actions from accommodation to quiet struggle.[7] Historians can document such actions only with much difficulty because intercepted letters, informants' reports, and other purported sources of peasant voices have been filtered by the party, government, and police.

In this light, the Soviet Union's agrarian economy appears more complex than allowed by the simplified model premised on Moscow's effective top-down command of the locales that Sovietologists considered accurate representations of practice. Although not acknowledging its existence, the authorities grappled with a labor market that kolkhozniks exploited by leaving the kolkhoz, most often to take up jobs in industry and construction. Under Stalin, wage labor in factories and on job sites mediated most economic life.[8] The kolkhoz remained one of the exceptions. By 1964, Khrushchev had spread wage labor from the few sovkhozes to the many kolkhozes. Facing foot-dragging by constituencies of officials steeped in the old ways, Khrushchev established price incentives, bonuses for exceeding planned yields, bonuses for raising specific crops and specific varieties, guaranteed wages, and monetary relationships between the kolkhozes and the kolkhozniks.

Labor policies and practices under Khrushchev responded to the legacies of the Stalin era, consisting primarily of threats of repression and appeals to abstract ideals. From 1928, the state had curbed peasants' mobility, controlled the fruits of their labor, and left them only a few kilograms of grain and the right to a personal plot. Although this regime grew even harsher after 1945, many kolkhozniks avoided compliance, little fearing the potential punishments for violations of labor discipline. Few kolkhozes voted to impose the sanctions, while extreme ones such as expulsion made employment off the farm possible, a way out of the oppression of the kolkhoz.

Stalin's successors recognized the failings of this coercive kolkhoz labor regime, which gave kolkhozniks little incentive to work diligently. As early as September 1953, Khrushchev recommended practices to address "low labor discipline" resulting from "violations of the principle of material incentives."[9] Echoing Khrushchev, officials invoked the maxims of Vladimir Lenin. Thus a party official in Lithuania cautioned that communism could not be built "'on the basis of enthusiasm alone, but with the help of enthusiasm birthed by the great revolution, and based on personal interest and incentives. There is no other way to achieve communism, there is no other way to guide the tens of millions of people to communism.'"[10]

Kolkhozniks confronted the consequences of Stalin's policies daily. Approved in 1953, rises in the prices kolkhozes received for mandatory sales of produce to the state increased kolkhozniks' income, but even in 1954 the average income for a standard 200-day working year still amounted to only 33 percent of an equivalent sovkhoz worker's pay.[11] Further reforms increased procurement prices, cut taxes, and scaled back compulsory deliveries, solidifying kolkhoz finances and raising output. Khrushchev and his rivals agreed on these initiatives to raise individual incomes and increase enthusiasm for work on the kolkhoz.

Despite improvements, the number of kolkhozniks working and the quality of their labor remained inadequate to kolkhozes' needs, which were rising in response to Khrushchev's demands for corn and the associated food products. Leaders soon comprehended that their initial reforms had not gone far enough.

In May 1955, the deputy minister of agriculture, Vladimir Matskevich, informed Khrushchev and the Central Committee about the paltry pay kolkhozniks received again in 1954. Nearly 25 percent of kolkhozes had allocated fewer than 500 grams of flour per labor-day. More than 10 percent paid as little as 300 grams, and 1.9 percent paid nothing at all.[12]

Miserly earnings correlated with low output, while small gains got significant results. This was the case on the farms served by the Reutov MTS in Moscow Oblast. In 1955, 15.7 percent of kolkhozniks there did not meet the labor-day minimum and, of those, 4.9 percent did not work at all. The kolkhozes harvested only 5.6 metric tons of corn silage per hectare, barely 20 percent of the goal of 25 tons, and they valued a labor-day at only 300 grams of flour.[13] Overall, kolkhozes' income had risen 27 percent, allowing kolkhozniks' average income from work there to increase from 1,195 rubles in 1950 and 1,589 in 1953 to 2,670 in 1955.[14] However limited, this improvement reversed the flow of migration. Matskevich reported with relief that, for the first time since 1948, new kolkhoz members outnumbered those who left the village.[15] Yet this proved to be only a brief pause in the long-term flow from farm to factory.

Fostering change after decades under Stalin, Khrushchev launched critiques that were supported by social scientists' studies. Between 1953 and 1955, officials developed a proposal to found the All-Union Research Institute for Agricultural Economics, whose staff of eighty soon began to analyze planning, organization, wages, productivity, and costs. Pursuing critical inquiry made possible by the Thaw, they soon started to influence Khrushchev and his advisers.[16] Economists, ethnographers, and statisticians contributed to the intellectual and evidentiary foundations for Khrushchev's subsequent policies.[17] Under Stalin, leaders had imagined "the peasant" as a socialist-realist subject, motivated by desire to construct socialism and possessing few interests distinct from those of the state. After 1953, socialist social scientists developed new concepts of peasants, whom they imagined as rational actors who saw to their own self-interest, characteristics strikingly similar to those assumed by liberal economists.[18]

Even before 1953, scholars Aleksandra Sanina and Vladimir Venzher had advocated for abolishing the MTS. Khrushchev and his advisers then developed the 1958 MTS reform with this new subject in mind. Expressing fears about class resistance and petit-bourgeois mentalities, the authorities who implemented it fell back on familiar mobilization mechanisms designed to motivate the socialist-realist subject.[19] Eliminating a coercive mechanism that had allowed Stalin to dictate to the kolkhozes, Khrushchev demonstrated innovative thinking and indicated a capacity for fundamental reform. Nonetheless, the breakneck pace and coercive methods made the reform universal and reckless, rather than tailored to individual kolkhozes carefully over several years. It harmed all farms, but particularly weaker ones taking on unbearable debt burdens, which local authorities then compelled them to repay rapidly at the expense of other obligations.[20]

Initially, kolkhozes lacked the capacity to take on the burdens of manually planting, cultivating, and harvesting corn. Recognizing that pay incentivized productivity, leaders redoubled programs to pay bonuses for output exceeding the plan. A 1955 policy promised to pay kolkhozniks whose work teams surpassed planned yields a percentage of the extra harvest. Even in locales where peasants had long grown corn, kolkhozes had ignored earlier directives mandating such bonuses. In Vinnytsia Oblast of southwestern Ukraine, only 40 of nearly 1,000 kolkhozes paid them.[21] Even on those farms doing so, barely 10 percent of kolkhozniks earned them.[22] Officials in Moscow noted, "With corn plantings expanding significantly, establishing monetary incentives takes on special importance because they better encourage each kolkhoznik to increase the area planted in corn and to grow a [large] harvest."[23] They called on kolkhozes to earmark for bonuses 15 percent of the corn harvested in excess of the plan, granting half to those who had earned labor-days working in that cornfield and distributing half to all kolkhozniks proportionally to the total number of labor-days earned. Promulgated in August 1955, the policy applied everywhere corn grew, remained in effect with modifications for the remainder of the corn crusade, and later applied to sovkhoz workers.[24] Moreover, the policy went into effect for most other crops in 1956.

Few kolkhozniks achieved the requisite harvests of corn or earned the bonuses defined in the directive from Moscow. Owing to the legal fiction of kolkhoz democracy, the policy, in the official terminology, made recommendations for kolkhozes to adopt at meetings of all members or, on larger farms, of their representatives. This central directive and decentralized procedure seemingly constrained local authorities' ability to meddle in kolkhoz affairs.[25] In reality, these practices limited Moscow's influence. In late 1955, the Stavropol newspaper published what appeared to be a letter from a kolkhoznik, Mezhniakova, who drew attention to her kolkhoz's lagging corn harvest. Appearing under the headline "When Are They Going to Pay Corn for Labor-days?" the letter explained that kolkhozniks needed corn to feed personal livestock and wood for heating, but received neither as payment. It blamed this on the farm's bosses, whose failure "destroyed kolkhozniks' belief" in earning what they deserved. If the managers had taken initiative by "implementing measures for increasing material incentives," the author wrote, "it is doubtful that members of my work team, Comrades Demenko, Brykalova, and Zhukova, would stay home to gather fuel."[26] The women refused to work because they doubted that the kolkhoz would pay them.

Even if a party official, rather than a rank-and-file kolkhoznik, wrote this overtly didactic story, its pointed criticisms pressured administrators to heed Moscow's call to pay kolkhozniks. Yet kolkhozes altered the spirit of such policies even as they formally adopted them. The policy had recommended not only bonuses, but also a penalty of 1 percent for each percentage point the corn harvest fell below the planned yield. In Stavropol Krai, kolkhozniks accrued bonuses for harvests larger than the plan, but few paid any penalty for those that fell short.[27]

Accompanying the rising income, increased labor-day minimums and new monthly quotas prevented kolkhozniks from concentrating on their personal plots during planting and harvesting, when kolkhozes needed them most. Kolkhozes also implemented these requirements unevenly. In Tambov Oblast, one kolkhoz's bosses had brazenly falsified meeting minutes. "The kolkhozniks affirmed," an inspector's report explains, "that the agenda item on the minimum was not discussed at that meeting. [The managers] added it later, only when they learned that there would be an inspection."[28]

Kolkhozes also responded to the reality that skilled workers, mostly men, earned several labor-days for one person-day worked, while women primarily milked, did manual field labor, and carried out other tasks worth few labor-days. Across the RSFSR, 20 percent of kolkhozes established differentiated norms for men and women.[29] They thus acknowledged that women shouldered responsibility for housekeeping, personal plots, and child care. As one local party authority concluded, "The cause of insufficient participation in collective labor on certain kolkhozes remains the lack of [measures] to increase material incentives for kolkhozniks, as well as conditions ensuring regular participation by women kolkhozniks who have young children."[30] A survey of 500 families in Stavropol Krai found that men earned an average of 2.2 labor-days per person-day whereas women earned nearly 20 percent fewer.[31] Moscow condemned these differentiated norms, most likely because they threatened to cut labor-force participation, but perhaps because they discredited claims about equality.[32]

Guaranteeing kolkhozniks some payment for their labor may not seem revolutionary, but it fundamentally altered policy while increasing workforce participation and productivity. A March 1956 directive recommended that financially sound kolkhozes fix a monthly wage, rather than pay a lump sum at year's end. A few exemplary farms set aside between 40 and 60 percent of income to paying wages each month.[33] Gosplan, the State Planning Agency, supported the proposal because it "increased kolkhozniks' material incentives to expand output."[34] In addition, the new policies raised annual incomes. That year, the aggregate income of kolkhozes in the RSFSR reached 37.3 billion rubles, a nearly fourfold increase since 1945. Of that, their wage bill amounted to 14.72 billion rubles, or 39.5 percent, yielding an average annual income of 3,022 rubles. This arithmetic mean, however, concealed a wide range: the lowest average for an administrative region, 1,142 rubles, fell far short of the comfortable 7,896 that distinguished the highest.[35] By 1958, the wage labor initiative had reached only 16 of 37,000 kolkhozes. Drawing conclusions from early experiments, leaders then insisted that kolkhozes convert conditional payments into reliable wages. In 1959, the figure grew to 2,222 out of 27,500 kolkhozes. By 1962, some 10,093 kolkhozes of 17,900 had made the change.[36]

This trend indicated the changing nature of kolkhoz incomes and of kolkhozniks' labor. To adequately compensate kolkhozniks for the value they produced, the kolkhozes first had to abandon the labor-day. The old system's tax burden, payments to the MTS, state procurements, and other obligations had pushed the wage fund to the lowest priority. Consequently, "*raspredelenie po trudodniam*," or distribution proportional to labor-days, often amounted to little or even nothing. After 1954, this practice gave way to new forms of payment for work, "*oplata truda*."

Theory and terminology changed slowly and irregularly. In 1957, the head of the Institute of Agricultural Economics, Academician Aleksandr Tulupnikov, criticized draft materials for the agricultural display at the 1958 Brussels World's Fair for improperly using the term "*oplata na trudodni*," or payment for labor-days. "It is necessary," Tulupnikov explained, "to convey the understanding that the kolkhoz is a cooperative enterprise distributing income [*dokhod*] among the members. Do not use terms such as wages for labor-days. Instead, favor the term 'distribution of income proportional to labor-days.'" Only that way could the text demonstrate that peasant income in the Soviet Union was rising faster than in capitalist countries. Tulupnikov concluded that this formulation would provide "eloquent proof of the advantages of the socialist economic system."[37] This official conception ignored the decades-long practice of repressing the kolkhozniks' income and consumption.

Even kolkhoz leaders understood that a conceptual change was occurring. As one explained in 1958, "On kolkhozes that function well, which have their income and their profit [*pribyl*], then it should be possible to distribute profit. That's a new bit of terminology.... But let it be 'distribution of profit,' and that way we can determine what percentage goes to the kolkhozniks to apportion as wages [*zarabotnaia plata*]."[38] The array of terms illustrates the issue's complexity, but nonetheless the chair sought to explain that kolkhozniks now deserved a fixed share of the kolkhoz's earnings, rather than the leftovers.

The advice agricultural economists offered to policymakers demonstrated such an evolution. In a 1958 report to the Central Committee, the Institute of Agricultural Economics outlined "progressive forms of remittance" and guidelines for establishing daily work norms for a given pay band. Even this document mixed terms by including both payment and distribution of income. Yet it presaged practices increasingly common over subsequent years by recommending that kolkhozes establish a wage fund assigned a high priority, rather than the lowest. From this fund, kolkhozniks received a guaranteed monthly wage in cash, or the equivalent in kind, amounting to one-twelfth of 70 percent of the expected annual pay. Paid at the end of the year, the remaining 30 percent was subject to bonuses or deductions appropriate to the fulfillment of the yield plan.[39] Now the kolkhoznik's effort was to determine only that final 30 percent, rather than the entirety of income, as before.

The proposed system thus addressed problems the institute's economists had isolated during their research. They listed "a lack of guaranteed pay," "different levels of pay for the same work on the same kolkhoz," "the multiplicity of extant forms of bonus pay, many of which depress the importance of base pay and do not facilitate holistic development of socialized enterprise," and a resulting "complexity and unwieldiness of accounting for the kolkhozniks' [pay]." Already tested on a few select farms, the new system offered "a method of establishing work norms which significantly increased the links between the personal incentives of the kolkhozniks and the output of the kolkhoz. Therefore it should be spread widely." Last, the system of differentiated bonus wages addressed the problem of leveling, in which each kolkhoznik received the same wages as her neighbors no matter what results she achieved, and therefore had little incentive to work more and better.[40]

Wage labor and incentives received authorities' approval and became linked to new management practices appropriate to industrial farming. The same year, a report by the institute to the Economic Commission of the USSR Supreme Soviet advocated *khozraschët*, a system of intraenterprise accounting for production costs and profit hitherto absent from the kolkhoz. The report was overly optimistic about the state of practices on all but the most well-equipped and efficiently managed kolkhozes, but its underlying principles are revealing. "Commodity relations are expanding on the kolkhozes, which are developing the real potential to decide for themselves the basic questions of their system of management," the institute explained. In light of the growing economic strength indicated by the MTS reform, the kolkhozes "realize their production at prices which cover production costs and support investment in collective economic activity." The key was "a system of remuneration for the labor of kolkhozniks, specialists, and managers dependent on their fulfillment of the gross production plan and lowering of production cost."[41]

Meager incentives had caused labor shortages restricting the number of workers available for planting, weeding, and harvesting corn, labor-intensive efforts that coincided with other difficult and time-sensitive tasks. Kolkhoz managers lacking Khrushchev's devotion to corn allocated the laborers they did have to tasks they considered more profitable. By thus ensuring that corn went unattended and yielded only small harvests, they confirmed their preconceptions.

In 1958, inspectors discovered a common scene in Krasnoiarsk Krai: the kolkhozes had fulfilled their corn-planting plan by selecting fields with the poorest soil and without applying needed fertilizer. They then had assigned few kolkhozniks to cultivate the crop, which those did "with considerable delay and low standards, with the result that the corn was fully overgrown with weeds." In one district, kolkhozes "underappreciated corn as the principal feed crop," declared the inspector. Only 37 percent of cornfields received fertilizer and only 57 percent benefited from moisture-saving plowing practices, resulting in yields of just 5.5 tons of silage per hectare.[42] Inspectors documented the same problems in Briansk Oblast. "None of the inspected kolkhozes," they reported, "have adopted any material incentives to reward corngrowers" or they simply did not pay the money when earned.[43]

Even in Krasnodar Krai, the southern region Khrushchev and Roswell Garst considered best suited for industrial corn growing, kolkhozes refused to pay bonuses. Inspectors concluded that only kolkhozes that implemented proper techniques harvested high yields. "Many farms undervalue corn," and therefore "it is often planted on poor, unfertilized plots and the required cultivation is not carried out."[44] Especially in the early years of the crusade, farmworkers' limited numbers and lack of motivation slowed the labor-intensive corn harvest. The country had few mechanical harvesters adapted to this laborious task, as well as to digging sugar beets and potatoes. To compensate, many regions brought in students and soldiers to do these jobs.[45] For many years, kolkhozes opted for two-part harvesting, in which laborers manually picked the cobs for use as nutrient- and calorie-dense feed for hogs. Others subsequently used machines to chop the plant itself into fodder suitable for cattle. Only around 1960 did mechanical corn harvesters become more common, but even those models functioned poorly and wasted much.

In practice, material incentives and wage labor remained irregular. Despite policy recommendations from Moscow, local authorities continued to assign kolkhozniks' pay the least priority. In 1958, one kolkhoz chairman, a certain Alperovich, denounced local bosses who often pressured him to "help the working class" by curbing his kolkhoz's wage fund and thereby holding down food prices. "How am I supposed to go and slash a kolkhoznik's pay?" he queried, recounting how his protests came to nothing. For example, his kolkhoz sold eggs at government-mandated prices below the price of feed for the hens, leaving him to pay those who tended the birds from some other source of income.[46]

Alperovich and other participants in meetings with officials from the ministry of agriculture decried how the outmoded labor-day's lack of intrinsic value left them at a disadvantage on the labor market, which was an unacknowledged fact of life. Alperovich's kolkhoz was located in Moscow's hinterland. Any farm in such an industrial zone had to compete with neighboring nonagricultural enterprises for laborers. It could not do so when offering irregular and conditional payment, often in kind. When Alperovich had become chairman in 1954, the kolkhozniks, earning next to nothing, had "voted with their feet by leaving the kolkhoz."[47] "We're all people," he explained, "and no one will work if they don't know what they'll earn." Another chairman countered that payment in kind was incentive enough: "If [kolkhozniks] receive an extra two tons of potatoes, that will make them happier than a wage." Incredulous, Alperovich questioned, "Then I ask you: why do the kolkhozniks leave to work in a factory knowing full well that there they'll earn only 300 rubles, instead of 700 on the kolkhoz?" When the other chairman countered that the higher figure was an incentive because the kolkhozniks could earn it "only if the yield is higher," Alperovich seized on the point: "No, everyone wants to avoid that 'if.'" Pay, he concluded, "must begin with the work, not with the harvest. They say that the kolkhozniks don't work because they don't get paid. They don't get paid because there is no harvest. It's a vicious cycle."[48]

To remedy their many woes, kolkhozes had to pay a steady wage that could incentivize work and keep kolkhozniks on the farm. Population statistics show that millions of kolkhozniks left their villages for good after World War II, but many moved from kolkhoz to nonagricultural employment and then back. In 1957, as many as 2.3 million of the 31.6 million kolkhozniks lived on the physical territory of the kolkhozes, but worked in industry, transportation, construction.[49] Disregarding restrictions on their movement and employment, working-age men in particular sought whatever occupation offered a living. Others, more often women, exclusively tended a personal plot and lived on earnings from selling unconsumed produce in legal markets.

Disdaining material incentives, local officials maintained authoritarian tactics familiar from the Stalin era, which only sowed disaffection among kolkhozniks. In isolated cases, authorities prosecuted and punished kolkhoz chairpersons for irresponsible, rude, aggressive, and even violent behavior.[50] When lodging complaints with higher authorities, kolkhozniks posed a threat to delinquent bosses. An anonymous letter to Moscow called attention to 1,134 hectares of potatoes and sugar beets in Ukraine's Chernihiv Oblast unharvested due to mismanagement. The district bosses "tried to hide this by falsifying data and compelling the people to work in inhumane conditions." As winter approached, they tyrannically threatened to assess a fine of 100 rubles to anyone, even the district's nonagricultural workers, who did not contribute to the harvest. The letter further complains that the oblast party committee knew about the illegal orders, but did nothing. "It is time to hold these people responsible," it concluded, relying on the idea corrupt local authorities might receive comeuppance from superiors, "*Look into this by sending an inspector* [emphasis in original]."[51] Moscow compelled the oblast party to launch an investigation, which bore out the charges. Lagging because of poor weather, work was also slowed "by the large volume of labor-intensive tasks, including digging sugar beets and potatoes, as well as harvesting corn." The oblast committee rescinded the fine, which the letter-writers had found so outrageous, before anyone had paid. To placate the kolkhozniks, the committee sacked a kolkhoz chairman and an MTS director, but only issued warnings to the district party secretaries.[52]

Kolkhoz chairs usually enjoyed the protection of local authorities, who shielded them from punishment if they broke laws and customs. This violated kolkhozniks' sense of fairness, driving them to write letters to newspapers, inspectorates, and party officials up to and including Khrushchev, a last-ditch effort to seek justice within the system. Not complete or systematic records, summaries of the letters compiled kolkhozniks' accusations about injustices. According to one complaint, a chairman had driven his kolkhoz into poverty, prevented it from providing a bathhouse and other communal necessities, and abused his access to an official automobile. "Velichko has set himself up like a dictator: whatever he says goes," a group of kolkhozniks charged, "He pays those who don't work and leaves nothing for those who do." Enjoying protection from district chiefs, Velichko punished those who protested. "And if [Vladimir] Lenin were to return, what would he say?" the kolkhozniks queried.[53] In another letter, a visitor to a kolkhoz in Ukraine's Kharkiv Oblast communicated kolkhozniks' wisdom: "Wherever they rule from Moscow, there is order. Wherever there is local power, it's better to stay quiet."[54] Although rarely sufficient to ensure firing, charges of having "suppressed criticism" or "violated the democratic basis of kolkhoz management" augmented the charges against those replaced during Khrushchev's initiatives to professionalize farm management.

Suffering abuse and receiving low pay, kolkhozniks stole from and refused to work for kolkhozes they regarded as alien. Such actions seemed a justified means to make up for unpaid work extracted from them. "Velichko has corrupted the people," the above complaint explained, "who have begun to refuse to work and to steal kolkhoz property."[55] Although common, theft involved quantities insignificant compared to overall output. In 1955, police officials reported on efforts to combat the "numerous cases of theft." As of October 5, some 3,292 legal actions against 4,229 individuals concerned the loss of 756 tons of grain.[56] Averaging a substantial 220 kilograms per prosecuted case, this total comprised an insignificant 0.00007 percent of the 103.7 million tons harvested.[57] Having detected, investigated, and reported these crimes, officials had not caught all thieves, but even if undetected cases had been many times more common, the losses of wheat, barley, corn, and other commodities would have remained inconsequential.

Although not economically ruinous, the prevalence and persistence of theft provide evidence about kolkhozniks' dissatisfaction with conditions. Most cases never reached the legal system because culprits took so little that local officials did not bother to report them. Corn was particularly attractive to pilfering because kolkhozniks could grab a handful on the way home from working in the field and immediately feed it to their personal livestock. An inspection in Ukraine's Rivne Oblast found that most theft occurred this way, leading authorities to complain about "weak organization of security and preventative measures." Striving to emphasize their own effectiveness, police and prosecutors soon caught more than forty culprits.[58]

Outside legal processes, kolkhoz officials sought to sanction those who stole amounts too trivial for the legal system. In neighboring Ternopil Oblast, they searched for punishments fitting the crimes, resorting to what the procuracy condemned as "administrative measures." One chairman explained, "I understand that this is unlawful but . . . must adopt these measures when bringing a criminal case would be ineffectual." Petty thefts of up to 5 kilograms of produce or of firewood chopped on kolkhoz land prompted fines of as much as 5 labor-days or 25 rubles.[59] In Mikolaiv Oblast, one kolkhoz chairman, Zhidkikh, struggled for years to tighten lax labor discipline. He complained that, when he arrived on the kolkhoz, it "had caused the corn harvest, as well as processing of hemp and rice, to end only in the early months of the following year." Moreover, it facilitated "the many instances of pilfered kolkhoz property." The threat of court action failed to deter theft, prompting the chairman to assess fines to those assigned to guard property that went missing. To compensate for some stolen hogs, Zhidkikh deducted labor-days from kolkhozniks absent from guard duty. They appealed to the district party committee, which overturned the judgment and ordered the matter taken to the courts. Exasperated, Zhidkikh wrote to Moscow asking for clarification of the law.[60] Authorities documented 52 similar cases in one Ukrainian district, and more than 3,000 in Belarus between 1955 and 1957.[61] Ultimately, Moscow ruled that chairpersons lacked the authority to mete out such punishments.[62]

The petty thefts testify to kolkhozniks' belief that the kolkhozes were unjust. Low pay compelled them to acquire corn for livestock feed, fuel for home heating, and other necessities by any means necessary. Inferring that low pay correlated with high rates of theft, a Ukrainian procuracy official noted that farms often failed to allot the feed kolkhozniks needed for their personal animals.[63] On one kolkhoz, yields of rye, wheat, and corn reached respectable figures, but poor management and lack of workers meant the last 37 hectares of corn came in only in February. By that time, more than half of it had spoiled or been stolen.[64] One kolkhoz in Ternopil Oblast experienced "widespread and largely unpunished" theft that was "especially common during planting and harvesting, when potatoes, corn, beets, grain, hay, and other crops are stolen." Amounts ranged from as little as 4 kilograms of wheat to as many as 30 kilograms of corn. Much like others, this kolkhoz initiated at most one court case per year. Even individuals caught stealing twice in a year received light punishment. In contrast, a neighboring kolkhoz had few members, giving it a tight-knit community that simplified managers' efforts to oversee labor in outlying fields, strengthened mutual surveillance among kolkhozniks, and ensured that officials punished thieves. Its losses to theft remained comparatively small.[65]

Alienated from the kolkhoz, kolkhozniks did not view theft as stealing from themselves, which it was in a strict sense. Earning low wages for work they considered imposed from above, they had little incentive to preserve kolkhoz property, as official jargon phrased it. Blind to kolkhozniks' motivations, officials investigating theft shed light primarily on their own attitudes. Incapable of imagining that kolkhozniks considered the kolkhozes exploitative, they saw only holdovers from a lower stage of development. In Western Ukraine, a procuracy official conjectured that kolkhozniks' habit of carrying home an armload of grain or cornstalks at the end of the workday survived from before 1940, when aristocratic landowners had dominated the locale, then part of interwar Poland.[66]

Kolkhoz officials also stole, thereby compromising any authority they might have had to combat theft. In one instance, a kolkhoz bought three loads of hay from a neighboring kolkhoz. The chairman then colluded to divert one to the farm's accountant, whom kolkhozniks readily denounced to visiting inspectors. Investigating further, inspectors found the farm's party and Komsomol organizations compromised and therefore unable to intervene. The Komsomol secretary did not speak out, for instance, because her sister recently had been convicted of theft.[67] On another farm, the agronomist, the Komsomol secretary, and six kolkhozniks were each caught with between 10 and 15 kilograms of straw.[68] Kolkhozniks might justifiably consider hypocrites any superiors who thieved while pursuing action against thieves.

In officials' eyes, alcohol abuse and idleness also harmed kolkhoz productivity. A 1959 investigation revealed so-called malignant parasites in rural communities in Ukraine. These required decisive "measures of social action," according to the report republic party boss Nikolai Podgornyi sent to Moscow, including the comrades' courts tasked with correcting behavior without custodial sentencing.[69]

In a sample of 373 villages, inspectors found many who avoided "socially beneficial labor," instead pursuing "moonshining, speculation, and . . . other paths to an easy living." They documented freelance farmworkers working seasonally for wages and those legally classified as kolkhozniks who exclusively worked their personal plots. Echoing language familiar from the 1948 expulsion campaign initiated by Khrushchev, Podgornyi concluded that those "leading antisocial, parasitic ways of life arouse the just resentment of honest kolkhozniks."[70] For his part, Khrushchev ordered Podgornyi's report circulated to the entire Central Committee and each administrative region, evidence of the problem's seriousness in the leader's eyes.[71]

Alcohol deserved blame for legal transgressions, interference with kolkhoz work, and property destruction. In 1958, the Central Committee mandated local party meetings to condemn drunkenness and moonshining, forums open for denouncing offenders. During spring planting, one kolkhoz chairman organized a night of binge drinking with an entire work brigade. In another instance, a kolkhoz official capped the May 1 celebrations of International Labor Day by commandeering one of the farm's automobiles, driving to the city of Saratov, going on a drinking binge, and wrecking the car.[72] In Belarus, drunkenness among the workers of one MTS disrupted work. "During work hours, they make the rounds on their tractors, going from village to village in search of vodka," officials reported. "Or they leave the tractor in the field and go get drunk." One fell from his machine and was seriously injured. By failing to respond, local leaders failed to prevent "some tractor drivers [from] moving on to criminal acts, including fraudulent work orders for jobs left incomplete."[73]

Theft and drunkenness directly contributed little to the underperformance of agriculture under Khrushchev. Instead, they provide evidence for kolkhozniks' continued disaffection with dismal pay and low status. The kolkhozniks' moral economy held that, in return for work on the kolkhoz, they deserved some of its produce and the money needed to pay taxes, buy manufactured goods, and build houses. Conditioned by experiences under Stalin, their habits of mind told them to expect little or no compensation for labor-days, and therefore to secure what they needed when opportunity arose. The low priority assigned to kolkhozniks' labor gave them little incentive to stay sober and work diligently.

Small thefts and occasional large-scale looting testify to limits on the power of the party and government in the countryside. Although the state possessed the preponderance of formal power, its top-down control did not reach deeply into rural areas and the everyday lives of their residents. Even village and kolkhoz authorities had neither the will nor the authority to prevent theft, being unable or unwilling to punish offenders.

The farms of Stavropol Krai experienced the daily realities of kolkhoz labor. Located in Goriachevodsk, a settlement (*stanitsa*) adjacent to the Caucasus Mountains spa town of Piatigorsk, the Lenin kolkhoz had much in common with the krai's many large, comparatively wealthy farms.[74] The cycle of winter preparations for growing corn, spring planting, summer cultivating, and fall harvesting structured each annual run of the newspaper the farm began to print in 1957. Published three or four times monthly in a print run of 1,000, *Kolkhoznaia zhizn* [Kolkhoz Life] declared that its mission was to mobilize and inform workers, acting as "operational auxiliary" to the managers and party committee.[75] Indicating the importance of corn, chairman Khadzhiet Agnaev urged a target yield necessary to soon "catch up to and surpass America," as Khrushchev's slogan required. Bringing in between 3 and 4 metric tons of grain and 20 and 25 tons of silage, the kolkhoz was to put itself among the must successful in the krai by "creating a stable feed supply for collectivized livestock," as a stock phrase put it.[76] The following May, headlines exhorted the kolkhozniks to plant faster and begin cultivation in timely fashion. They announced one work team's pledge to "genuinely struggle for a high yield," and reminded farmworkers to "carefully tend the corn, not breaking the rules of agronomy."[77]

As spring turned into early summer, the farm took up the task of removing weeds from the developing corn. By publicly shaming those who shirked work, the newspaper applied pressure and brought problems to managers' attention. "The [female] kolkhozniks of work team no. 1 care for their designated areas, but six have not yet begun," a brigade leader explained. "One member of the work team, M. Boiko, has not been seen in the fields for three weeks. The brigade members hope that the work team of Comrade Miasoeva will not cause the whole brigade lag behind the vanguard in tending corn and other crops."[78]

Using naming and shaming tactics to police violations beyond the cornfields, *Kolkhoznaia zhizn* singled out individuals and groups violating norms at work and at home. Condemning those guilty of public drunkenness, it publicized punishments imposed by the kolkhoz management. Difficult to supervise because of their mobility, the farm's drivers had opportunities to steal, falsify work orders, and commit other misdemeanors. After repeated drinking bouts, one driver was fired from the job, another reassigned to different work, and a third docked five labor-days.[79] Publicizing such punishments, the newspaper aimed to prevent recidivism and warn others of the consequences of bad behavior.

Authorities referred more serious crimes to the criminal-justice system. They did so when several drivers collaborated to steal and sell 158 kilograms of sunflower seeds, using the proceeds to fund a bender.[80] Like other farms, the kolkhoz suffered from thefts, even if individual quantities remained small. During the harvest, mature corn enticed those looking to nourish personal livestock. An August

1957 story denounced and demanded punishment for a dairy worker who stole feed. "On the night of July 31, S. Lutsenko was caught in the cornfield planted for silage, where he was harvesting corn for his personal cow," it explained, "People say that this was not the first instance of Lutsenko 'procuring' feed in this manner."[81] In September, the kolkhoz adopted a rule that any kolkhoznik caught taking any corn would become ineligible for bonus pay for the year.[82]

Kolkhoznaia zhizn held up several kolkhozniks for exemplary working and living, making them local vanguard workers. One, Nikita Kaplun, led the work team that grew the farm's largest corn harvest and spearheaded efforts in local corngrowing competitions against a nearby vanguard kolkhoz, Will of the Proletariat. The local party secretary praised Kaplun's orderly personal life, careful approach to work, and attentiveness to fellow party members and kolkhozniks.[83] In other instances, the newspaper praised the examples of Kaplun and Evdokiia Ulianik, the star corngrower of the neighboring kolkhoz.[84] Kaplun appeared on the Honor Roll, a list of exemplary workers published periodically and posted on a billboard in a public place for all to see.[85]

Kaplun, Ulianik, and their peers contrasted with those admonished for flagging enthusiasm. In June 1957, the newspaper questioned, "One must ask brigade leaders Zozuli and Morgatyi ... when they will organize an actual socialist competition among the work teams to raise high yields of corn and other crops. The kolkhozniks do not want to lag behind in this important initiative by the work teams of Kaplun and Ulianik."[86] The farm also participated in competitions pitting farms across the krai against each other.[87] In fact, Kaplun became the object of wider attention when a team from the North Caucasus Documentary Film Studio came to the Lenin kolkhoz to shoot footage about his work and life.[88]

Large, profitable, and winning awards from the krai leadership, the Lenin kolkhoz continued to combat theft and drunkenness. At a 1959 meeting of kolkhoz's management committee, speakers singled out the head of the drivers for failing to discipline his subordinates. In particular, one had crashed one of the farm's Moskvich compact cars while drunk driving.[89] A vanguard farmworker and head of one of the corngrowing work teams complained, "Much has been said about thieves, drunks, and moochers who are of no use to the kolkhoz." She asked, "Why do they not have any cases of theft on Will of the Proletariat?" "Because there, all members of the kolkhoz look after collective property," she answered. Moreover, she continued, to remedy the problem the Lenin kolkhoz "must follow the example ... of the Will of the Proletariat kolkhoz and establish ... oversight such that no one gets in the habit of carrying it off."[90] Given the ubiquity of petty theft and alcohol abuse, the neighboring kolkhoz likely had its own problems, but at a distance they were unseen, allowing it to serve as an example for goading violators into compliance and encouraging vigilance among the rest.

In the late 1950s, the Lenin kolkhoz joined others in transitioning from using manual labor to machines to plant, cultivate, and harvest corn. Corn plantings consequently expanded rapidly toward their 1962 peak. In Stavropol Krai in 1958, at least 774 work teams composed of between four and ten kolkhozniks pledged to cultivate 377,000 hectares of corn, or an average of 487 hectares per team.[91] On February 19, 1959, the first in a series of articles in *Kolkhoznaia zhizn* raised awareness of the innovations, declaring the Lenin Kolkhoz "On the march for the corn harvest!"[92] Brigade representatives met to discuss mechanized corngrowing, part of the nationwide campaign spearheaded by Aleksandr Gitalov, Nikolai Manukovskii, and other vanguard workers. The farm participated in a krai campaign by forming twelve mechanized work teams.[93] By June, however, the cornfields had become overgrown with weeds, as the first brigade had fallen behind the fourth and the sixth. One team of tractor operators had weeded the corn especially poorly, which the team leader blamed on machinery in poor repair. Notwithstanding "many requests to the chief machinist to replace [dull cultivator blades], Comrade Prutkov remains deaf to our appeals," he explained, "Thus we save a few pennies and lose hundreds of rubles" because of widespread "instances of shoddy work."[94] Employing martial language typical of the corn campaign, *Kolkhoznaia zhizn* attempted to mobilize workers for manual weeding: "Comrade kolkhozniks! The battle for a high yield of row crops has entered the decisive phase. . . . Rainy weather had caused weeds to grow quickly, so we must devote everything to eliminating them!"[95]

Even when farms had machinery, they continued to use it poorly. "In addition to the vanguard farms," the Stavropol Krai party committee explained in 1961, "there are also serious shortcomings in the planting of corn in the necessary timeframes." Officials blamed district authorities for leaving machines idle by organizing only one shift, not two. Farmworkers often waited while others repaired machines or completed prerequisite jobs. On one kolkhoz, inspectors found planters out of repair. Having not learned how to operate a new model, kolkhozniks left it unused. Local bosses did not visit the farm or call attention to problems, an approach condemned as "formalistic and cautious."[96] To achieve the necessary results, the krai committee demanded that "corn cultivation be at the center of the district party committee's attention," so that the subordinates would "take measures for the mobilization of the entire able-bodied population for carrying out the cultivation of row crops."[97] These descriptions echo the opening phases of Khrushchev's crusade: the equipment was new, but haphazard efforts to use it remained little changed from those of eight years before.

To put industrial farming into action by using modern farming methods (Figure 6.1), the Lenin kolkhoz and other farms in Stavropol krai had to pay kolkhozniks cash wages to incentivize more efficient labor, a step on the path

FIGURE 6.1 A 1963 press photo shows workers on the Staromarevskii sovkhoz in the Shpakhovsk Production Administration, Stavropol Krai, RSFSR, as they apply agricultural chemicals to newly germinated corn.
Photo by V. Mikhalev (Photokhronika/TASS), courtesy Russian State Archive of Cinema and Photo Documentation, ID#1-77444

toward more sophisticated financial calculation. In late 1957, the Stavropol Krai party committee secretary, Fëdor Kulakov, and soviet chairman, Evgenii Krotkov, wrote to Moscow to advocate for implementing the new system of pay, accounting, and planning on kolkhozes of their administrative region. They proposed implementing *khozraschët* to track expenditures on production, for the first time learning which commodities brought net income and why. This promised to allow a kolkhoz to plan to produce the most at the least cost within the requirements set by government procurement orders.[98]

Kolkhozes had hitherto been unable to calculate production costs because the money value of a labor-day varied according to the volume of goods and cash left over at year's end. In 1958, one kolkhoz in Novo-Aleksandrov District attempted to apply *khozraschët*, but the labor-day, with its fluctuating value, could not provide "brigade leaders a clear indication of production costs, as well as of ways to decrease them."[99] A comparison required payments to kolkhozniks in units constant from farm to farm and year to year. Kulakov and Krotkov appealed, therefore, "to organize [kolkhozes'] finances based on full implementation of *khozraschët*, allowing calculation of production costs, financial results of production, and net income."[100] Furthermore, "a transition to cash payments to kolkhozniks will increase material incentives, bringing closer the pay systems of kolkhozes and sovkhozes." The krai's leaders proposed implementing the system first on Will of the Proletariat and eleven other wealthy kolkhozes over the next several years, and only subsequently on the rest.[101]

Even the basic process of calculating profit by subtracting production expenditures from income was an innovation. Since their founding, the kolkhozes had not determined the cost of a ton of wheat or a liter of milk. By establishing the cash value of a person-day worked, even if it was paid in kind, the kolkhozes learned which activities brought a profit and which a loss. Only then could they plan to increase profits by producing more value with less expenditure. Before, when they sold produce to the state below production cost, increases in production brought only further losses. By contrast, accounting prioritized efficiency and savings of fuel, seed, labor, and other inputs. After a few experiments, the practice spread alongside the wage system that removed the uncertainty and arbitrariness of the labor-day.

Although integrated into the planned economy, *khozraschët* was a market-like financial mechanism in its own right. Under the New Economic Policy in the 1920s, the state trusts managing the textile and other light industries had operated according to the principle to encourage rationalization, efficiency, and budgetary discipline.[102] For kolkhozes, the government set purchase prices for most marketable output, enmeshing them in the state economy. In theory, *khozraschët* made them self-financing and increasingly self-managed. The system simply extended the accounting practices used when meeting government quotas, paying taxes, and repaying loans to actions within the kolkhoz. As a 1958 pamphlet explained to local officials and kolkhozniks, "The innovation is in ... raising the brigade's responsibility for the results of its economic activities and in requiring efforts from each kolkhoznik to economize labor, feed, fuel, etc." *Khozraschët* first shed light on real financial outcomes, including missed targets and excessive expenditures, preventing "poorly working brigades [from] hiding behind the broad shoulders of the whole kolkhoz."[103]

On paper, the system made production costs for crops a primary indicator of an enterprise's success. Cheaper corn promised meat, milk, and wool at prices affordable to consumers. In principle, *khozraschët* also prevented kolkhoz bosses from arbitrarily altering daily work norms, as they had under the labor-day system. Within certain boundaries, individual kolkhozes fixed a table of six classes of work, with appropriate daily norms and wages fixed at a level appropriate to its finances. This change pressured kolkhozniks to accept higher work norms in exchange for a pay raise and a guaranteed monthly wage, with outcome determining only the final 30 percent and any applicable bonuses.[104]

Their cautious initial timeframe notwithstanding, Stavropol leaders directed a hurried campaign. In May 1959, Krotkov reported that 141 of the 146 kolkhozes had begun to pay guaranteed cash wages. Backing this "extremely important measure," he explained the obsolescence of the labor-day, which "had played a positive role in the past, but now we're saying 'so long' to it."[105] In practice, mistakes plagued the initiative. The party committee official in charge, Sergei Maniakin, reported that almost every kolkhoz achieved "positive results," including "improved organization, labor productivity, and material incentives, ... as well as decreased unproductive expenditures." Nonetheless, he wrote, "spot inspections demonstrated that on certain kolkhozes ... serious shortcomings occur."[106]

These shortcomings highlight how kolkhozniks made demands on managers. Kolkhozniks maintained well-grounded fears that the new system exploited them, much as the old one had for decades. Voicing concerns about the wage table, meetings of kolkhozniks acknowledged that women often did low-paid manual labor in the fields and on dairy farms. Some proposed raising the wage associated with the lowest two grades to mitigate the effects of this gendered wage gap on such "underqualified workers."[107] A 1959 inspection showed that the kolkhozes that had made the transition had little heeded such concerns, maintaining inequalities. On poor and wealthy kolkhozes alike, the wage for the first category remained half of the sixth, the rate for tractor drivers and specialists.[108]

Especially given past experiences, kolkhozniks quite reasonably felt cheated when the new wages seemed to offer no advantage. A 1959 inspection found that pay amounted to 42.5 percent of kolkhozes' total income.[109] On some farms, however, accountants distributed monthly payments in cash and kind totaling a full one-twelfth share, rather than 70 percent of it. The vagaries of weather, crop disease, pests, and other ills rendered kolkhozes' cash flow irregular. Having paid full wages in February, the farm could not pay them in November. In other cases, paymasters recorded the full 100 percent of the monthly share in kolkhozniks pay books, but kolkhozniks received only 70 percent of that figure and a promise of the remainder later. Exasperated, Maniakin reported that such occurrences "elicit unneeded rumors and doubts about the guaranteed nature of the kolkhozniks' pay." Applied on "several" kolkhozes, this method nullified the advantages of the new wage. In other cases, it became "impossible to implement this differentiated ... pay when the kolkhozniks are assigned a sum for the whole year's pay."[110] In other instances, officials fretted that kolkhozniks' wages might rise too high, even exceeding those earned by state-employed sovkhoz workers.[111]

As a rule, kolkhozniks preferred a safe wage based on collective achievements, rather than one predicated on uncertain individual ones. Officials emphasized that the new policy eliminated the labor-day system's leveling tendencies, meaning that better work earned higher wages.[112] The labor-day inherited from Stalin had ensured that earnings were uniform, if low.[113] Kolkhozniks cultivating better land, operating advanced machines, or assigned to grow cotton, sugar beets, or other crops fetching high procurement prices stood to gain under a system of differentiated wages. Those who lacked such advantages, however, likely anticipated little benefit. Justifiably unwilling to work hard for the kolkhoz, kolkhozniks indicated that they understood it as a mechanism for extracting grain and other commodities from the countryside. Long experience under Stalin had taught kolkhozniks to prioritize growing reliable sustenance on their personal plots. Preferring to preserve time for working there, they limited their commitments to the kolkhoz, doing just enough to receive the grain that they could not grow themselves. Many therefore cared little for incentives encouraging more intense labor for the kolkhoz.

The change from shares to wages had to overcome those decades of experience. Kolkhozniks quickly rejected guaranteed wages if they failed to actually improve life. Characteristically, Krotkov blamed local officials for mismanaging the initiative, rather than policies, old or new. Denouncing those who "did not change their approach" to the issue, he cited a case where a kolkhoz had distributed less than one-half of the planned wage fund for January and nothing at all for February and March. "What kind of guaranteed wage is that?" Krotkov asked indignantly. The kolkhoz chairman and accountant "claimed there was no money," he reported, "However, sometimes we must not only trust comrades, but also verify [their actions]." Inspectors did find that farms sometimes suffered such shortages. In this case, a call to the bank revealed that the kolkhoz had received the money, but its managers had spent it on operating costs. Nothing had changed for the unpaid kolkhozniks, on whose behalf Krotkov grumbled, "'However much they cheated us before, now [they] are still cheating us.'"[114] Even when claims of cash shortfalls proved true, kolkhozniks likely still felt swindled. Nor was this an isolated incident: in one district, the *majority* of kolkhozes fell at least one month behind.

Kolkhozniks sensed duplicity when bosses altered the new wage scales and work norms soon after the kolkhoz had formally adopted them. In 1959, unusually heavy early summer rains fell in Stavropol Krai, harming the corn, winter wheat, and other crops. Krai and district authorities pressured kolkhoz managers to alter pay scales, balancing the books by offsetting lower-than-expected earnings from smaller-than-planned harvests. On aggregate, kolkhozes had earmarked 39 percent of their income for pay and guaranteed monthly advances of 78.7 percent of that figure. Actual income fell 13.5 percent short of the anticipated 2.5 billion rubles. By midsummer, however, kolkhozes had overspent *original* wage budgets by some 112 million rubles, a figure 189.6 million above budgets revised to reflect lower incomes. Krai officials condemned this as "incorrect management."[115]

Kolkhoz bosses did not flout superiors' orders lightly, but because they had to persuade kolkhozniks to work at tending and harvesting corn, wheat, and other crops. Managers prioritized wages, whose importance had increased because the new policy had raised kolkhozniks' expectations. To compensate, kolkhoz managers declared their enterprises in financial difficulty, requesting short-term credit and temporary relief from debt repayments increased considerably by the 1958 MTS reform. Decrying the situation, krai officials indignantly rejected these petitions as "completely baseless."[116]

Even after the wage reform, higher officials viewed kolkhozniks' pay as a shock absorber allowing farms to cushion the budgetary consequences of drought, flooding, disease, or other unanticipated events without reducing deliveries. Under any circumstances, the state received its produce, but kolkhozniks saw their wages fluctuate wildly, quickly coming to mistrust the seemingly more advantageous system as much as the obviously exploitative old one. Brigade and kolkhoz leaders then had to negotiate with them so that they would turn out to work, including by paying wages even in a year of low yields and falling incomes. Weighing the prospects, some farm managers disregarded financial consequences, raising wages in hopes of fulfilling output plans. Krai officials expressed exasperation with those who privileged workforce morale over the budget, bemoaning the strategy as incorrect.[117]

On January 1, 1959, the Lenin kolkhoz joined others in the krai in transitioning from labor-days to guaranteed wages. An account of a public party meeting described how "material incentives" promised to improve discipline. Extolling the farm's economic strength, Chairman Agnaev cited figures showing that milk production had increased 42 percent since 1955. Not coincidentally, corn plantings had grown from 1,570 to 2,435 hectares and yields from 2.08 tons to 5.63, a figure comparing favorably to American averages of the time. Agnaev outlined the new six-tiered table of wages and work norms. The work once earning a labor-day now paid 13 rubles, the second grade. Those operating machines earned 26 rubles, the top bracket, and could earn bonuses up to 13 more. Finally, Agnaev outlined rules for production bonuses, declaring, "To reward corngrowers' achievements, a brigade or work team overfulfilling its planned yield will earn 10 rubles more for every tenth of a metric ton of corn over 5 tons."[118] In 1959, the Lenin kolkhoz made the most of the difficult weather, enacting special one-time bonuses for higher yields of corn.[119] Like most other kolkhozes, the Lenin kolkhoz also had to quickly repay loans for MTS equipment acquired in 1958. Its purchase of 56 tractors, 32 combines, 300 implements of various sorts, 60 trucks, and 7 cars indicates the size and productive capacity of the farm. Additionally, the kolkhoz had to finance construction of an electric power station, irrigation systems, and other means of production, all signals of the modernization and industrialization of the kolkhoz.[120]

Kolkhozniks soon grumbled about the new wages. Within six months, some charged that the new wages shortchanged them much as the old labor-day. The farm's senior bookkeeper took to the pages of *Kolkhoznaia zhizn* to respond to "those who say that pay is too low." Wages guaranteed working kolkhozniks cash income allowing them to purchase flour and other commodities from the farm at the low price paid by the state for procurements. Addressing complaints, the accountant examined the case of one kolkhoznik, not by chance a woman performing low-wage manual labor. Having worked nineteen days, she had earned an average of 23 rubles per day, or 423 rubles for the month. If she had worked the expected twenty-six days that month, the total wage of 579 rubles would have amounted to livable sum, the accountant explained.[121] Regardless, the wages achieved some results. In Stavropol Krai, average annual household income increased 27 percent, from 4,749 rubles in 1957 to 6,019 rubles in 1959. Average individual income similarly grew 40 percent from 2,315 rubles to 3,350 over the same period.[122]

Having little formal input into policy, kolkhozniks demonstrated displeasure by refusing to work in the fields, forcing leaders to respond. In one district in 1960, bosses pressured kolkhozes to abandon wages. "The district committee changed the kolkhoz, among the strongest financially," a report explained, "back to labor-days in the spring. Within two days, the kolkhoz managers were ordered to revert to cash wages because the kolkhozniks refused to return to the old pay system."[123] Some kolkhozniks left altogether in response. In one district, conditions "did not allow settling accounts with the kolkhozniks on time, causing labor discipline to fall and the departure of a substantial proportion of kolkhozniks."[124]

The krai authorities criticized the bosses of another kolkhoz for their "dependent attitude," suggesting that they expected the government to make up financial losses incurred by paying high wages. In 1959, kolkhozes in the krai's Karachai-Cherkess Autonomous Oblast had budgeted 50.9 million rubles for wages, but income had fallen from an expected 149 million rubles to 112.9 million because of the rains. Despite a 24.2 percent decrease in income, guaranteed wages had *risen* 17 percent to 59.5 million rubles. This was because "kolkhoz managers and primary party organizations did not honor the kolkhoz charter and do not follow the [*khozraschët*] plans as if they were law."[125] Without regular wages, kolkhozniks stayed home. Local bosses had concluded that, by failing to pay the kolkhozniks, kolkhozes would suffer even more from the extreme weather because no work would get done at all.

The wage system had other glitches that provide evidence that kolkhozniks indirectly influenced wages by holding views of fairness distinct from those implicit in the policy. Because of intermittent "mistakes" in planning to make payments, "wages were distributed to kolkhozniks at the same level," the krai party authorities concluded, "irrespective of the individuals' fulfillment of production plans. A leveling occurred."[126] Officials fretted that kolkhozniks would become dissatisfied with variation among pay scales, which caused wages for same job to diverge among kolkhozes. In some cases, this gap reached 37 percent across a single district.[127]

The wage policy sought to incentivize participation, but only partially succeeded. On the Lenin kolkhoz, the fifth brigade had long completed jobs rapidly and had even helped others that fell behind. By July 1959, Chairman Agnaev criticized it for lagging in cultivating corn and cutting hay, even with the aid of new machines designed for these jobs. He blamed "falling labor discipline," a result of party members and brigade leaders becoming, in a boilerplate phrase, "complacent." They allowed kolkhozniks to dodge work, meaning that only 70 or 80 of the 259 members typically assigned to fieldwork turned up regularly. The remainder, "a certain undisciplined element," caused progress to slow, while simultaneously complaining about low incomes.[128] As late as 1963, officials continued to have difficulty matching kolkhozniks' pay to their output. A krai conference reminded authorities about policies for paying bonuses above guaranteed wages. The percentages of such payments in money or in kind differed: between 15 and 25 percent for milk, eggs, and meat and other products. Those for raising corn, on the other hand, reached 50 percent.[129]

The new pay, labor, and accounting policies illuminate the pressures kolkhozniks, managers, and district officials faced. An analysis of the Lenin kolkhoz's finances illustrates how *khozraschët* worked, but also failures in translating the principles into practices. "The brigades themselves formulate the production plans," as standard procedure required. "However, the plan figures for important products are distributed from above by the kolkhoz management," a practice counter to the spirit and letter of policy, but nonetheless widespread.[130] The 1955 planning reform explicitly required plans to begin at the bottom, but in reality this was only for show. "After formulating the production plans, the brigades submit them to the kolkhoz management for review and approval. The kolkhoz managers then adjust these brigade plans," the report concludes. Having shaped the plan at the beginning, managers then unilaterally altered it. The process repeated itself at district and krai levels.[131] Krai officials denounced one district committee's orders to kolkhozes to trim expenditures on schools, clinics, and other social services. While necessary, such nonproductive expenditures had to wait until the kolkhoz had financed its own investments in production capacity, an achievement unlikely because of the government's constantly rising demands for produce.[132] In the last step of the planning procedure, "on the basis of the brigade production plans, confirmed by the managers, *khozraschët* targets are formulated, approved, and distributed to the brigades." Here, authorities noted managers' failure—or unwillingness—to implement a differential system of pay that allowed bonuses to increase the pay of those individuals and brigades with above-average results in measures such as economizing fuel.[133]

Notwithstanding some kolkhozniks' discontent and refusal to work, the reforms made most appreciably more enthusiastic about work on the kolkhoz. In the summer of 1959, Stavropol Krai officials documented that a higher proportion of kolkhozniks worked a greater number person-days. In quarters one and two, 11,800 more kolkhozniks had worked at least once, an annual increase of 7 percent. In total, kolkhozniks had worked 2,153,100 more person-days, or 11 percent more than the same period in 1958, amounting to 4.5 percent more per farmworker. In light of labor shortages, this significant improvement facilitated increases in output, including 22 percent more milk and eggs.[134] By 1962, production on the Lenin kolkhoz had risen still more. The expected corn yield (3.5 tons per hectare), as well as output of 1,468 tons of meat (2.5 times the 1955 figure) and 5,430 tons of milk (2.4 times the 1955 figure), proved to party officials the effectiveness of Agnaev's management, which the district secretary characterized as "honest" and "conscientious."[135]

As late as the summer of 1960, the krai committee received complaints about *khozraschët*, its implementation, and financial burdens imposed by concurrent campaigns. In one district, only two of eight kolkhozes had implemented it fully, and one had not even begun the process. Admonishing them, the district committee achieved little. One kolkhoz's bookkeeper defended himself by declaring that for six months the targets "achieved nothing positive and, therefore, the brigades and departments refused to continue *khozraschët*." A neighboring kolkhoz, by contrast, had smoothly distributed plans, met them, improved efficiency, and lowered production costs.[136]

In August, one kolkhoz chairman bemoaned the many external financial burdens inhibiting *khozraschët*. The farm had paid kolkhozniks for work only through April, leaving it more than three months and 2.7 million rubles behind. Citing low rates of capital investment in the past, the chairman further lamented year's poor grain harvest, which depressed income from a planned 7.1 million rubles to only 1.8 million. In addition to back wages, the farm owed 4.7 million rubles to the government for the equipment it was forced to purchase in the 1958 MTS reform. Officials required these repayments not over many years, as Khrushchev's policy envisioned, but in only a few. Widespread, this problem was proof that "it was premature to seek the full cost of the machines purchased," because it prevented the kolkhozes from building up a wage fund or paying wages required for current production.[137]

The kolkhozes also had to expand their livestock herds as part of Khrushchev's drive "to catch up to and surpass America." Under pressure, they bought kolkhozniks' personal animals, putting simultaneous burdens on kolkhoz finances and on supplies of corn and other fodder. One kolkhoz had borrowed 2 million rubles of the 2.85 million needed to buy 767 cows. This coercive campaign went forward behind a thin screen provided by claims that kolkhozniks, now happy with wages they earned, willingly sold personal livestock. Using their cash incomes, they supposedly purchased milk in state shops. As one chairman put it, "Because kolkhozniks desired and requested to give up personal cattle, the management purchased them on a strictly voluntary basis."[138] In the first quarter of 1960, the cost of these purchases across the krai exceeded kolkhozes' savings, eventually requiring a grant of 148 million rubles in aid. When the abuses came to light, the krai authorities shifted responsibility onto district subordinates.[139]

Becoming almost universal after 1964, wage labor spread most rapidly in Stavropol Krai. Paying comparatively high wages, most of its kolkhozes quickly transitioned to cash.[140] By 1963, 138 of 141 kolkhozes paid wages, a proportion higher than any other region. Other leaders included Orenburg Oblast (304 of 398), Leningrad Oblast (82 of 90), and Kostroma Oblast (177 of 209). Yet these practices achieved only so much. A survey that year of sixteen locales in the RSFSR found that kolkhozes still underutilized their workforce even under the new system. The average kolkhoznik worked only 170 person-days per year. Average wages fluctuated wildly from region to region, not to mention from kolkhoz to kolkhoz.[141]

Similarly, the problems prevalent in Stavropol Krai cropped up across the country. One 1959 inspection in Ukraine and Moldova found "significant increases in the productivity of kolkhoz labor" and concomitant boosts in crop yields and livestock productivity. Yet it also uncovered "serious shortcomings." Kolkhozes paid bonuses for all production, not just that above planned yields. Field hands, most of them women, earned few bonuses regardless of the size of their harvest.

Their bonuses rising, managers and technical specialists came to view them as integral to their pay. A few manipulated the system to earn most of the bonuses paid out. Inspectors found cases where a few individuals received more than 50 percent of the total. In an extreme case, the five specialists on a kolkhoz in Lithuania collected 88,900 rubles out of 136,000 rubles paid in bonuses, or 7.5 percent of the kolkhoz's money income. The chairman's bonus of 36,400 rubles alone was equal to 147 percent of his basic pay, while for the others this percentage was even higher.[142] Because wages and finances were poorly supervised, kolkhoz bosses had little difficulty circumventing the formal need for approval by the kolkhozniks, whose sense of fairness would not have allowed such exorbitant payments.

In Ukraine, inspectors judged the wages as "a progressive method of remunerating labor" despite "the lack of regular payment of the monthly wage." Thus during the first three quarters of 1959, kolkhozes countrywide had received cash totaling 28.4 billion rubles to make the wage payments, but had actually paid out only 19 billion. In some districts, fewer than 20 percent of kolkhozes had done so.[143] As an executive summary sent to the USSR Council of Ministers dryly concluded, "The remaining sum of 9.4 billion rubles was disbursed by the kolkhozes toward other ends."[144]

In March 1963, Kulakov spoke to a Stavropol Krai conference on agricultural development, outlining continuing efforts to achieve Khrushchev's goals. The local secretary declared that the herculean task of "constructing communism" required not incremental development, but revolutionary change. "Now we must double [or even] triple output, and not in forty years, but in just a few," Kulakov explained. "In organizing production, in labor and pay, and in management methods we retain much that is outdated, backward, and conservative, useful for extensive use of the land," which is to say the opposite of intensive practices of industrial farming. "All of this holds back productive forces [and] the rapid development of agriculture," he concluded.[145]

Labor and pay remained paramount. Kulakov denounced new and, in his view, inflated expectations not matched by concomitant gains in output. He chided, "Each farm [must] pay its workers in relation to the quantity and quality of their output."[146] Other speakers concluded that, from the beginning of efforts to replace labor-days with wages, authorities imposed norms and plans from above. These were therefore "not based on accurate, rigorous information, but [were] calculated imprecisely." Another warned the audience, "The system of pay is determined by the character of production. We must remember that production influences pay, but pay, in its turn, makes labor more productive." Only persistent exertions, another official concluded, could complete the transition to wages, ensuring policies were "based on the effort to expand material incentives for agricultural workers to boost agricultural output."[147] Additionally, speakers stressed specializing land management and altering crop rotations to include more corn and other row crops, thereby reinforcing connections among technology, labor, and productivity.

Describing the circumstances facing rural consumers, Kulakov raised concerns that seemed to reach a critical state by 1964. When Khrushchev's former comrades removed him from power, they charged that corn and slogans to catch up to America had not improved living standards, especially those of kolkhozniks. Blaming him for lagging agricultural output, his critics attacked pay policies in particular. "The question of farmworkers' material incentives has not been answered," they declared. "Comrade Khrushchev has delivered many speeches and signed numerous memoranda, but the results have been insignificant." In 1958, kolkhozniks earned an average of 1.56 (new) rubles a day. By 1963, that figure had risen to 1.89 rubles.[148] Concluding that an increase of 20 percent in five years was "insignificant," Khrushchev's opponents tacitly acknowledged how depressed wages had been before 1958. Moreover, their charge did not take into account that wages had become to some degree regular payments in kind or, increasingly, in cash.

In fact, frustrated expectations had become a considerable problem. In 1963, Kulakov stated, "We know that supplying bread and other foods is an important material incentive for kolkhozniks because they are in the habit of stockpiling grain, but also because even a ruble earned cannot buy all needed goods in the state trade network."[149] Peasants acted according to their sense of justice. They worked neither to increase production for production's sake, nor to advance abstract ideals of socialism, as authorities sometimes hoped. Instead, they responded to material incentives because they hoped to earn the money and procure the goods they did not produce for themselves.

The poor harvest of 1963 had raised the specter of tightened belts and low wages, leaving kolkhozniks apparently fearing the disappearance even of grain or flour from kolkhoz stores. During the August harvest, police reported to the Central Committee several unconnected "disorders" and "mass refusals to work" more accurately described as strikes by farmworkers. In Ukraine's Poltava Oblast, eighty kolkhozniks confronted a kolkhoz's chairwoman, refusing to bring in the harvest until she resolved their pay grievances. In another instance, groups of hundreds of kolkhozniks gathered on successive days to make demands. On the first day, they demanded that the district party secretary order payment of 1 kilogram of grain or flour. Rebuffed, they returned on the second day to challenge the Moldovan republic party leaders, a scene marked by angry shouts and so-called anti-Soviet oaths. As soon as the bosses met their increased demand for 1.2 kilos, the kolkhozniks quickly dispersed and returned to work.[150]

Kolkhozniks produced meat, milk, and other foods, but could not buy them in the local shop, a fact that violated their sense of justice. A letter from Stavropol Krai to Khrushchev described how life was improving in one rural area. "In the stores, you can buy whatever manufactured goods you want," the writer explained, "there's even bread and flour." On their small personal plots, kolkhozniks concentrated on growing vegetables, but acquired grain as an in-kind payment from the kolkhoz or purchased it with their increased income. Milk and dairy products were another matter because rural stores never had any. "It's not good that every year our community ships a large quantity of milk to the towns, but for some reason dairy products do not make it back to our shops," the writer lamented. Inquiries to the authorities revealed that no one had ever sanctioned sale of dairy products in rural areas, likely on the assumption that kolkhozniks produced them at home. In the aftermath of the campaign forcing them to sell personal livestock, the man wrote, "I think that this is not entirely just... that [kolkhozes] sell [milk] to the state and only enough to raise calves remains on the farm. Nothing remains for use on the kolkhoz, not even for the nursery." "Nikita Sergeevich!" he concluded, "I ask you to tell us how to solve the problem of supplying the people with milk products."[151]

Overall, kolkhozniks produced more and earned more by 1964. The views of Khrushchev's critics notwithstanding, this period initiated a trend of rising earnings and thus standards of living. Combined household income from kolkhoz and personal plot doubled between 1953 and 1967, even as the proportion of income from working on the kolkhoz increased.[152] As late as 1971, however, kolkhozniks still had poorer access to consumer goods, electricity, public transportation, public health services, and education than an unskilled urban worker. Moreover, they earned approximately one-half the income.[153] The pull of urban jobs and wages remained strong.

As Khrushchev acquired authority over agriculture after 1953, new initiatives curbed old abuses and augmented previously meager material incentives. The kolkhozes struggled because kolkhozniks had little avenue for personal initiative and received little income, even if they put greater effort into growing corn or any other crop.[154] At first, kolkhozes lacked machines to grow corn, making labor the most important contributor to the success of corn. Anticipating rates of growth outpacing the rate of investments, Khrushchev assumed that kolkhozes only needed to bring latent capacity into production. Only higher productivity could raise farmworker income without either forcing higher consumer prices or increased government subsidies.[155]

Implementing the reforms associated with *khozraschët*, authorities intended to learn what it cost to produce a ton of corn feed, a hectare of wheat, or a kilogram of pork, information vital to improving efficiency. These practices pointed the way to cautious reforms in industry undertaken in 1965 and to the presence of *khozraschët* in the reform program of Mikhail Gorbachev after 1985. The leadership also considered reliable wages superior compensation for kolkhozniks' labor, which it undoubtedly was in comparison to the old system of arbitrary disbursements from the leftovers at the end of the year.

The significant changes in pay, accounting, and technology all pointed toward industrial agriculture. A guaranteed wage and financial accounting signaled a transformation of the relationship between kolkhozniks and kolkhozes established during collectivization. They formed part of a long-term, gradual transformation of kolkhozniks' social status, one that in the 1960s granted passports, pensions, and the further benefits of citizenship that other groups had enjoyed for decades. Machines, chemicals, and other technologies promised to make labor more productive, providing more surplus product from each person-day worked. Producing more commodities at lower per-unit cost, kolkhozes could cover higher wages from their profits, which in theory were no longer to be appropriated through low state procurement prices. Although the terms kolkhoz and kolkhoznik had remained the same, changes in the agrarian political economy had fundamentally altered their practical meaning.

Even though labor-saving technologies began to become more common by the 1960s, labor productivity lagged. Those with skills and training left at the first opportunity because of wages that remained lower than any other sector, rugged living conditions, and the general poverty of opportunity in rural areas. It was a vicious cycle, which could only be solved by improving social and economic conditions.

Khrushchev's labor reforms fell short of expectations because of poor management by local leaders who misunderstood the kolkhozniks and their moral economy. Labor policies in action provide glimpses into the kolkhoz's workings and political economy. The advent of regular wages changed kolkhozniks' lives and work, but did not necessarily make them efficient corngrowers. Despite legal constraints on their ability to move and change jobs, a market for labor allowed them to exploit the labor shortage by seeking off-farm employment if conditions on the kolkhoz became too harsh. In other, more common instances, they used the threat of doing so to achieve goals while remaining collective members and, therefore, retaining a prized personal plot. Kolkhozniks' experiences conditioned them to seek regular and dependable sources of produce and income while spurning attempts to coax them into working for the kolkhoz. Learned under Stalin, this lesson shaped kolkhozniks' understanding of the kolkhoz and presented Khrushchev's reforms with a challenge.

Kolkhozes could not discipline laborers by the threat of redundancy or by actual firing, as commercial farms in other economic formations might. Whatever the costs, they needed vast workforces to make up for shortages of labor-saving machines, particularly at harvest time. Instead, they faced shortages, which were exacerbated by the demands of corn. In fact, kolkhozes augmented the agrarian workforce at these moments by bringing in students and workers from nearby towns to dig potatoes or pick corn. Kolkhozes thus imitated industrial enterprises by hoarding of laborers, who sat idle for long periods. Only in the vital moment could workers, including all the extra workers made necessary by technical and supply problems, throw themselves into high gear, storming production to meet the quota—or bring in the harvest—just in time.

7 AMERICAN TECHNOLOGY, SOVIET PRACTICE

Seeing his ambitions for corn and industrial agriculture hindered, Nikita Khrushchev developed policies confronting the bureaucracy, whom he denounced in speeches amplified by press campaigns. This antibureaucracy rhetoric employed epithets that shed light on existing practices and officials' unease at losing their position. Beginning early in the decade, public and secret documents alike utilized *po shablonu* (a *shablon* is a template) to denounce administration by formula with little thought for outcomes. At intervals, they condemned *formalizm* (concern for bureaucratic form rather than results), *ochkovtiratelstvo* (duplicity, deceit), *pripiski* (adding fictional production to reports or forging work orders), *obman gosudarstva* (cheating the government), *biurokratizm* (runaway bureaucracy), and the related *volokita* (red tape). Although it is not exhaustive, this list does not include bribery, a practice that seems to have been prevalent, but which did not number among these antibureaucracy slogans.

Khrushchev, the Central Committee Presidium, and the USSR Council of Ministers established formal policies, but the formidable administrative apparatus did not automatically transform policy into action. Inherited from Iosif Stalin, the apparently rigid and self-serving bureaucracies impeded Khrushchev's efforts to transform ambitions into agricultural commodities.[1] The leader mandated technological innovations to facilitate the rapid growth of corn plantings, but these failed to ensure that officials responsible for construction, manufacturing, and procurements executed these policies.

Khrushchev struggled to discipline administrative organizations into a functional management apparatus. Launching the full-scale crusade for corn in 1955, Khrushchev ordered factories to build specialized harvesters, planters, and cultivators, but the answerable ministries delivered slowly. In early 1957, Khrushchev and his rivals together initiated a reform to break up vertically integrated economic ministries, an effort to transform the bureaucracies into responsive tools for implementing policy.² In subsequent years, the government made an investment in hybrid corn seed by importing American technologies and methods. Achieving partial successes, these programs shed light on how economic management functioned. Moreover, the seeds challenged orthodoxies about genetics associated with Trofim Lysenko. Winning a battle, proponents of hybrid corn secured official backing to ensure that, by 1960, the country had made substantial advances in seed technology in spite of Lysenko. Over the course of a decade, Khrushchev proved able to maintain a corn crusade of such scope and intensity because he inherited a model of top-down authority from Stalin. Unwilling to threaten violent repression, Khrushchev and his ambitions collided with the functionaries whose practices encumbered efforts to implement policy.

Particularly in their archival records, bureaucracies portray themselves as orderly hierarchies. The Soviet Union's government and the Communist Party were no different, but behind that façade, a mishmash of formal procedures, unofficial practices, personal relationships, stopgap measures, and outright falsehoods permitted the system to function while constraining its capacity for disciplined and effective action. Studies of informal relationships in the Soviet Union have concluded that practices were often "self-subversive," meaning that they permitted individuals to achieve ends but, in greasing the wheels, undermined formal procedures and rendered systematic operation impossible.³ From ministries in Moscow to enterprises far from the capital, officials impaired policies through inertia and subterfuge. They did so not out of particular malice for Khrushchev, or even for corn. Instead, they had to respond to pressure from above, secure personal authority against demands from below, and fortify job security by at least appearing to get results. In short, they saw to personal interests first as they mediated between many competing incentives.

Authorities uncovered these practices by dispatching inspectors. Their reports traveled upward outside normal party and government channels, which tended to trumpet success while concealing failure. Records from the USSR Ministry of Government Oversight and its local subordinates, if taken in isolation, overemphasize breakdowns and underappreciate normal operations.⁴ Inspectors, moreover, did not quantify the scope of fraud, waste, and inefficiency, but the prevalence of these phenomena in the archival record and contemporary public discussion suggests that they were common.

The scale of inspections illustrates the limits of inspectors' activities. Each administrative region had an inspectorate, which only coordinated with republic and union counterparts on a few large sweeps each year. In 1959, for instance, the USSR Commission for Government Oversight carried out 23 multiregion inspections to verify fulfillment of resolutions about agriculture passed by the Twenty-first Party Congress, directives of the December 1958

Central Committee plenum, and joint orders of the Central Committee and USSR Council of Ministers. Local personnel worked alongside those sent from Moscow. After an inspection, the officials of an enterprise, government office, or party committee had to promise action to remedy any problems uncovered. The commission collated reports about enterprises, districts, oblasts, and republics into a summary for the whole country, which it dispatched to party and government authorities. Inspectors covered nearly 2,000 sites related to agriculture across 70 administrative regions, including 829 kolkhozes and 510 sovkhozes, as well as procurement facilities, construction projects, and research institutes. This resulted in punishment of only 115 individuals, of whom 27 were fired and 25 given "strict reprimands," a warning leaving them on the verge of dismissal.[5] The number of kolkhozes inspected amounted to 3 percent of the total, an inconsequential number. Even multiplied by the number of independent regional inspections, the process likely had little deterrent effect, but its records do provide insight into day-to-day operations.

Inspectors left a glimpse of the heart of the Soviet Union's manufacturing sector. Comprising a tangle of organizations, each ministry controlled factories, defended its own resources, and prioritized its own interests. In 1955, a crash project to assemble machines for the corn crusade stretched their resources. The results only partially met Khrushchev's demands, which began as early as November 1954. "We have not yet valued corn," he stated, "which is a giant among crops." To take advantage of its potential, farms needed proper equipment, such as silage harvesters (Figure 7.1), which required investment. "We must adopt the silage harvester at full speed," he ordered. "It is as necessary as the grain combine, perhaps even more so. This machine will facilitate rising milk yields because the kolkhozes will not be able to cope [with harvesting] . . . without these machines."[6] During the January 1955 plenum, Khrushchev reprimanded Stepan Akopov, the minister of automobile, tractor, and agricultural machine-building. Akopov admitted that domestic machines were poorer in quality and fewer in number than American counterparts. "Unfortunately," he conceded, "We use foreign technology very poorly, study foreign know-how poorly. There is much we do not know." The number of factories manufacturing harvesters had *fallen* from eight in 1951 to just three because the others had been retooled. In 1954, the Soviet Union had produced 291,000 tractors, or 22 percent more than in 1953, but output lagged far behind the 400,000 produced in the United States, not even taking into account the latter's much smaller planted area.[7]

During the early seasons of the corn crusade, shortages of specialized machines checked progress. The chronic shortage demanded that machine-tractor stations (MTSs) and sovkhozes to use the machines they did have on hand as efficiently as possible. They did not. In spring 1955, Stavropol Krai officials ordered

FIGURE 7.1 Workers on the Kulundinskii sovkhoz in Altai Krai, RSFSR, use a mechanical silage harvester and heavy trucks to reap, chop up, and transport the whole corn plant for use as animal feed.

Photo by A. Agapov (Sovinformbiuro), courtesy Russian State Archive of Cinema and Photo Documentation, ID#1-50443

the MTSs to redistribute their 1,911 planters. Spread evenly among the planned 280,000 hectares of corn, the burden on each planter amounted to a daunting 147 hectares. Some MTSs had many and some had few, however, meaning that actual averages ranged from only 36 hectares to more than 1,000 per planter. Krai authorities ordered stations with many planters to ship some to those with few, a common practice. Later, inspectors found that "many MTS did not carry out the order, . . . while others transferred [planters] in disrepair."[8]

Anticipating the shortage, Khrushchev ordered advisers to remedy expected equipment shortages. Akopov's ministry had manufactured only 2,000 silage harvesters in 1954. The initial 1955 plan called for only 8,000 more, as well as 2,360 grain harvesters for corn, a 36 percent increase.⁹ In a March 1955 private meeting with advisers, Khrushchev gave informal orders spurring action, a demonstration of his authority. "I would ask," he politely commanded, "that you do something about wheeled tractors [for cultivating row crops], implements specialized [for corn], and silage harvesters. Concentrate on that, consulting with Comrade [Iosif] Kuzmin."¹⁰ This verbal instruction set in motion the Central Committee's department responsible for industry and transportation, whose head, Kuzmin, served as a liaison to the ministries managing manufacturing.¹¹

Khrushchev's informal directive and the resulting formal policy required ministries to divert resources. In May 1955, the USSR Council of Ministers allocated funds and ordered ministries to manufacture the harvesters in time for the fall harvest. They promised to alleviate expected extra burdens on the farm workforce caused by the simultaneous maturing of corn and other labor-intensive crops. The directive called on factories belonging to ten separate ministries to produce not the 8,000 harvesters Akopov had promised in January, but 40,000. Their number included the Ministries of Heavy Machine-Building, Transportation Machine-Building, Agricultural Machine-Building, General Shipbuilding, and more. The jumble of bureaucracies hints at the sisyphean task of coordinating an investment of some 300 million rubles.¹²

In oral instructions and speeches, Khrushchev stressed machines' importance, ensuring the government passed a policy. He could not, however, ensure that they were manufactured on schedule. Even before the directive, ministers tasked with producing the harvesters protested that they lacked necessary materials. The minister of construction, Nikolai Dygai, and minister of ferrous metallurgy, Aleksandr Sheremetev, objected that existing projects had already depleted reserves, preventing their respective ministries from filling the new orders.¹³ By early August, investigators had proven those claims false, discovering that the ministers had claimed a lack of materials while simultaneously ordering subordinates to find and allocate them.¹⁴ The ministers had attempted to avoid the new tasks, which complicated their existing production plans. Having little chance of doing so, they pleaded inability likely in hopes of securing additional resources or other concessions. In light of political leaders' skeptical attitudes toward the ministries, the ministers pressed these claims too insistently. The USSR Council of Ministers concluded that the ministers had shirked obligations to fill state orders, favoring parochial priorities over those established by party and government leaders. Sheremetev and Dygai received reprimands, while other ministers received warnings for lesser infractions.¹⁵

Subsequently, all ministries failed to meet the established deadlines, substantiating leaders' charges that the vertically integrated industrial ministries functioned primarily to stockpile resources, perpetuating inefficiency. Such outcomes strengthened a rising consensus favoring reform. By early August, inspectors concluded that the ministries were far behind the schedule requiring the harvesters in time for use in late August and September. Writing to the minister of government oversight, a senior official at the ministry of agriculture reported that the ministry had not received scheduled deliveries. In fact, the manufacturers' record was dismal: the directive mandated completion of approximately 35,000 harvesters by August 10. The most successful of the ten ministries, the Ministry of Heavy Machine Building, had assembled 600 of 10,000, or 6 percent. The agricultural official could only note wryly, "Such unsatisfactory manufacturing of these implements threatens to prevent delivery in time for the harvest."[16] Much like an interim report issued on June 20, 1955, this report went directly to Khrushchev, a sign of hands-on involvement in the ministries' activities despite his lack of a formal government post, illustrating the party's growing role in guiding economic policy.

Shortages of machinery and the slow pace of manufacturing new ones remained a constant refrain, joined by grievances about poor quality and a chronic lack of spare parts. In 1956, the Council of Ministers enacted a similar supplementary program, earmarking for it a sum of 2.3 billion rubles.[17] Ministries again fell behind, delivering the machines late. Inspectors repeated concerns familiar from 1955. In the end, the ministries completed the program by November 1956.[18]

That year, a group of combine operators working at an MTS in Altai Krai wrote to the quality-control inspectorate in Moscow to complain about the defects in new equipment received from the factory in nearby Barnaul. Factory workers painted harvesters improperly and assembled components so haphazardly that they fell apart during transport. Because the machines were "shipped barbarically," they reached their destinations covered in rust, having been stored in the open air, exposed to rain and snow. Nuts, bolts, and other basic parts were missing. "Why should we pay 30,000 rubles for a combine?" the writers asked, indignantly, "And what's more—for this junk?" Closing their letter, they pleaded for inspectors to "bring some order to the situation."[19] Constant complaints in the press established the fact that manufacturers allowed many defects, which purchasers had to fix, wasting time and resources. The need to repair new equipment illustrates the infamous problem, which had considerable cumulative effects as defective goods circulated through the economy.[20]

Responding to findings by experts sent to the United States in 1955, the government launched a program to produce more and better equipment. The delegation's engineer, Aleksandr Ezhevskii, explained that the number of tractors in the United States had increased from 1.5 million in 1950 to more than 4.5 million in 1954, while the number of farmworkers had decreased from

13.5 million to 8.5 million. This meant that "the degree of mechanization in the United States is very high." Although Ezhevskii did not bluntly state that the Soviet Union lagged behind, his findings contrasted with public claims about the resources available to sovkhozes and kolkhozes. Moreover, Ezhevskii concluded, farmers in the United States utilized "many diverse machines enabling complete mechanization of various productive and auxiliary tasks."[21] Each one permitted an American farmer to produce more at less cost, a primary objective of industrial farming.

Ezhevskii also found his country's machines poorer in quality. Americans' light, maneuverable tractors excelled in cultivating corn and other row crops. Most of the Soviet Union's tractors were propelled by tracks better suited to heavy plowing and other tasks requiring more horsepower. Compared to American models, Soviet wheeled tractors were heavy and had lower power-to-weight ratios. The common Belarus model weighed 3.25 metric tons and delivered 37 horsepower. The International Harvester model Ezhevskii selected for comparison produced 49.5 horsepower while weighing just 2.83 tons.[22] These findings spurred a redesign.

Ezhevskii considered not only the American machines, but also practices of the companies manufacturing them. He therefore proposed to retool assembly lines to produce not whole farm machines, but standardized components. Other factories might then assemble the engines built on one line and the transmissions manufactured on another into complete machines of a wider range of designs at lower cost. Describing operations of the International Harvester Company, Deere & Company, and other manufacturers, Ezhevskii argued that this measure promised productivity and efficiency.[23] Equally, the individual components also served as spare parts for repairing existing machines, the shortage of which was the subject of constant complaint in the Soviet Union. Flaws in incentive systems encouraged factories to construct few spare parts because they counted less than whole machines in measuring plan fulfillment.[24]

Khrushchev praised Ezhevskii's proposal for its interpretation of American practices, but expressed skepticism about its viability in the Soviet Union. "This is an illustration of what I was just talking about," Khrushchev explained, referencing a discussion of the spare-parts shortage. "Comrade Ezhevskii visited America and is drawing his own conclusions. What kind of conclusions? He is drawing Russian conclusions. . . . This is our Soviet American."[25] Dating from 1920s debates about models for industry, the term connoted someone conversant in foreign technological expertise, but also able to critically evaluate it. Instead of blindly copying machines and methods, the Soviet American accounted for the principles of the new society, molding foreign achievements into something suitable for socialist society.[26]

Khrushchev also voiced exasperation with the ministries managing the economy, whom he accused Ezhevskii of antagonizing. Each ministry fiercely defended its labor, raw materials, factories, and other resources. Khrushchev concluded that Ezhevskii wanted to take over other ministries' factories in the process of specializing factories. If granted, this request would cause the other ministers to demand new capital investments to replace factories lost to Ezhevskii. "Then we'll have to build another batch of as many factories as we already have, and only then will we be able to raise production," Khrushchev grumbled, "It's a vicious cycle." When Ezhevskii responded, "We will put specialization of our factories first," Khrushchev interrupted, "Not first. Only that," suggesting that each ministry should specialize its own factories separately. "That is extra capital investment," he continued, "I want you to consider how to more rationally use your own productive capacity. You're trying to take capacity from others."[27] Expecting to get the most from existing resources, Khrushchev countenanced heedless spending on neither production of agricultural machines nor industry in general.

Insufficient in 1955, manufacturing thereafter never became adequate to meeting Khrushchev's plans to mechanize farm work. In 1956, authorities began experiments using American methods for growing corn without manual labor, cutting the time to cultivate a hectare of corn to as little as 14 percent of that required by then-standard methods employing intensive manual labor. In 1957, kolkhozes and sovkhozes grew some 60,000 hectares in this new way.[28] New policies painstakingly built machine-building capacity, but even these improvements did not make up for long-standing shortfalls. In 1957, the leaders of Kostroma Oblast petitioned Moscow for tractors, harvesters, 37,000 tons of synthetic fertilizer, and 14,700 tons of improved wheat, oat, barley, and corn seeds. Moscow refused their request, indicating their inability to support those designs.[29]

In subsequent years, the manufacturing ministries widened the shortfalls rather than solving the problem. After peaking in 1958, output of combines, trucks, and tractors decreased in 1959 and 1960.[30] In 1962, Stavropol Krai leaders pled for more implements necessary to grow the vast plantings of corn they planned. They calculated a need for 2,000 more planters to make good on shortfalls resulting from having received *none* in 1960 and 1961. They had requested 1,900 through the standard allocation process. Moscow earmarked 220, and scheduled those for delivery in the third quarter, after the planting season. Similarly, a shortage of cultivators threatened to allow corn plantings to be overgrown by weeds. The krai bosses petitioned that the number earmarked for delivery during the first half of 1962 increase from 650 to 1,000. They claimed an additional need for 290 trucks, but received an allocation of only 80.[31] Their special plea gained a supplemental allocation of 30 trucks and 540 planters, accompanied by the unhelpful assurance that Moscow had no additional resources.[32]

Stavropol was hardly alone. Neighboring Dagestan ASSR, which was known for high yields of corn, petitioned to replace corn with wheat in its state procurements. Its farms had harvested less corn than planned, a failure republic leaders blamed on a lack of machines. Without planters and cultivators, farms used unapproved methods to economize manual labor, but the resulting delays in work made corn susceptible to dry midsummer weather.[33]

Khrushchev had ambitions for complete mechanization, which was to trigger "a revolutionary breakthrough in the technology of corn cultivation," as one of his expert advisers called it.[34] In the late 1950s early 1960s, designers developed new machines "taking into account the best practices of domestic and foreign technologies," in the words of a 1960 directive. It listed efforts to develop 377 new labor-saving designs "necessary for the complete mechanization of fundamental agricultural labor." The policy produced results only slowly. In late 1962, an interim report found that 38 of designs had been abandoned, 32 had not yet reached prototype stage, 132 were still in testing, and only 169 were moving toward manufacture.[35] Substantial investments increased mechanization, but it nonetheless bore limited fruit more slowly than authorities envisioned.

Machines, however, were but one component of this program. The 1955 delegation reported on the double-cross hybrid corn seeds becoming dominant in the United States. The Soviet Union subsequently adopted the seeds, a sign that Trofim Lysenko's influence over crop genetics had waned and proof that it had come to be seen as a value-neutral technology.

Since the 1930s, Lysenko had advocated intervarietal hybrids created by crossing two stable varieties developed when specialists improve on landraces with desirable characteristics, themselves products of prescientific selection of open-pollinating cultivars. These intervarietal hybrids offer improved yields and resistance to adverse conditions. Double-cross hybrids offer still larger yields and greater resistance, but require delicate and labor-intensive research and production. To create them, scientists manually self-pollinate several generations of corn plants of a variety exhibiting desirable traits, producing inbred lines reliably expressing those traits. However, these lines have little practical value because they produce low yields. Breeders overcome this by hybridizing two lines to create single-cross hybrids or, more commonly in the 1950s, four distinct lines to produce double-cross hybrids. Genetically uniform and expressing the parental lines' advantages, these seeds benefit from heterosis, or hybrid vigor, to yield not only more than the parental lines, but also up to 30 percent more than the best landraces, varieties, or intervarietal hybrids. The yield benefits of heterosis persist for only a single generation, however, requiring annual repetition of laborious process of crossing the parental lines.[36]

Genetics was an epicenter of the contentious history of science under Stalin. Advocates for the classical genetics founded in the nineteenth century by Gregor Mendel struggled for two decades against Lysenko. In the 1930s, Lysenko defeated and purged his foes, most notably Nikolai Vavilov, the specialist in the natural history of crop domestication, who died in prison.[37] In 1948, Stalin had elevated Lysenko's ideas to the status of orthodoxy, granting him control over important institutions.

While double-cross hybrids were conquering American cornfields in the 1930s, Lysenko denounced them as inferior to intervarietal hybrids and rejected as fabricated the data showing otherwise.[38] He dismissed American advances because they required inbreeding, which his theories rejected along with the idea that fixed genes were altered primarily by random mutations. Instead, he believed that change occurred through inheritance of acquired characteristics influenced by environment. More important, his campaign tapped into rising antiforeign sentiments, which peaked in the late 1940s with denunciations of all manner of "bourgeois" knowledge. The Soviet Union therefore rejected double-cross hybrids that had become essential to American industrial farming since the 1930s.[39]

After 1953, Lysenko's influence first ebbed and then strengthened anew, only to collapse completely when Khrushchev fell from power in 1964. The relationship between the two was complex. Khrushchev is remembered as a defender of Lysenko because he at times lent support and, in 1964, viciously attacked the pseudobiologist's critics. Yet Khrushchev knocked Lysenko down a peg on more than one occasion.

As soon as Stalin died in 1953, Lysenko's opponents looked to the dictator's successors for aid. They wrote letters to Georgii Malenkov and the Central Committee accusing Lysenko of favoritism in his capacity as a member of the Stalin Prize committee and of abusing his powers as editor of academic journals.[40] As president of the Lenin All-Union Academy of Agricultural Sciences, he had failed to implement the Central Committee's 1952 orders to formalize internal departmental organization and to regularly convene meetings of its presidium.[41] The most evocative, if not the most effective, letter denounced "the mistaken positions of Academician Lysenko and his group in the sphere of agricultural development and agronomy." "Using his theories," the writer, the head of a regional research station, fumed, "Academician Lysenko throws into confusion practical and research workers in their *struggle with nature* for high yields" [emphasis added].[42]

The petitioners soon got some results. In early January 1955, a commission of inquiry reported on the influx of letters about agricultural science, education, and Lysenko. Its findings cataloged "serious shortcomings" in teaching biology in higher education, "backwardness" in scientific research in biology and agronomy, "the incorrect recommendations of Academician Lysenko," "a lack of free discussion of research questions," and "the one-sided publicizing of open and pressing questions in biology and agricultural practice." Yet for all this, the commission apparently lacked the authority to reverse the 1948 holding that, in support of Lysenko, had established "a progressive, materialist tendency" and subjected "reactionary-idealistic" ones to criticism.[43]

Some letters gathered by the commission called attention to corn. One underscored that in the United States "hybrid corn seed had revolutionized the use of the crop" by allowing farmers to increase yields enough to produce annually up to 25 million tons of grain more. The writer noted that work in the Soviet Union on these hybrids and the underlying genetics had begun under Vavilov, continuing on the margins since his downfall.[44]

Each of a cluster of letters received in March 1954 argued that hybridization and heterosis in corn promised gains in productivity. Letters from at least three different advocates touched on the theme, suggesting coordination. At the very least, each sought to exploit to Khrushchev's evident interest in corn in light of the visible increase in his authority.[45] One writer, Anton Zhebrak, explicitly linked corn and genetics to industrial farming. "Lysenko's adventurism" had caused scientists to abandon research on corn genetics, whereas by 1950 farmers in the United States planted these hybrid seeds on 75 percent of their fields. Intriguingly, Zhebrak called attention to the hybrids' further spread: Italian farmers had used the seeds to achieve a 30 percent increase in their harvests since World War II. Since 1943, the seeds had spread in Mexico, allowing the country to stop importing grain—which it had done for 35 years—in 1948, feeding 28 million more people. Although Zhebrak could not yet discern the pattern that later became known as the Green Revolution, he pointed to these developments as evidence that methods suppressed by Lysenko actually had highly practical applications.[46] Double-cross hybrid corn had demonstrably increased American yields, forming a component of the global farming revolution.[47]

At the January 1955 Central Committee plenum, Lysenko sailed with the prevailing winds. His speech praised corn as livestock feed, decrying the traditional view of it as a grain for human consumption. It lionized Khrushchev's remarks for elucidating "the scientifically accurate basis for the crop's potential." "Corn can grow wonderfully in Siberia," Lysenko affirmed, echoing Khrushchev's enthusiasm for spreading the crop far and wide. "But it is important to ensure that the locals don't harvest the corn until the beginning of September," he continued, "when the weather is often very favorable."[48] Under intensifying attacks from opponents, Lysenko was silent about hybrid corn.

The commission investigating the charges against Lysenko soon made a new report on "substantial shortcomings and mistakes" in the work of both the man and the academy. It faulted him for dictatorial control over the editorial policy of journals under his control, as well as over biology instruction in higher education, which failed to "elucidate the achievements of Soviet and *foreign* researchers in the field of biology" [emphasis added].[49] This charge stands out particularly strikingly for its contrast to the campaign of the late 1940s to denounce foreign scientists and their ideas as bourgeois and formalist, in contrast to the labels of materialist and Marxist-Leninist that Lysenko attached to his own theories.

Khrushchev had spoken favorably of hybrid corn before the October 1955 report by the delegation to America, but not about double-cross hybrids specifically. The delegation devoted considerable attention to studying American mastery of the science behind the hybrids, their production, and their use. Presenting their findings, they demonstrated to Khrushchev that the Soviet Union could no longer ignore the technology because double-cross hybrids outperformed the intervarietal hybrids previously favored at Lysenko's insistence. In both written and oral reports, Vladimir Matskevich and Boris Sokolov, the delegation's corn-breeding expert, evaluated research and applied science, emphasizing that American researchers investigated only double-cross hybrids. On this basis, they appealed to Khrushchev for official support and funding for producing these seeds at the expense of continued spending on intervarietal hybrids.[50] The choice to send Sokolov in particular is revealing: he had been a pupil of Vavilov and had known Khrushchev when the latter had been party boss in Ukraine.[51]

Together, Sokolov and Matskevich appealed to Khrushchev based on the extra output corn expected to result from the research, rather than on the value of genetics as a theoretical framework. Emphasizing this point, Sokolov speculated that double-cross hybrids had accounted for some 75 percent of the extraordinary growth of American corn yields and harvests in the preceding decades. He argued on these practical grounds to reverse the decades of hostility by sanctioning the investments needed to turn theoretical knowledge into production hybrids. Having worked on hybrid corn since 1930, he had seen the Soviet Union fall behind the Americans, but also knew that the country possessed the requisite knowhow. "It is incorrect," Sokolov asserted, "to concede hybridization to the Americans." The rival's advantage was in practical application of the technology allowing the country to "exploit this biological phenomenon [heterosis] by organizing very large hybrid-seed firms, which sell only the hybrid seeds to farmers."[52] In fact, research in the Soviet Union used the same methods and had even begun with genetic lines imported from America.[53]

Because the Soviet Union applied existing knowledge poorly and concentrated on intervarietal hybrids, the country had sidelined double-cross hybrids and the researchers who developed them. Sokolov carefully but forthrightly confronted the ideas of Lysenko, who had long claimed authority by linking his theories with production, setting them in opposition to lab-based research into fundamental questions. Working with cows and wheat, Lysenko taunted his adversaries among the geneticists for conducting laboratory research on fruit flies. When the geneticists proved that their work led to better hybrid corn, they won Khrushchev's backing.

Sokolov's version of this history is telling for its conciliatory tone but firm message. Real problems with producing and putting double-cross hybrids to work had facilitated Lysenko's attacks. "How do we explain," he asked during his presentation to the Presidium of the Central Committee, "the fact that such a beneficial measure has been adopted so poorly?" He emphasized disagreements among experts during the 1930s. In cautious and conciliatory terms, he outlined how Lysenko had asserted that intervarietal hybrids suited conditions on the kolkhozes because they were easier to produce and transmitted their benefits to second and subsequent generations.[54] Lysenko's charges that the superior yields of double-cross hybrids in the United States were fictive did not stand up to scrutiny, Sokolov explained. But managers of kolkhozes and sovkhozes favored those who spoke for the simpler intervarietal hybrids and against the double-cross alternatives, which therefore were ignored.[55]

Khrushchev satisfied Sokolov's requests by sanctioning the research and providing funding. Opponents of Lysenko received millions of rubles in investment to put their corn-breeding ideas into practice.[56] Responding to Sokolov's appeal, Khrushchev endorsed the geneticists, at least insofar as their work concerned hybrid corn. Possessing the required knowledge, the specialists needed funds and institutional resources that only the leader's patronage could grant. "I am convinced," Khrushchev responded, "that 99 percent of what Comrade Sokolov reported here, he knew prior to the trip to America. The benefit of the visit is that he personally saw [the technologies in use] there and became troubled by the fact that we have not developed them, even though we had the knowledge." He praised the geneticist, adding, "Comrade Sokolov has spoken well and drawn correct conclusions. Now we must set this matter in motion with his help."[57]

Khrushchev ordered aides to prepare proposals specifying where and how to produce the hybrids. They soon returned with a policy designating research institutions, selection stations, and sovkhozes to carry out the necessary work.[58] Once Lysenko's power base, the Academy of Agricultural Sciences became home to a special corn institute and many research stations advancing work on corn and double-cross hybrids.[59]

In 1956, Khrushchev ordered the party to curb Lysenko's administrative and editorial duties, combating his brazen abuses of power. Nonetheless, no one formally revoked his authority to define orthodoxy.[60] Investment in American-style hybrids signaled disfavor for Lysenko, forcing him and his supporters to read the shifting currents and cease opposing the double-cross hybrids. By 1958, Lysenko was back to proclaiming sharp distinctions between his own theories and the bourgeois alternatives, but hybrid corn was out of bounds.[61] He reclaimed editorship of journals and attempted to resurrect former methods of demonization, but never equaled the triumphs he achieved under Stalin. Thus on the basis of the practical improvements in corngrowing—not fruit flies—the geneticists scored a victory.[62] Sokolov soon became the face of double-cross hybrids in the official press, and eventually earned awards, including a Lenin Prize.[63]

Embracing double-cross hybrids, the Soviet Union invested in required infrastructure. As in the 1920s and 1930s, it imported technology. The 1955 delegation preached to leaders the importance of seed processing and of the specialized companies—Pioneer, DeKalb, Garst & Thomas, and others—that carried out the component tasks of producing, harvesting, drying, sorting, treating, packaging, and distributing the seeds.[64] Eyes open to potential profit, Roswell Garst sold the technology in part to contribute to global stability by guaranteeing food security. In early 1956, Garst arranged sales of double-cross hybrid seeds, the parental lines for growing more, and the machinery to process hybrid seeds produced in the Soviet Union. For their part, the Soviet Union's leaders sought technology facilitating higher productivity to overcome the limitations that rendered early stages of the corn crusade extravagantly labor intensive.

Purchasing technology from Garst, the Soviet Union's leaders planned to master the process they called calibration. Thus they invested in machinery to economize on labor in accordance with principles of industrial agriculture. In this process, a factory used specialized equipment to preserve the raw seeds by drying them, to treat them with fungicide, and to sort them according to size and shape (Figure 7.2). This facilitated square-cluster planting by permitting the mechanical planter to distribute the seeds not in a continuous row, but in evenly spaced clusters of two or three plants. The seeds had to be uniform in size and shape so that at regular intervals the same number passed through a mechanism in the planter—a metal disk with small holes—and into each cluster. Tellingly, while Khrushchev was stipulating this method, American farmers had begun to abandon it in favor of rows because increasingly common chemical herbicides made the extra expenditure on squares and accompanying mechanical cultivation superfluous.

The Soviet Union's hybrid-seed program achieved significant success despite difficulties and delays. Ministry officials and local authorities demonstrated an

American Technology, Soviet Practice • 181

FIGURE 7.2 In February 1957, workers manually inspect the output of the sorting machine at the calibration plant constructed in Novomoskovsk, Dnipropetrovsk Oblast, Ukraine. These machines comprised one of the three complete processing plants sold to the Soviet Union by Roswell Garst under a 1956 agreement. The name of the American manufacturer of the equipment, Finco, Inc., of Aurora, Illinois, is visible on several components.
Photo by Viltman, courtesy Russian State Archive of Cinema and Photo Documentation, ID#0-249470

unresponsiveness that, although seemingly a result of inertia, reflected their weighing of the risks and rewards of carrying out orders. Tantalized by yields up to 30 percent higher and improved resistance to pests and drought, central authorities pursued a crash program to produce the hybrids. Having bought the genetic lines needed to augment existing domestic ones, they established procedures for growing the hybrid seed, harvesting the grain, storing the raw seed, transporting it, and processing it in time for use the following spring. The workings of this elongated supply chain left a wider paper trail in the archives than the corn earmarked for animal feed, which appeared primarily in statistical reports subject to shoddy record keeping and outright falsification. By contrast, Khrushchev and his aides tasked party committees, government personnel, and inspectors with fully documenting the processes of producing vital hybrid seeds.

By making substantial capital investments, leaders intended to supply enough hybrid seed for all plantings by the end of the 1950s. After Khrushchev endorsed double-cross hybrids in October 1955, the government moved swiftly to build capacity. The resulting plan, "On the Transition by Kolkhozes and Sovkhozes to Planting Hybrid Corn Seed," established targets of 169,000 tons for 1956 and of

300,600 tons for 1960.⁶⁵ In 1956, this effort involved 600 sovkhozes and 1,400 kolkhozes across many southern administrative regions. Officials had to coordinate time-sensitive and technically precise processes spaced throughout the year. Failing to execute them would render the grain useless as seed. To illustrate, that year Ukraine exceeded its plan of 120,000 tons by harvesting 141,300 tons. Faults in planting, detasseling, harvesting, and storing the corn, however, ensured that 34 percent was unsuitable for use as seed. The republic therefore actually delivered less than 100,000 tons.⁶⁶

Central authorities first had to ensure that farms planted the seeds at all. To function smoothly, bureaucracies require written orders, but officials often issued and executed informal verbal instructions. Having received informal authorization, a subordinate might then face questions about missing documentation. On April 27, 1956, the chairman of the Will of the Proletariat kolkhoz in Stavropol Krai petitioned district authorities to release his farm from an obligation to plant 150 hectares of seed corn on top of standing orders for seed potatoes. He pleaded that this new mandate overburdened his kolkhoz. Over the telephone, he received permission from the krai agricultural department to plant only 50 hectares of the genetic lines necessary to produce the prized VIR-42 double-cross hybrid, named for the All-Union Institute for Plant Breeding where it was developed. The necessary written documents, however, never arrived.⁶⁷ Relayed by the kolkhoz's deputy chairman, the story of this unwritten order failed to placate the inspectors who soon arrived.

From a distance, the kolkhoz apparently refused to implement a directive. In a standard written response to the inspectors' findings, the kolkhoz managers cited this oral order. A subsequent report by the inspector to the USSR Ministry of Government Oversight does not mention the claim, providing no reason at all for the kolkhoz's decision to plant only 33 percent of its assignment. To a reader in Moscow, the farm had simply refused. Informal authorizations unsupported by written confirmation rarely appear in the archival record, an unsurprising fact given the fleeting nature of telephone conversations. Common occurrences, such telephoned orders had two possible outcomes: they either subsequently gained the official stamp provided by a written directive or, if not, entered the documentary record as an unexplained failure to follow orders. This kolkhoz's managers had understandable reasons to seek a change in their assignment. In fact, the farm received commendations as one of the best in the krai, as the assignment to produce various kinds of seeds attests. Far from resisting, the kolkhoz's bosses calculated that they should concentrate on familiar tasks they considered more useful. In so doing, they failed only to follow bureaucratic procedure.

Inspectors revealed other shortcomings, as they often termed them, in hybrid-seed production. In the Moldavian SSR, they found that "many kolkhoz and

MTS managers undervalue the importance of raising high-yielding hybrid corn seeds." Kolkhoz bosses planted far fewer hectares than ordered, but reported that they had planted the full amount. In the severest case of six cited in the inspectors' report, a kolkhoz had documented planting the 74 hectares required. "The inspection revealed that this contradicted the actual situation," the inspector's report dryly noted. "In fact [the kolkhoz] planted a total of 4 hectares."[68] Despite the priority assigned to the program at the top, local counterparts did not follow through for reasons the archival record leaves indiscernible. Much like the leaders of Will of the Proletariat, these managers seem likely to have calculated that labor, land, machines, and other inputs necessary to carry out this plan might instead advance other efforts they deemed more important. Having made this decision, they may have then lied to superiors to maintain the appearance that they had complied.

Even farms making good-faith efforts to plant the prescribed number of hectares sometimes failed. In 1956, inspectors dispatched to the Donetsk sovkhoz in Russia's Kamensk Oblast discovered significant flaws in the plantings of hybrid corn seed. Its director and agronomist justifiably protested that they were not to blame. First, in May the oblast agricultural administration had sent a telegram with instructions that reversed the names of paternal form of the VIR-42 hybrid to retain its tassels and the maternal form requiring detasseling. Second, the sovkhoz then received shipments of the seeds in a proportion matching the incorrect formula. The sovkhoz's managers estimated that the fields would yield approximately 80 metric tons of a nonstandard hybrid without the desirable characteristics.[69] Furthermore, the inspectors later reported that the shortages of the necessary parental lines had made the problem, magnified by the agricultural administration's incorrect instructions, an oblast-wide concern.[70]

Although Moscow's inspectors unearthed many mistakes and coverups, they occasionally described positive results. The one responsible for Kamensk Oblast concluded that despite these "serious shortcomings" on the Donetsk sovkhoz and others, "the inspection showed that many MTSs and kolkhozes have endeavored to carry out party and government directives, having organized work fairly well."[71] In comparison to the typical reproving tone, this was a glowing review.

Having planted the seeds, the farms faced a second task: detasseling. They had to organize scarce manual labor for a technically demanding and time-sensitive job, made even more pressing because, in southern regions suitable for growing seeds, the crop faced threats from dry midsummer weather. Laborers had to cut off or pull out the pollen-producing tassel, the topmost part of the plant, from the rows containing plants of the maternal line before they dispersed their pollen. This ensured their pollination by the paternal line in a neighboring row, resulting in seeds combining genetic material from each line.

Reports from 1956 confirm that the sovkhozes tasked with raising the double-cross hybrids had few available farmworkers in July, forcing a scramble to assemble enough to manually detassel the millions of individual plants constituting thousands of hectares of corn. One person could detassel 2 hectares during the two-week window. In Stavropol Krai, farms planted 7,325 hectares and therefore required 3,670 workers.[72] In mid-July, the krai agricultural administration requested that heads of local secondary, postsecondary, and technical schools recruit students to do the job during summer recess. The farms pledged to instruct the students in the science and benefits of hybrid seeds, while offering room, board, and pay.[73] The farms completed the process, but in some cases the poor quality of the work forced officials to abandon hope of harvesting hybrid seeds.[74] In the Kyrgyz SSR, farmworkers detasseled the wrong plants and worked carelessly, causing approximately 33 percent of the plantings to have no value as seed. In Moldova, that figure was 9.3 percent of 16,346 hectares.[75]

In 1956, the laborious harvest also proceeded slowly. Many regions experienced a comparatively late spring and early frost, making timely picking even more important than usual. Shortages of labor and its poor organization caused many farms to lag.[76] Beginning on September 27, inspectors in Krasnodar Krai and Stavropol Krai, as well as Belgorod, Kamensk, Voronezh, and Kursk oblasts, took stock of the situation. On October 5, they reported to Moscow that farms had harvested only 527 hectares, or 1.7 percent of the total area, a figure only 2.7 percent of the target for that date. As late as November 10, progress remained "extremely unsatisfactory," with only 70 percent of the crop picked.[77]

Among the local considerations contributing to delays, kolkhozniks and farm bosses had little incentive to concentrate on hybrid seeds because that work coincided with many other pressing autumn tasks. In the absence of the price incentives instituted in subsequent years, farmworkers dug potatoes and tended personal plots. On one farm in Krasnodar Krai, 94 hectares of prized VIR-42 hybrid seeds awaited harvest while 700 kolkhozniks devoted their energies to fields of regular corn, likely because they received fodder for personal livestock as pay for the latter job, a stronger incentive. Additionally, some 20 percent of kolkhozniks did not turn out for work at all. Even when farmworkers turned up for work, the harvest might proceed in a "disorganized" manner as they unknowingly or uncaringly picked the separate rows together, mixing the double-cross hybrids with less-useful grain from the neighboring rows.[78]

Once harvested, the seeds might be stolen in transit. In one of many cases, kolkhozniks in the Kyrgyz SSR harvested 100 metric tons, but by the time the grain reached the government collection point, only 41.8 tons remained. As a report on the incident laconically noted, "The lack of necessary protection [means that] corn in the field is fed to livestock, carried off, or spoiled."[79] Even the seed that did reach procurement points did so only slowly. In late November 1956, Matskevich reported that only 19.3 percent of the 178,000 tons harvested had been collected, a situation he described as "exceptionally worrying."[80] The longer transport took, the more time thieves had to steal it, managers to divert it to their own uses, and its high moisture content to cause spoilage.

Often, local officials made self-congratulatory assertions of success even when the seeds were nowhere to be found, making higher authorities wary of inflated claims. A final report on a countrywide inspection noted that "soviet, party, and agricultural organizations have carried out major [sic] efforts to ensure effective hybrid seed-corn production. At the same time, the inspection demonstrated that substantial shortcomings exist." The document reproduced a common formula, contrasting major efforts with only minor shortcomings. One of the readers, most likely the minister of government oversight, to whom it was addressed, underlined "major" in pencil and wrote, "Really?" in the margin.[81]

Once farms harvested the seeds, obstacles remained. State enterprises had to store, transport, and distribute them. In 1956, inspectors visiting collection points operated by local subsidiaries of the USSR Ministry of Grain Procurement revealed additional serious flaws.[82] The managers of these facilities had to construct new structures and maintain old ones needed to dry and store seed corn. Inspectors found that these jobs remained incomplete and far behind schedule.

Enterprise managers lagged for comprehensible reasons, especially a lack of construction materials, but their subsequent actions require careful interpretation. Puzzlingly, they occasionally reported that they had completed jobs that they had not even begun. In July 1956, scrutiny of the Nevinnomysk grain elevator in Stavropol Krai discovered irregularities in repairs and preparations for the harvest scheduled for completion on July 1. Managers had submitted routine paperwork documenting that the projects were finished and given quality control ratings of "good" or even "excellent." Inspectors from Moscow found a different scene altogether. "In fact," they concluded, "on July 13, 1956, . . . work was not complete on certain [storage] bins, as had been reported in the fraudulent documents." Those also indicated that 625 meters of border fence had been repaired. In fact, no one had begun that task. On May 23, the managers had reported that a gate in this fence had been installed. In July, inspectors found that it had not yet been mounted.[83] As part of the inspection procedure, heads of organizations under scrutiny had to justify themselves. In this case, the director could only plead a lack of building materials—chronically in shortage—and of specialized drying and ventilation equipment.[84] This explained the incomplete work, but not the fraudulent documents claiming that it was complete. A similar state of affairs existed at three of the five other inspected sites in the krai, including the elevators at Bogoslov, Urakov, and Eren-Shakhar.[85] Inspectors frequently identified such machinations in other administrative regions across the country.[86]

Beyond missing deadlines and exceeding budgets, these organizations' bosses lied to superiors by declaring tasks complete when, in fact, they had not yet begun the work. They did so likely to create the appearance that they had fulfilled the plan, a result of an incentive structure favoring appearances over outcomes. They incurred punishment for submitting false documents only if caught, a potential sanction less threatening than the reprimand sure to follow an admitted failure to complete production plans and repair schedules on time. Inability to meet plans was certain to lead to extra scrutiny, loss of bonuses, official censure, and—if frequent—firing. On the other hand, superiors might not notice fraudulent reports, or might turn a blind eye, in both cases leaving the culprit unpunished. Officials acted in ways reflecting their weighing of the lower risk of punishment for actions that were inefficient and disruptive to the economy against the certain reproof for allowing plans to go unfulfilled.[87]

Even when procurement agencies acquired, dried, and stored the seeds, they still had to distribute them, a considerable logistical challenge. Khrushchev required farms across the Soviet Union to plant corn, but most could not produce seeds themselves because of short, cool growing seasons or hot, dry summers. Specialization required farms in a narrow band stretching from Moldova through parts of southern Ukraine, the North Caucasus, and irrigated lands of Central Asia to produce the hybrid seeds. These locales then had to supply all others, requiring substantial efforts to transport the seeds each winter. Committees in recipient oblasts insistently inquired about shipments, hoping to guarantee their arrival in time for planting. In 1956, the schedule called for transport primarily in March and April, but already in February the party committee in distant Chita Oblast began to dispatch telegrams to Kyiv, urging the Ukrainian Central Committee to expedite the process.[88] Many other regions did likewise.[89]

Once the sought-after seeds arrived, they still had to be collected by the farms, a task many neglected. In Moscow Oblast, inspectors revealed that many had received their allotments, but stored them improperly and allowed them to spoil. Questioned before the oblast soviet, the responsible district officials tried to blame producers and shippers by stating that the seeds had already rotted by the time their district had received them.[90] These unsupported claims notwithstanding, the officials received a "strong reprimand" for failing "to demonstrate the necessary care for storing seed corn prior to planting."[91]

Ministries, enterprises, and farms employed a mix of formal and informal practices when running their economic activities. These tactics broke through the procedural barriers of bureaucratic organization, but introduced their own inefficiencies. Inspectors subjected double-cross hybrid seeds to heightened scrutiny because they entered the state procurement system. Because the seeds offered a substantial increase in yields, they became policymakers' priority, attracting attention that made hybrid-seed production an unrepresentative sample of all output of corn. Inspectors were in position to counter local authorities' claims to party and government superiors to have made major efforts with only a few difficulties. Inspectors thereby documented how the system functioned—or malfunctioned—by revealing informal practices otherwise difficult to discern in the archival record.

Once grown and purchased from the farms, the double-cross hybrids traveled to specialized calibration factories. The Soviet Union imported three factories under a contract with Garst that resulted from contacts with the 1955 delegation. Leaders planned to install the machines in time to process the 1956 harvest for planting in 1957. Top officials in Moscow supported the projects, dispatching inspectors with wide powers to identify delays, speed progress, and single out those responsible for slowdowns. Efforts to complete the three factories, as well as the many constructed in subsequent years using Soviet-made copies of the equipment, demonstrate characteristic shortages of competent managers, construction materials, and skilled workers. In 1956, authorities chose two sites in Ukraine, Novomoskovsk in Dnipropetrovsk Oblast and Buialik in Odesa Oblast, with annual capacities of 2,500 and 5,000 metric tons, respectively. They sited a third plant at Ust-Labinsk in Krasnodar Krai, consisting of a cluster of factories with a combined capacity of 12,500 tons. In early 1956, plans called for completing them by December.[92] Party officials pressured local authorities and workers themselves to speed construction, indicating the priority of these projects and their overambitious schedules.

Delays threatened almost from the start. On August 17, the director of the building trust—equivalent to a general contractor—responsible for the two sites in Ukraine alerted authorities in Kyiv and Moscow to such delays. Although officials typically minimized failures when reporting to superiors, this director bluntly declared, "Conditions on the construction site threaten to disrupt the timetable established by the USSR Ministry of Grain Procurements in order No. 315 of May 31." Defending himself, he blamed a lack of workers trained in necessary trades, probably in hopes of securing aid in finding skilled workers.[93] A month later, the republic central committee in Kyiv dispatched envoys to Dnipropetrovsk and Odesa, and assigned an official in Kyiv to verify progress at regular intervals. The republic party authorities then ordered oblast and district subordinates to designate someone to take "personal responsibility" for each site, exercising "strict oversight . . . over all aspects of construction" and submitting weekly reports up the chain of command.[94]

The delays continued through the autumn. In September, Kyiv authorities moved to secure the needed skilled workers to organize a second shift.⁹⁵ They also found that irregular deliveries caused further delays. For instance, the machines reached the country from the United States in August, but neither the machines nor blueprints were initially available on the construction sites. First, they had been shipped to Moscow, where engineers studied them to aid in reverse-engineering their own models.⁹⁶ This left the construction trust facing a tight schedule. In early September, the Novomoskovsk firm had spent only 1.3 million rubles, or 26 percent, of its budget of 5 million rubles. Similarly, the Buialik trust had spent only 19 percent of its 4 million ruble budget.⁹⁷

The regular reports to Kyiv described improvements, but also ongoing challenges. By October 13, 1956, Odesa Oblast authorities had commandeered students, kolkhozniks, and others to organize a second shift, a practice reminiscent of "storming" at the end of the month to meet production quotas. The oblast party committee and the construction trust lacked the sway to acquire electrical equipment and structural steel needed, so they asked officials in Kyiv to use their influence.⁹⁸ Late in October, reports to Moscow confirmed that the delays continued.⁹⁹ The republic's newspapers, the Russian-language *Pravda Ukrainy* and Ukrainian-language *Radianska Ukraïna* [Soviet Ukraine] emphasized the importance of the project, shedding light on the slowdowns.¹⁰⁰ They provided a forum for complaints that forced the Kyiv authorities and the oblast party committees to put further pressure on local subordinates. That month, during Roswell Garst's second visit to the country, officials from each construction trust gathered at Ust-Labinsk for consultations with him.¹⁰¹ The reports on the actual situation contrasted with rosy press coverage, which described the project and signaled its importance. One story noted the arrival of the Americans who, it claimed, praised the speed and quality of construction, which "was increasing with each passing day."¹⁰² Unsurprisingly given the numerous delays, the November 1 deadline for launching the production lines passed. New orders established December 30 as the new target date.¹⁰³ In December, the operation was delayed until January 25, 1957.

Khrushchev's personal attention demonstrated the importance assigned to the plants. Ukrainian party authorities routinely dispatched reports on the two construction sites to Moscow. In turn, Khrushchev pressured his protégé Aleksei Kirichenko, republic first secretary, to speed things along. A December 21, 1956, report by the republic party committee's agricultural department made its way from Kyiv to Moscow, and from Moscow back to Kyiv to the central committee's files, having acquired along the way a note Khrushchev wrote in the margin to Kirichenko.¹⁰⁴ The former demanded that his subordinate appraise and respond to the update. An accompanying note indicates that Kirichenko read it, but not any corresponding orders he gave. The report itself described delays caused by unsatisfactory management by ministries, local party committees, and government organizations. In the end, the factories went into production late, in piecemeal fashion, and without permanent worker housing and other secondary but necessary structures (Figure 7.3).[105]

A 1958 cost-benefit analysis concluded that the factories had generally achieved positive results, encouraging additional plants using domestic adaptations of the technology. The largest, at Ust-Labinsk in Krasnodar Krai, proved costly and

FIGURE 7.3 A factory conveyor belt transfers hybrid corn from the drying chambers to the threshing shop at a corn processing plant in Buialik, Odesa Oblast, Ukraine, on February 7, 1957, soon after the calibration plant went into operation. The large crane visible in the background suggests that construction work continued at that time.

Photo by A. Fateev, courtesy Russian State Archive of Cinema and Photo Documentation, ID#0-249471

impractical because surrounding farms produced insufficient suitable raw seed to fully utilize its considerable capacity. Officials therefore did not contemplate building any more on that scale. More usefully, seeds processed at factories with an annual capacity of 5,000 metric tons cost 2,770 rubles per ton, while the 2,500-ton factories came in at 4,099 rubles per ton. Like the largest, both types also suffered from shortages of raw seeds. Officials bemoaned the rarity of appropriate sites near an existing grain elevator on a railroad line and connected to adjacent farms by asphalt roads. The analysis therefore recommended building primarily small-capacity plants, despite higher construction and operating costs. It furthermore suggested more rigorously managing seed production and processing.[106] Far from plunging ahead with the corn crusade blind to the challenges, authorities adopted and adapted best practices from abroad. In so doing, they consciously considered practical questions about economies of scale and the peculiarities of the Soviet Union's countryside.

After the factories using American-made machinery entered production in early 1957, the government embarked on an expansive program to construct more using domestic versions of the equipment, establishing a goal of 156 by 1962.[107] On December 4, 1956, the USSR Council of Ministers adopted a resolution designating sites in Ukraine, Moldova, Rostov Oblast, Krasnodar Krai, North Ossetia ASSR, and Stavropol Krai. In 1958, the program expanded by a further nineteen large-capacity and six smaller facilities.[108] Even after design improvements cut costs, each represented a capital investment in industrial corngrowing of several million rubles.

In March 1958, inspectors from Moscow discovered serious delays building two factories in Stavropol Krai, at the settlements of Rasshevatka and Bogoslovsk. At Bogoslovsk, the first of two 2,500-ton production lines started up at the end of 1957, but operated inefficiently due to poor construction. Workers had installed equipment behind schedule, a problem also evident on the line slated for completion in time to process the 1958 harvest. Lacking machines to move earth and raise structures, construction trust managers substituted large numbers of manual laborers. Nonetheless, inspectors blamed inadequate leadership and political agitation for allowing workforce turnover of 54 percent during the year's first quarter alone. They also found similar conditions at the Rasshevatka site.[109] These problems were common to sites across the country. Much like the plants constructed in Ukraine in 1956, many lacked bricks, timber, and other basic building materials.[110] Already in May 1958, inspectors pressured the Stavropol Krai soviet to pass a resolution demanding that hitherto "unsatisfactory progress" improve.[111]

Work continued slowly at Rasshevatka and Bogoslovsk because of labor force turnover and shortages. Inspectors noted that one site had no boss because the previous one had been fired "for poor organization, unsatisfactory management, low-quality results, and consequent cost overruns."[112] In seven months of 1958, 175 workers at the Bogoslovsk site had left, whereas only 103 new ones arrived. Turnover became so pressing that the construction trust contracted with a nearby corrective-labor camp for a contingent of 50 manual laborers, including some skilled construction workers, but on average the camp sent only 15 daily.[113] The need for these measures demonstrates how acute the labor shortage had become amid constant demands for progress. Authorities could not prevent mass turnover due to poor working conditions because the workers were relatively free to move about. Under Stalin, policies attempted to address substantial turnover by imposing a severe labor code. Reforms under Khrushchev first allowed penalties for unauthorized job changing to fall into disuse, and then eliminated them altogether.[114]

By August 1958, the head of the Stavropol Krai grain procurement office and the official responsible, a certain Samokhval, reported to party secretary Ivan Lebedev about continued slow progress. Typical of such documents, its first two pages list dry budgetary and technical details. Two more shift responsibility from Samokhval and his organization to others. The construction trust had poorly organized the workers and permitted the labor shortage to continue, resulting in slow progress installing equipment, completing buildings, and other laborious tasks. Components arrived late from manufacturers located outside the krai, beyond the control of local authorities.[115] An inspector from Moscow corroborated parts of Samokhval's evaluation, detailing delayed deliveries and an inadequate workforce, but also singled out careless onsite management. The construction trust had appointed a foreman who later proved "an uneducated con-man, who mostly took bribes and lowered output norms."[116] A common tactic, reducing labor quotas padded the pockets of workers and managers by easing plan fulfillment. This provided opportunities to overfulfill plans without extraordinary efforts, making for easier bonuses, happier workers, and the appearance of effective leadership.

Delays in deliveries resulted from the industrial-management reforms that created a *sovnarkhoz*, or council of the national economy, in each administrative region. Designed to counter overcentralization, the councils actually fostered parochialism. In 1953, Khrushchev's rival Malenkov initially moved to strengthen the ministries, but movement in the opposite direction soon followed. A year later, Khrushchev and other critics denounced these ministries for mismanagement, waste, and dishonesty. Enjoying broad support, the countervailing trend peaked in early 1957 with the reform establishing the councils.

Already in 1958, issues arising from localism had become evident at Rasshevatka and Bogoslovsk. There, the inspector proposed that the Stavropol

Krai party committee pressure the local council to entreat other councils to expedite production and delivery of needed equipment by manufacturers in their territories.[117] In effect, the reform replaced parochial ministerial bureaucracies in Moscow with parochial bureaucracies in 105 regions and republics. The councils had few incentives to coordinate and cooperate among themselves, and therefore placed narrow interests ahead of efficiency and national priorities. Because the councils complicated coordination, quiet moves toward recentralization gathered momentum after 1959, to be later codified in a 1962 reform.[118]

Inspectors' fears proved well founded. On December 30, 1958, long after the schedule called for the seed-corn production lines to run, the krai party committee passed a resolution condemning delays. In late February 1959, inspectors found that the plants had finally processed their first seeds, but many critical jobs remained incomplete. Having failed to clear storage capacity, managers had left unprocessed seeds exposed to the wind and rain, causing them to spoil. Although the factories had chemical fungicide on hand, they had not yet treated the seeds. Local party committees had failed to implement directives to expedite production while conserving fuel and electricity. Furthermore, they had not conducted the expected "cultural-educational work" with the workers designed to encourage the best workers to create so-called communist labor brigades.[119]

Inspectors identified the same problems at sites in Ukraine. The one at the town of Lozova in Kharkiv Oblast stood out as the most troubled. In several documents, inspectors and party authorities alike condemned the work of the local construction trust and its on-site personnel as "unsatisfactory." The first set of managers had "wasted" resources, resulting in a recommendation "to hold them accountable" by reprimanding or firing them.[120] They were fired in February 1958 for their "irresponsible approach to the job."[121] In April, inspectors found that new managers had failed to organize the labor force or increase the pace. Workers did such a poor job that inspectors labeled supposedly complete parts of the plant "defective goods," while many individual elements suffered from "amateurish" work.[122] Bosses denied that the flaws existed at all, or blamed them on irregular supply and poor quality of materials.[123] Holding a meeting of the site's personnel, a visiting inspector ordered workers to fix the defects and recommended formal reprimands for the second group of foremen.[124] Even after this second intervention, the oblast party committee found continued failures as late as June 1958. Despite considerable effort, "construction continues to lag behind and is unsatisfactory" because the proper machinery and materials were not on hand. Word of continued delays caused the republic's Ministry of Grain Procurements to issue orders to solve the shortages, delays, and shoddy construction rampant at Lozova, as well as at sites in Odesa, Khmelnytskiy, Chernivitsi, and other oblasts.[125]

Although word of these failures reached the Central Committee in Moscow, it could not ensure that on-site officials got results. A June 9, 1958, report alerted the USSR Council of Ministers and Aleksei Kirichenko, hitherto head of the Ukrainian party but by then the Central Committee secretary responsible for agriculture.[126] Summarizing the failures outlined above, it liberally assigned blame to Ukrainian and RSFSR ministries of grain procurements, their local subordinates, and the oblast soviet and party authorities. No one had "organized timely and complete construction." Republic governments had not overseen progress. All of this meant that construction plans were unfulfilled and the work continued to lag. Senior Central Committee secretary Frol Kozlov sent copies of this report to the heads of the respective councils of ministers, as well as to the USSR Ministry of Agriculture and of Grain Procurements.[127] His demand for additional action to remedy these failures testifies of the importance assigned to the matter, even if it did not achieve efficient construction of the factories.

Even after the calibration plants began to process seed corn for planting, day-to-day operations remained disorganized. In early 1959, inspections in Stavropol Krai revealed that managers of grain elevators often violated standard procedures. First, rather than housed in well-ventilated buildings, large quantities of grain designated for processing lay in the open, merely covered with tarpaulins offering insufficient protection against rain, snow, wind, and sun.[128] This common problem was caused by shortages of storage space brought on by rising output and insufficient capital investment in new capacity. By 1961, plans calling for more corn and more grain encouraged renewed state investments in storage, processing, and transportation capacities amounting to some 80 million rubles.[129] Nonetheless, inspectors found the same practices at the same sites in 1961 and 1962.[130]

The problems these inspections revealed, moreover, illustrate the bureaucratic inefficiency plaguing the local system. As of January 15, 1959, inspectors found that workers were still cleaning and drying seeds, the first stages of the process. Local procurement official Samokhval again explained delays by citing mitigating circumstances, acknowledging to the krai party committee only that progress had been "unsatisfactory."[131] None of this obscured the fact that plans remained unfulfilled. The annual plan called for Samokhval's organization to ship 16,000 tons of hybrid seeds to farms within the krai and a further 64,000 tons to other regions. On March 1, 1959, as spring planting loomed, only 3,766 tons were sorted and 2,000 tons ready for distribution, far short of the 42,000 tons planned for that date. Samokhval received the bulk of the blame for this failure. The krai party committee concluded that orders to "achieve the systematic functioning of all sections and machines, the Grain Procurement Administration, Comrade Samokhval, and the plant directors are being unsatisfactorily fulfilled." Parts of the production line did not function because machinery had been installed improperly. Demands that Samokhval report personally to the krai party committee on February 27 and again on March 3 achieved little effect. The party committee later expressed exasperation with Samokhval, but instead of reprimanding him, only required him to report to the committee a third time.[132]

Although the seed processing plants' everyday operation improved during 1959, a corruption scandal seems to have finally caused the downfall of Samokhval and associates. In September 1959, inspectors found that the Bogoslovsk grain elevator and the attached calibration plant regularly fulfilled its quotas.[133] They also revealed, however, irregularities in accounting for the grain that was the plant's raw material and for the calibrated seed corn. The krai procurement authority had declared lots totaling 178 metric tons unsuitable for use as seed, reclassifying them as livestock feed. Samokhval claimed that the calibration machine intermixed irregularly shaped kernels with the useful seed, making the output of substandard quality.[134]

This incident proved the culmination of a series of shady dealings that apparently led superiors to fire Samokhval.[135] A summary of "abuses of authority" by procurement authorities documents that this was only the latest of a number of cases over a period of years. Between 1957 and 1959, Samokhval and his deputy had sold grain to a kolkhoz in Leningrad Oblast, allegedly using the income to buy construction timber for the calibration plants. In fact, they had bought lumber of such low quality that it was nearly useless. Samokhval and his subordinates had profited illegally from the exchange, most likely by embezzling the difference between the sale price of the grain and the low price paid for the substandard timber.[136]

As in past cases, Samokhval "attempted to shirk responsibility," but his actions in 1959 reached the breaking point. He designated as unsuitable 697 tons from a lot containing 1,071 tons of corn, sending it to be refined into oil, starch, or industrial alcohol. The government had paid kolkhozes bonuses of nearly 600,000 rubles for producing the hybrid seed. When Samokhval diverted it, that expenditure went to waste. Insisting that all of the corn had been substandard, he produced documentation to that effect for only 236 tons. Other officials attested that between 86 and 88 percent of the seeds had been of good quality when they shipped the grain to the plant. The inspector's report detailed two similar incidents, and labeled the three together *prioizvol*, a term commonly used to characterize bureaucratic arbitrariness motivated by expediency or personal profit.[137] A search of the krai archives uncovered no documentation of Samokhval's firing, but his name does not appear on any relevant files from subsequent dates.

Despite hindrances, delays, poor quality, and spoilage, the program to introduce advanced double-cross hybrids achieved noticeable successes. By 1957, authorities utilized new incentives policies to offer kolkhozes higher prices for raw seed of appropriate quality. The kolkhozes in turn offered bonuses to kolkhozniks for growing and harvesting them.[138] Advocating larger bonuses for the 1960 harvest, Matskevich argued, "The present purchase prices for seeds of early- and average-maturing corn hybrids and varieties do not create a material incentive among kolkhozes and sovkhozes."[139] In particular, these types, well suited to growing further north, yielded less than longer-maturing varieties in the southern regions where seed had to be produced.[140] By 1961, plans called for farms to produce 1 million metric tons of a range of hybrids suited to various climate zones, with a rising proportion of advanced double-cross hybrids.[141] The endeavor involved 50 research institutes as well as more than 2,000 sovkhozes and kolkhozes.[142] Output reached 1.25 million tons of seeds, with the result that the state paid out bonuses of 188 million rubles to farms for the hybrids most in demand.[143] Even this colossal effort proved insufficient to meeting sovkhozes' and kolkhozes' requests for hybrid seeds, which expanded rapidly as planted area climbed toward its 1962 peak of 37 million hectares.[144]

When combined, these findings about bureaucratic malfunction provide a new understanding of how the Soviet Union's infamously ponderous administrative apparatus functioned. With American-style industrial agriculture providing the precedent, hybrid seeds and the associated technologies worked in many environments and socioeconomic formations around the world. In the Soviet Union, however, they did not fulfill their potential even in climates favorable to corn. Far from keeping the country functioning in spite of ill-conceived directives from Moscow, the bureaucracy stifled innovations, including potentially beneficial ones.[145] Khrushchev needed the functionaries, but they disregarded even directives with potential in pursuit of their own visions of orderly operation.[146]

Khrushchev wrangled perpetually with the bureaucracy in efforts to overcome obstacles standing between him and his goals. Efforts to fight ministries' parochialism by decentralizing policy implementation did not alter ingrained practices. Officials in charge of farms, districts, oblasts, procurement enterprises, and industrial concerns responded to the Kremlin's directives not with iron discipline, but instead with more regard for appearance than for substantive results. Incentives to fulfill plans and disincentives to admit failure combined to make the risk that lies might be discovered acceptable. This was not resistance, a conscious or programmatic effort to reverse central policy. Common tactics and a attitudes among officials curbed Khrushchev's authority to implement policy by granting local functionaries, in their own minds, the leeway to discount orders they considered impractical or to pursue a given end through deceit and other unauthorized means.[147] Local party authorities and economic bureaucracies concealed inability or unwillingness to follow orders from above, hoping to create at least the appearance of compliance.

Officials from the top down thus demonstrated doubt about corn that they could not express openly. In the mid 1950s, these actions coincided with the antibureaucracy campaign Khrushchev championed, an effort to make organizations carrying out economic policy more responsive to Moscow's orders. By increasing the Communist Party's duties to implement and oversee economic policies, he did not eliminate—and barely limited—the influence of local agricultural departments, construction trusts, district bosses, and farm authorities over policy outcomes.

Khrushchev used his authority to foster industrial farming methods and technologies needed to realize his vision of modern agriculture. Between 1955 and 1959, efforts to transform ideals of industrial corngrowing into practice concentrated on introducing more machines, new hybrid corn varieties, and related technologies. These policies were insufficient to realize the vision without additional heroic efforts to turn policy into practice. Local authorities sidestepped expectations, seeking to create a façade of apparent compliance. Obstacles grew after 1958, as officials began to recognize that the corn crusade was a long-term strategy, and not a one-off campaign, and therefore adapted to Khrushchev's insistent demands to put industrial farming ideals to work. Local authorities' strategies evolved in parallel with Khrushchev's campaign, which began to discard customary crops he denounced as mere grasses in favor of industrially farmed corn and other row crops.

8

BATTLES OVER CORN

"Comrades!" Nikita Khrushchev thundered while visiting Voronezh Oblast in February 1961, "We must punish strictly charlatans who try to embellish [successes] and hide mistakes they've made." He then described a whistleblower's letter. Anticipating Khrushchev's anger at the sight of unharvested corn, the bosses of kolkhozes and sovkhozes along his planned route in the region ordered workers to requisition a rail from a nearby depot, attach it to a tractor, and use it to knock down corn still standing in the fields. They thereby hoped to disguise the incomplete harvest, which should have ended months earlier. Confirming the story, an investigating *Pravda* correspondent faced pressure from oblast authorities to suppress his findings. When confronted by Khrushchev, party boss Stepan Khitrov claimed that this was standard practice for gathering corn for use as feed once the grain had been harvested manually. "I will soon be sixty-seven years old," Khrushchev countered, "and I don't believe such fairy tales. . . . In reality, this was deceit. Why did they do this? They wanted to deceive me."[1]

This mischief by Khitrov and his subordinates sheds light on center-periphery relations and policy implementation by local party organizations. Khrushchev's attempts to realize his vision of corn-based industrial farming confronted existing relationships among local officials, who felt pressure to appear to achieve results. Moscow's demands could transform policy prescriptions into frenzied campaigns, which hindered orderly management and economic calculation. Operating within local networks, party leaders built façades seeming to show increases in meat and dairy output achieved by pursuing policies that Khrushchev's programs of corn and industrial farming required.

Khrushchev's attitude toward local party organizations evolved over time. Before the scandals of 1960 and 1961, he promoted strong regional leaders who appeared to get results. After learning that apparently successful party secretaries had abused power and deceived

Moscow, central authorities moved to curb local bosses' power, a motivation for the administrative reorganizations implemented between 1961 and 1964. These regional party organizations comprised the regional leaders, who fit into categories of dictator, weak, and compromise, and the networks of officials appropriate to each type.² A functioning regional party network required party officials to trust one another to fulfill promises and not betray illegal actions to higher authorities, as someone in Khitrov's network had done by alerting Khrushchev.³ This typology and the concept of trust can aid in teasing out regional leaders' responses to policies related to corn and industrial agriculture, explaining how they implemented—or did not implement—policy.⁴

As early as September 1953, Moscow pressured regions to implement a lengthening list of agrarian reforms. Objections might provoke censure or firing, although the consequences were no longer potentially deadly, as they had been under Iosif Stalin in the 1930s. In early 1954, Khrushchev ousted the leaders of the Kazakh SSR because they did not embrace his Virgin Lands program.⁵ When launching the corn crusade in 1955, Khrushchev denounced those who merely sloganeered about corn, but did not mobilize local efforts to realize the directives. Choosing potentially strong, even dictatorial secretaries, the leader fostered local hierarchies capable of carrying out directives. Weak secretaries became targets because they failed to do so.⁶ Those who gained power and seats in the Central Committee, moreover, owed their positions to Khrushchev, making them disposed to support him against his rivals in the Presidium in June 1957.

Supported by top leaders, mass media campaigns decried lethargy and ineffectiveness in the regions. On March 27, 1955, *Pravda* published a drawing that linked this message to the new corn campaign. The text accompanying a sketch depicting two cobs read, "In the struggle for high yields of corn we must ensure that there are more cobs like [the one on the left of the sketch, depicting corn] and fewer like this one! [on the right]." The latter so-called cob consisted not of kernels, but of bureaucrats huddled around a conference table. Crowded around, they churned out resolutions keeping subordinates busy, but lacking any practical purpose. This and similar cartoons and stories condemned those who failed to execute orders and meet objectives, instead concentrating on formal documentation.⁷ Another drawing called attention to a new variety of corn. Named in the manner of hybrids associated with their place of origin, such as Krasnodar, Bukovina, or Voronezh, this *Kantseliarskaia*, or Office, variety grew from a filing cabinet as a bureaucrat looked on proudly. Alarmingly thin due to lack of actual fodder, a cow pushes her head through the window in an attempt to eat some of the voluminous pile of paper on the official's desk.⁸

Although under heavy criticism, officials' compulsion to document everything for superiors was ingrained into styles of governance inherited from Stalin. Thus, in April 1955 the Central Committee dispatched an order underscoring "major shortcomings" in preparations for spring corn planting, sternly reminding local committees that they were responsible. Instead of documentation, Khrushchev wanted hands-on leadership of the style he preferred, and condemned those who did not exhibit it. The order censured the authorities in Moldova for "managing from the office; visiting kolkhozes, sovkhozes, and machine-tractor stations [MTSs] extremely rarely and [therefore] knowing little about the state of affairs." Short on knowledge that might be gained only from on-site observation, it explained, "they do not make specific proposals needed to develop agriculture on the basis of local capacities." Only this sort of hands-off approach could explain a move by officials in the republic's Tiraspol District to *decrease* the corn-planting plan by 100 hectares when other locales were increasing their commitments tenfold or more.[9] The message was simple: good leaders made sure that farms in their domains planted more corn. Bad ones remained in their office issuing orders and demanding progress reports.

The point was made clearer by a condemnation of endless documentation for which farm personnel had to tabulate results, fill in forms, write reports, and complete other bureaucratic tasks, none of which directly improved output. In 1952, the Vologda Oblast agricultural department had sent subordinates more than nine directives, orders, and telegrams of various sorts per day and, in 1953, more than twelve. Local authorities governed from the top down through such demands for reports and forms, which amounted to an ineffectual despotism of paperwork.[10]

Between 1953 and 1957, Khrushchev used his powers as head of the Central Committee Secretariat to replace at least half of all regional secretaries. He sacked with particular alacrity those failing to get results in agriculture. In January 1954, Briansk Oblast secretary Aleksei Bondarenko lost his post because the Central Committee singled him out for his slow, ineffective response to directives of the September 1953 plenum. Officials from Moscow reported that the oblast's farms had prepared poorly for spring planting, in particular that of corn. At a local party meeting, Moscow's representatives opened the floor to criticism of Bondarenko. Seizing the moment, the secretary's subordinates "sharply criticized members of the oblast committee bureau for rarely venturing into the districts; for seldom speaking with kolkhozniks, MTS workers, and district party activists; for uncritically evaluating the state of affairs in agriculture; and for accepting the serious shortcomings of kolkhozes, MTSs, and sovkhozes."[11] With Moscow's blessing, the plenum replaced Bondarenko with Aleksandr Petukhov, a functionary of the Central Committee apparat, which was a reserve of potential regional bosses.[12]

As deputy chief of the Central Committee Department for Party Organizations, Iakov Storozhev supervised similar transfers of power in Iaroslavl,

Tula, Smolensk, and Kalinin oblasts. Summarizing that work, Storozhev characterized the scale of the problem: "Local organizational and political work... is weak.... Party activists do not struggle to develop agriculture." Linking these failures to the campaign against red tape, Storozhev concluded that the agricultural reforms "drown in a flood of directives."[13]

Khrushchev censured and pressured regional authorities he deemed insufficiently active. At the January 1955 plenum, he expressed displeasure with his successor as head of Moscow Oblast, Ivan Kapitonov. Acknowledging that the previous year had witnessed "very serious shortcomings in cultivating corn, for which Nikita Sergeevich Khrushchev justly criticized us in his report," Kapitonov outlined new measures designed to ensure successes.[14] On the plenum's final day, Khrushchev again criticized bureaucratic routine, which hindered effective management. "There are so many [officials], but the work is a failure," he lamented. "Why?" he asked rhetorically, "Because, comrades, there are many windy speeches made up of stock slogans." "They repeat [them]," he continued, "but they don't know how to plant [corn] and care for it. . . . My fellow Muscovites, for example, [have] . . . and a propagandist for every hectare, but their corn has failed." Khrushchev continued his broadside: "Why? Because, Comrade Kapitonov, very many speeches were made, but there was very little comprehension. This is the only way to explain it. They blathered and blathered, but at the end of the year there was nothing to harvest."[15]

Whether secretaries lost their positions, like Bondarenko, or they did not, like Kapitonov, all were vulnerable to Khrushchev's charges of failing to join the corn crusade with sufficient zeal. Consequently, the Moscow committee redoubled efforts by holding conferences and meetings on corn throughout 1955. In June, an oblast plenum featured speakers declaring it "a crop of decisive significance" in increasing meat and dairy output.[16] "It is impossible to say that . . . measures for fulfilling the directives of the January Central Committee plenum went smoothly, without mistakes," a district party committee secretary conceded. The campaign against red tape meant that officials had to acknowledge management failures: "For this, we were justifiably criticized in the regional newspaper . . . in an article titled 'Without Leaving the Office.'"[17] Each subsequent speaker recapitulated what his district or farm had done to grow corn and, thereby, to meet goals for meat and dairy output.

Success remained elusive, even as pressure form Khrushchev forced Moscow Oblast authorities to sustain their efforts. In July 1955, the oblast agriculture department found the state of corn cultivation "unsatisfactory." Some farms, MTSs, and districts had expended some effort on weeding plantings, but many had done little. The kolkhozes served by Mytishchi MTS had weeded 96.8 percent of corn at least once, and 65 percent a second time. By contrast, those under

Podolsk MTS had weeded only 39.7 percent the first time and none the second. The department therefore reprimanded officials, even firing one MTS director.[18] In January 1956, the authorities held another conference to evaluate the year's results. Political authorities and experts offered advice on how to select the correct fields, plant, cultivate, and harvest. Although many recommendations echoed Khrushchev's favored methods, some did not, such as that to plant not two or three grains in a cluster, but six or even eight.[19] Officials frankly conceded "widespread misfortunes with the corn crop." The oblast's plantings had expanded fivefold to reach 91,000 hectares, 19 percent of grain plantings, but only a few farms brought in a fruitful harvest.[20] Oblast party leaders blamed district subordinates who "gave less attention, ... did not demonstrate sufficient care, and allowed ... disorganized and untimely execution of work."[21]

Regional leaders faced compulsion to plant corn, procure grain, and execute Moscow's other directives. In public, such pressure began with Khrushchev. Behind the scenes, powerful officials such as Vladimir Mylarshchikov applied still more.[22] Head of the Central Committee's RSFSR Agricultural Department from 1954 to 1959, he managed information flowing to the Presidium and worked as one of the first secretary's agricultural advisers. Proximity to Khrushchev gave Mylarshchikov authority exceeding his modest formal powers, which he used to enforce Moscow's policies in the republic's regions. Respected and feared, he embodied the rude, brusque, and pugnacious figure characterized as a little Stalin, which novelist Ilya Ehrenburg had captured in factory director Ivan Zhuravlev, the antagonist of the epochal novel *The Thaw*.

Earning a formidable reputation as a troubleshooter, Mylarshchikov worked to ensure that oblasts planted corn and met state grain procurement quotas. In December 1956, Mikhail Karpenko wrote to Khrushchev outlining a case against the Central Committee operative. First, Karpenko related events that occurred during his earlier posting in Siberia's Krasnoiarsk Krai, where the party secretary, Nikolai Organov, had gained fame in 1955 for overfulfilling grain procurement quotas. Karpenko bemoaned how Organov had intimidated the kolkhozes, even forcing them to sell grain set aside for seed and kolkhozniks' pay. Having listened to Karpenko's protests, Organov had explained that he understood that those actions had disastrous effects. "Alluding to compulsion by Comrade Mylarshchikov," he had explained that he had to obey orders from the Central Committee to procure grain at any cost. To meet those demands, the leaders of the krai took actions they considered "irresponsible" and "counter to their party conscience." Anyone who spoke out, however, faced firing and blacklisting.[23]

Karpenko furthermore observed that Mylarshchikov's detrimental influence, and the abuses it engendered, became more evident once he moved to Moscow to become deputy to the RSFSR minister of agriculture. From this vantage point, he learned of inspections in Krasnodar Krai revealing irregularities in procurements. Local authorities had completed paperwork attesting to delivery of 10,000 metric tons of corn, which in reality stood unharvested in the fields. This ploy ensured that annual procurement plans seemed to have been achieved. Karpenko described the corrupting influence of Khrushchev's own aide thus:

> Who spreads this antigovernment practice around the country? Why is this done? Does it not happen in the wake of the one considered a practical, competent organizer of grain procurement? I have mentioned Comrade Mylarshchikov's name many times. I have the impression that it has some kind of magic power. People speak [of him] as an omnipotent figure, one whom they especially fear.[24]

In early 1957, investigators found that local officials had tampered with the data. The chairman of the krai soviet, Boris Petukhov, and an official in the local procurement department, the inspectors reported, had issued unwritten orders to cover up the shortfalls in corn deliveries by accepting other grains, while recording that kolkhozes had delivered corn. This cost the kolkhozes some 3.6 million rubles. Using the same terminology as Karpenko, the inspectors termed these actions "antigovernment behavior," and demanded that Petukhov appear before authorities in Moscow to account for himself.[25]

According to Karpenko, Mylarshchikov backed up flagrant abuses of power with threats and oaths. Resisting Mylarshchikov's illegal orders, Karpenko had been met with "all manner of insults." His description of the dictatorial department head as "rude, haughty, irascible, and vindictive" is corroborated by others' testimony.[26] In a later memoir, veteran party secretary Fëdor Loshchenkov characterized Mylarshchikov as "a rude man, considerate of no one." Oblast and ministerial officials' protests to Khrushchev achieved nothing, as Mylarshchikov "mercilessly forced [oblasts] to expand corn plantings."[27] This dictatorial behavior served as a model for subordinates, who employed the aggressive approach apparently required to execute Moscow's orders. Karpenko concluded that Mylarshchikov's actions in procurement campaigns "taught people [to act] in such a way that the grain [so procured] becomes bittersweet."[28] The complaints of Karpenko, Loshchenkov, and others failed to rein in Mylarshchikov. Khrushchev noted in the margin of Karpenko's letter only that he had read it. Mylarshchikov remained in his post for more than two years afterward.

The same year, an anonymous kolkhoz chairman wrote to premier Nikolai Bulganin to describe this pressure to sell grain. The writing of someone with a formal education, the evocative letter begged Bulganin to dispatch investigators to uncover abuses in Stavropol krai. He pleaded:

Please get to the bottom of this. Send people from the Central Committee. Only let them be judicious, not like those from the bureau of the krai party committee, who threaten the kolkhoz chairmen: "If you don't meet wool procurements, we're going to shear you and fulfill the plan that way." I came to agriculture from industry, and have fallen in love with it despite all of its unbelievable difficulties. I will say that [the bosses] do not value, do not like, and want neither to hear nor to understand the people who work these dusty steppes in the rain and snow.

He then described the pressure on kolkhozes to sell grain even at the expense of their operations and kolkhozniks' well-being. He did not understand "why the Central Committee's published resolutions, even after Stalin's death, diverge from the actual state of affairs." Officials in Moscow and the krai administration "continued to sugar-coat things." They forced the farms to write long-term development plans, but "when it comes to the root of the matter, we forget about them." Under these conditions, his kolkhoz had nothing to pay kolkhozniks, who justifiably refused to work until they could earn wages permitting them to purchase flour and other needed goods. Finally, the chairman condemned the krai bosses: "They holler, 'Give up the grain!' and the krai party committee secretary, Comrade [Ivan] Lebedev, declares that the kolkhozniks instead 'can pig out on corn.'"[29] The phrase captures Lebedev's equal contempt for farmers and for corn: a feed fit only for animals, a judgment reinforced by the verb "*zhrat*," which connotes not eating, but messily consuming in an animal-like fashion.

Pressure to take responsibility for policies incentivized local leaders and their networks to push initiatives, including the corn crusade, beyond logical limits. Khrushchev and party leaders purposefully strengthened regions to make them tools to achieve practical results. A new cohort of strong secretaries brought subordinates under control, becoming secretary-dictators able to carry out Moscow's policies. Having proved themselves capable by getting results, the secretary-dictators could call on powerful backers in Moscow to intervene on their behalf, solving conflicts and shielding their regions from suspicion.[30] This self-reinforcing process helped Aleksei Larionov become so prominent amid rising pressure on regions to outdo each other. Khrushchev catalyzed new efforts with the 1957 campaign surrounding the slogan "catch up to and surpass America," which triggered a runaway chain reaction. In Riazan, the secretary-dictator Larionov plunged into Moscow's campaigns, using any means available to appear even more successful and achieve even more acclaim.[31] Such secretaries made outlandish promises to double and triple deliveries of meat and dairy products, backing them up with fraud to create an image of successes. Those in charge of kolkhozes, districts, oblasts, and republics cast themselves in the best light, amplifying chronic data inflation.[32]

Through 1959, events seemed to bear out Khrushchev's belief in the revolutionary leaps forward in output apparently made possible by new technology and by thorough use of latent productive capacities already present in the countryside. Larionov foremost among them, secretary-dictators apparently provided leadership, the missing piece seemingly needed to realize the promise of an industrial farming revolution.

Under Larionov's leadership, Riazan Oblast worked a so-called miracle that became first a farce and then a tragedy. In 1958, he boasted about rapid rises in milk output. In August of that year, he complained in a letter to Khrushchev about inspectors from Moscow, who scrutinized his region more frequently than others only because it was nearby.[33] Larionov tried to use his influence with those at the top to deflect attention at a time when thorough investigation might have uncovered systematic fraud underway in milk procurements.[34] Having pulled off the 1958 scam, Larionov pledged that his oblast's farms would deliver outlandish quantities of meat in 1959. As Larionov seemed to produce food on the cheap, his actions concealed the hand of Moscow in inspiring Riazan's claims.[35] Larionov and other local leaders set a target of 75,000 metric tons of meat, 150 percent of the year's government-mandated quota. The day before a conference called to formally adopt the pledge, Mylarshchikov arrived armed with orders from Moscow: Larionov named not the initial ambitious target, but one twice as high.[36] Praise for Riazan skyrocketed as the oblast reported delivery of first 100,000 and then 150,000 tons. Leaders received prestigious awards, and the oblast garnered the Order of Lenin. The press made them a model for the others by trumpeting these apparent successes on the front page of *Pravda*.

In fact, Riazan was a model for all. Similar frauds were perpetrated everywhere, if nowhere else in such a concentrated and spectacular form. Farms slaughtered animals of all ages and types, including dairy cows and calves. They bought individuals' livestock and sold it to the state as their own production. When these methods secured only the first 100,000 tons, representatives traveled to neighboring oblasts to buy animals. Pressed to meet soaring demands, individuals and enterprises bought food in stores, selling it back to the state as if it were new output. At first only whispers, rumors of fraud grew into a roar as 1960 wore on. When the scandal broke, Larionov took his own life. Damaging the legitimacy of Khrushchev's reforms, the debacle left the oblast's exhausted farms incapable of meeting even the modest quotas characteristic of earlier years.[37] In late 1959 or early 1960, Mylarshchikov left the Central Committee to head a consortium of sovkhozes near Moscow. Although possibly a demotion resulting from his role in the still-hidden affair, this seems unlikely, especially because in 1963 he was appointed RSFSR Minister of Agricultural Production and Procurements.

By 1961, the Central Committee began to take a hard line on regional leaders and their party organizations, initiating a campaign to stamp out such schemes. Those secretaries who fell victim had responded to Khrushchev's calls to boost production and compete with the United States, but had substituted appearance for substance.[38] The scandals went farthest where a secretary-dictator of long tenure headed a stable network of subordinates whose mutual trust was sufficient to launch and sustain such schemes for a time.[39] Larionov had indeed been a model. His fall triggered discovery of other scandals. When Khrushchev had his subordinates clean the slate in 1961, the purge harmed his prestige and that of the party.

Local officials had enjoyed the protection of the Moscow authorities who pressured them. Ties such as those between Mylarshchikov and Larionov encouraged collusion in cover-ups. Gaining prominence thanks to apparent successes under their supervision, officials arranged credits, cajoled subordinates, and marshaled needed resources. Top officials tacitly approved of the fraud, even if they did not order it. Aleksandr Shelepin, chairman of the Committee for State Security (KGB), and Pëtr Doroshenko, head of the Central Committee department for agriculture of union republics, investigated reports from Riazan before the scandal broke. They stopped when Central Committee Secretary Nikolai Ignatov stepped in to chastise them.[40]

Letters arriving in Moscow complained that officials sent to inspect the oblast ignored legal and ethical violations, confirming an established pattern. A young Central Committee functionary, Georgii Smirnov, discovered various frauds during a 1959 trip to Kazakhstan's Pavlodar Oblast. Agriculture lay outside his responsibility, so he dutifully reported his findings to superiors, who warned him not to pursue the issue.[41] Above all, the post-Riazan campaign aimed to break ties shielding the regions.[42] In the end, Moscow could not ignore the flood of reports. By 1961, it swelled hundreds of letters a month addressed to Khrushchev alone: 866 in January, 1,571 in February, 2,172 in March, and 1,808 in April. Most carried accusations of fraud, cheating on wages, drunkenness, theft, and other malfeasance on the part of kolkhoz chairs and other local authorities.[43]

Many deceived regarding deliveries of meat, milk, and eggs to the state, but they sometimes inflated claims about corn plantings and harvests. An anonymous letter to Moscow charged that Mikhail Stakhurskii, party secretary of Ukraine's Zhytomyr Oblast, had required subordinates to submit deceptive records about their corn crop. It claimed that these measures caused actual production to fall, reversing gains made between 1953 and 1958. The letter writer explained how Stakhurskii ordered efforts "'to improve the situation' any way possible," and thereby make the oblast appear more productive. He and his subordinates, the letter concluded, "embarked on thorough but poorly disguised fraud about yields, gross harvest, and output of meat and dairy products." Kolkhozes and sovkhozes planted many more hectares of corn than the plan required. They then claimed the full amount harvested while reporting that they had planted only the number of hectares planned, artificially doubling the apparent average yield from 12 or 13 metric tons.[44] In 1961, the USSR Statistical Administration informed the Central Committee that this practice had been widespread.[45]

Stakhurskii had attempted to rule as a dictator but did not possess the necessary authority and trust among his subordinates. The anonymous denunciation detailed how Stakhurskii's henchmen had made district secretaries force kolkhozes into such actions, suggesting that the writer was subject to this pressure. Describing Stakhurskii's behavior, the letter underscores his obstinacy, unwillingness to listen to others, and inclination to pressure subordinates, all marks of an aspiring secretary-dictator.[46] The Ukrainian Central committee reported to Moscow on Stakhurskii's removal for that leadership style, dodging the question of corn-planting fraud altogether.[47] In reality a weak secretary, Stakhurskii had established neither sufficient trust nor fear to prevent a subordinate from blowing the whistle.

This was also the position of Andrei Naumenko, secretary of Ukraine's Sumy Oblast. In early 1961, a letter to Kyiv had claimed that crops remained unharvested and Naumenko was rude and abusive. It was purportedly—but not actually—written by a local Supreme Soviet deputy, a claim lending the charges credibility.[48] Some of the charges proved false, but the letter succeeded in bringing investigators to the oblast, illustrating why secretaries often "suppressed criticism" to present the best possible face to outsiders. Something had spurred a disgruntled member of a party network to write a letter to Moscow revealing what everyone knew or, at least, what was rumored. In this case, Central Committee secretary Frol Kozlov ordered Ukrainian party chief Nikolai Podgornyi "to administer harsh justice" to the perpetrators.[49] Unable to keep the fraud hidden and lacking the authority to force compliance, such secretaries failed to carry out Moscow's directives or conceal fraud, ensuring their eventual removal.[50]

Cases like these proved widespread. USSR chief procurator Roman Rudenko reported to the Central Committee that, in the first half of 1961 alone, authorities had begun prosecutions for fraud in nearly 300 cases, mostly in agriculture.[51] Party committees everywhere held meetings on the issue, revealing still more charges. On one kolkhoz in Ukraine's Vinnytsia Oblast, kolkhozniks denounced the chairman for false reporting about the sale of 100 hogs, as well as about the planting of grains, including 628 hectares of corn.[52] Other cases from the RSFSR and Ukraine involved falsified records regarding fields as small as 30 hectares of corn up to as many as 164 hectares.⁵³ Come harvest time, the Central Committee learned of a district in Odesa Oblast where officials reported the corn harvest complete, when in fact nearly 10 percent remained in the fields.⁵⁴ Oblast and krai statistics offices also made mistakes or, at least as likely, purposefully recorded inaccurate data.⁵⁵ The Central Committee concluded that a "formalistic approach" to the issue had allowed "instances of deceit to continue to exist now, when the question of the struggle against *pripiski* has already been deliberated."⁵⁶

The wave of scandals sweeping the country included false accounting for planting, harvesting, and yields of corn. More than making corn a footnote in the larger, better-known story of fraud in the meat and dairy schemes of Riazan, this fact demonstrates divergence between efforts to maintain appearance and those to produce substantive results. Only an unlucky few were denounced and discovered, meaning that information on cheating remains fragmentary. Any estimate of the scope, scale, and time frame of fraud must conclude that it was substantial. Taken together, this indicates that summary figures for corn planting should not be taken at face value, even though they are the only statistics available. A single exemplary kolkhoz in Ukraine serves as a case illustrating the phenomenon.

The leaders of the Red Plowman kolkhoz in Kyiv Oblast perpetrated fraud resulting in a scandal typical for the period, an episode shedding light on participants' motivations and the atmosphere of the time. In a book published in the 1990s, Petro Shelest recounted serving as oblast party secretary in 1961. Rising from aviation engineer to secretaryships in district, city, oblast, and republic party organizations, Shelest benefited from Podgornyi's patronage to become head of the oblast in the mid-1950s. Naturally taking pains to portray himself in the best light, he wrote of the Khrushchev era in all its complexities, including the era of Riazan-style scandals. He mixed dismay at the turmoil with a nostalgia for the era that is not surprising in light of later experiences as first secretary of the Ukrainian party until a 1971 falling out with Leonid Brezhnev.

In late 1954, Shelest became the oblast's second secretary. In support of his appointment, Kyiv authorities characterized his prior work in the city committee and as a factory director by calling him "energetic," "experienced in party and economic affairs," and "possessing organizing talents, initiative, and determination," all boilerplate terminology.⁵⁷ Shelest later wrote of his superior, Grigorii Grishko, as a respected colleague but bluntly described his ineffective leadership. Shouting at subordinates, Grishko nonetheless demanded too little of them. Grishko soon fell ill, leaving Shelest to act as first secretary.

Initially, Shelest struggled to establish his own authority, control district committees, and turn around struggling kolkhozes. He achieved this goal by emulating Khrushchev's leadership style. "There was no corner of the oblast, its enterprises, farms, and fields I did not personally visit," he later wrote, "speaking with the people; observing what they did in the fields; and listening to their advice, suggestions, and requirements." He cultivated relationships with individuals, rather than relying on the oblast committee's administrative apparatus. Shelest describes an informal meeting with a kolkhoz business manager who informed him about life on the farm. Shelest thus learned about kolkhozniks who secretly distilled moonshine, news that was unlikely to reach him through official channels. He visited kolkhozes unannounced, often discovering that district officials had forced kolkhozniks to overfulfill procurement plans or to replace popular chairpersons with candidates more amenable to arbitrary orders from above. Getting his hands dirty, Shelest devoted by his own estimate 75 or even 80 percent of his time to managing agriculture, despite training in engineering and a preference for industry.[58]

Having been on the front lines of the scandals, Shelest downplayed his own involvement. Disgusted by subordinates' actions but acting to protect them, he blamed Khrushchev for creating the conditions that drove otherwise good communists to fraud. News from Riazan of spectacular successes agitated Shelest. An entry for February 21, 1960, describes discomfort with the campaign that resulted when Khrushchev and the mass media harangued other oblasts for failing to equal that supposed success, requiring that each party committee reproduce it on a local farm or in a district but giving them no time to actually increase output. "Deep down, neither Podgornyi, with whom I have spoken several times, nor I agree with such methods," Shelest wrote. "Shameful phenomena, lying and double-dealing," he termed them in another entry from that year.[59] Yet the feverish atmosphere of the campaign forced him to join all others in selecting a kolkhoz to mimic the purported production miracles. Shelest had previously proven himself willing to protect subordinates. In December 1959, he had punished the secretaries of the Berezansk district party committee for fraud and related offences only after an article denouncing them appeared in the Moscow newspaper *Izvestiia*.[60]

The exemplary farm in Kyiv Oblast was the Red Plowman kolkhoz of Obukhov District. Needing to promote a model, Shelest formulated a plan with its chair, Ivan Kabanets, and its party secretary. Shelest described Kabanets as an honest and scrupulous leader, one whom Khrushchev had met and praised for his businesslike approach and practical knowledge. Kabanets had received accolades and had given a speech at the December 1959 Central Committee plenum, a rare honor.

In response to Riazan, the kolkhoz's leaders redoubled their efforts, promising to deliver even more meat and milk. Large investments furthered legitimate efforts, but mischief soon followed. Shelest lamented to his diary, "This is double-dealing and adventurism! But there is nowhere to run, and we cannot lag behind.

I reported to Podgorny and he, as much as I have, 'gave his approval.'"⁶¹ The tone of the passage indicates trepidation, signaled by the quotation marks surrounding the phrase. More likely, however, it also reflects Shelest's retrospective point of view. Although it is possible that he had harbored reservations, he had remained silent because of the reigning campaign mentality. He thereby at least tacitly condoned the actions necessary to meet Moscow's demands. Outlandish pledges and their fulfillment by any means, including fraud, proved the only viable strategy that in the short term could provide a shield from pressure from above.

At first, the Red Plowman kolkhoz appeared astoundingly successful, garnering praise for Kabanets and Shelest alike. In 1960, the kolkhoz produced more and paid its kolkhozniks well. The capstone of the chair's twenty-five years of leading the kolkhoz, success received glowing publicity in Moscow newspapers and earned Kabanets a citation as a Hero of Socialist Labor. In January 1961, Shelest glowingly reported about the kolkhoz to a Central Committee plenum, staking his own prestige and that of his patron, Podgornyi, on its reputation.⁶²

Rumors soon circulated hinting at prohibited practices, falsified reports, fraudulent sales to the state, and violations of the kolkhoz charter. In March 1961, inspectors from the Central Committee in Moscow began to investigate these charges and other "charming things," as Shelest sardonically termed them in private. He wrote that all of these "unpleasant affairs, if they are substantiated, are bad; but even if they are not confirmed, they cast a black shadow."⁶³ In early April, inspectors authenticated the allegations and forced local officials to mete out punishments.⁶⁴ The managers of the Red Plowman kolkhoz had systematically purchased milk, meat, and other commodities from kolkhozniks and other private individuals, disguised them as the farm's output, and resold them to the government, common if expressly condemned practices. "Comrade Kabanets," the inspectors concluded, "transformed the kolkhoz into a procurements office, engaged not in producing agricultural goods, but in double-dealing." In 1960, the scheme had provided more than half of the farm's purported production. The inspectors also found that the oblast committee had scrutinized the kolkhoz's operations in March 1960, but had uncovered infractions sufficient to earn Kabanets and the party secretary only a minor reprimand and a warning. Moscow's investigators condemned this as "superficial" in light of the new discoveries.⁶⁵

The investigation found additional infractions. To organize the purchases, Kabanets had hired non-kolkhozniks who served as a function similar to industrial enterprises' *tolkach*, or pusher, a word literally meaning a simple tool used to apply force to another object. Sometimes known as expeditors, such individuals used connections, barter, and bribes to smooth out irregularities in the supply system, locating necessary materials in time to fulfill production plans.

The rural expeditor used similar means to achieve slightly different ends. Tolerating such agents because they completed jobs, sold surplus produce, and acquired goods, officials apparently reported them only when they behaved badly.[66] A 1959 inspection of six districts and more than thirty kolkhozes in Kyiv Oblast alone, although not Obukhov District, discovered cases of purchasing agents who disbursed sums ranging up to 540,000 rubles to facilitate sales of meat to the state at the cost of significant financial harm to the kolkhozes. Inspectors furthermore uncovered purchases from "private individuals" of spare parts, timber, tires, and other necessary goods at "speculative prices," with total expenditures reaching as high as 65,000 rubles by one farm.[67]

The Red Plowman's buyers, each a resident of the nearby city of Kyiv, had spent over 4.5 million rubles buying livestock for resale to the state. This substantially exceeded the 3 million rubles in credit Shelest had arranged for expanding the farm's production capacity. Each of the three buyers had earned for his efforts some 44,000 rubles, a substantial sum when a kolkhoznik might receive 10 or 20 rubles for a day's work.[68] Their given names—Ziama, Avram, and Khaim— and family names furthermore suggest that they were of Jewish ancestry. Given common stereotypes about such individuals' inclination toward economic activities the state deemed illicit, those who complied and read the report may have lent this identity significance. Documents show only that the local authorities turned the three over to the criminal justice system.[69]

The investigation also revealed fraud in the Red Plowman kolkhoz's accounting for corn. Managers falsified production figures for meat and milk, but also inflated data on the corn harvest. Official records indicated that the kolkhoz had planted 916 hectares and had grown in impressive yield of just over 40 tons of silage per hectare. In reality, it had planted barely one-third as much and harvested 37.9 tons per hectare.[70] The ruse played a decisive role in the larger fraud. If the farm had not reported harvests of fodder sufficient to feed the livestock it purportedly raised but actually bought from individuals, its output would have appeared suspicious. Documents attesting to the nonexistent corn lent credibility to the façade of productivity the kolkhoz had constructed, while concurrently conforming to the demands of Khrushchev's crusade.

Shelest and the oblast committee had to censure the Red Plowman kolkhoz's leaders, a necessity he later judged "difficult." He wrote, "They were condemned because they, according to the Riazanites' example, purchased livestock for producing meat. And now we hold them responsible? But for what?"[71] The dealings that landed Kabanets and the Red Plowman kolkhoz in trouble were in fact common in Kyiv Oblast and across the Soviet Union. On March 24, 1961, the bureau of the Ukrainian Central committee handed down reprimands to district party secretaries in the oblast, including those of Obukhov District.[72]

Local officials began to understand the agrarian reforms only after the frauds came to light. Regional bosses had created showpieces. By investing only in the best farms, they exacerbated existing patterns of underinvestment, ensuring that the productivity of the majority lagged. Local bosses had pledged a great deal and created model farms while allowing business-as-usual to continue out of the spotlight. They thought Khrushchev's initiatives only a form of window dressing, one designed to substantiate for domestic and foreign audiences alike the leader's pledges to "catch up to and surpass America."[73]

For his part, Khrushchev demanded real production because he believed in the rightness of his causes: industrial agriculture and corn as foundations for abundance. When the first secretary became aware of the leaders' deception, he expected their regions in short order to actually produce what they had long only claimed, a task beyond the capacity of sovkhozes and kolkhozes due to the exhaustion resulting from the recent campaigns. Khrushchev boldly reversed course, breaking the regional networks headed by dictatorial leaders of the Larionov type, whom he had previously fostered. He soon instituted the territorial production administrations and then the scheme to divide party committees into agricultural and industrial branches, which were to specialize in each sphere. Alienating many, Khrushchev removed the dictators, paving the way for the norm-bound, compromise leaders and networks that became the common model under Leonid Brezhnev.[74] One such secretary, Grigorii Zolotoukhin, led Tambov Oblast for a decade before being moved to the larger Krasnodar Krai in 1967. Encouraging comradely behavior among party officials, never shouting at or insulting them, he respected subordinates and secured substantial authority.[75]

A champion of industrial farming, Khrushchev overturned long-standing land-management policies in most corners of the Soviet Union, an effort that attests to the reach of his authority in the regions, but also its limits. His convictions began to solidify while in Ukraine after the war, when he contested Moscow's one-size-fits-all policy. The grassfield system of crop rotations, or *travopole*, had become Stalin's dogma even before the 1939 death of its originator, Vasilii Viliams. It remained the basis for land management in 1953. As early as February 1954, Khrushchev criticized it while making his case for corn. He reviled farms in Rostov, Kursk, Krasnodar, and Stavropol for following Viliams system at the expense of corn. A voice from the audience offered an explanation: "They are afraid of row crops!" These required weeding between the rows and, therefore, more labor. In reply, Khrushchev called attention to several farms where corn provided the largest harvest of livestock feed, while local authorities favored less-labor-intensive crops. He charged that these sovkhozes used their land "incorrectly." "Why do we run things so poorly?" he asked. "Because the people who run these farms have lost their sense of responsibility for their assignments.

There is no communist left within such people," he angrily continued, "We must steadfastly wage war against this evil." Alarmingly employing a Stalinist trope, he concluded, "We must unmask such people."[76]

At the Central Committee plenum in January 1955, Khrushchev raised the issue again while attacking Aleksei Kozlov, minister of sovkhozes and an ally of Georgii Malenkov. Previously chided for his ministry's practices, Kozlov mounted the podium and began an anodyne speech about the 1954 harvest and plans for 1955. The Presidium members pounced. Nikolai Bulganin charged, "Your policy actually contradicts the party line. At the last plenum, they warned you about your work methods—clerical-bureaucratic work methods."[77] Khrushchev and Lazar Kaganovich joined in, reprimanding Kozlov for breaking rules on land usage and failing to ensure that sovkhozes planted corn. Pressing the issue, they raised mistakes dating from before Stalin's death, when Kozlov had headed the Central Committee Agricultural Department. In that post, he had forced implementation of the grassfield system. Khrushchev fulminated that Kozlov, "as department head [had] determined agricultural policy," meaning that he "was primarily guilty for the current chaos." "You are like a goose who pops out [of the water], flaps its wings, and goes along its way, clean and dry," he continued, indicating that Kozlov would not escape responsibility so easily.[78] Attacks such as these could only serve to silence open criticism of Khrushchev's corn policies.

Khrushchev considered the grassfield system antithetical to modern, progressive farming. He derided its advocates as grassfielders (*travopolshchiki*), but could not so easily eradicate the crop rotations and perennial grasses. By 1960, he had spent years promoting corn, but grasses and fallow remained widespread. Khrushchev acknowledged that shortages of planters, cultivators, and harvesters suitable for row crops hampered the effort. Yet he also pointed the finger at "peasant conservatism" and officials who followed the system as a rigid doctrine. Redoubling his attempt to overturn it, he advocated for a system based on synthetic fertilizers, machine power, herbicides, pesticides, and related innovations, even though they all were in short supply.

Outwardly, local leaders followed orders, but the reality appears far more complicated. The anti-grassfield campaign peaked after a Central Committee plenum in late 1961, where Khrushchev again advocated for replacing pastures with corn, sugar beets, sunflowers, and other intensively farmed row crops, likely as a way to make up for the shortfalls revealed by the recent widespread frauds. In December, he railed against grassfielders, officials who "rob the kolkhozes and sovkhozes of the ability to rationally utilize the land's abundance." At their insistence, farms planted not high-yielding corn and other crops suited to industrial farming, but low-yielding grasses. He indignantly charged that of the 220 million hectares of arable land, some 29 percent grew annual grasses, perennial grasses, hay, and oats or—worst of all—were fallow, idle while nutrients reaccumulated in the soil. Industrial agriculture rejected these, especially the last, in favor of synthetic fertilizer.[79]

The Soviet Union was to learn from the practices in the North America and Western Europe to realize Khrushchev's dream of catching up. Having quoted Lenin to that effect, Khrushchev offered an anecdote from Russia's more remote past:

> We must learn and implement every useful thing, as Peter the Great did after the Swedes defeated him at Narva [in 1700]. He responded, "Thank you for the lesson!" and set about learning how to make war. He mastered that, and defeated them at Poltava [in 1709]. He handed them such a defeat that afterward the Swedes never again tried any military campaigns.[80]

Khrushchev's message was clear: the capitalists had bested the Soviet Union in farming technology in the postwar period. Now the time had come to master those technologies.

At times crude and imperious, Khrushchev censured the leaders of republics, oblasts, krais, districts, and even of individual farms for insufficient zeal for the offensive against pastures. Earlier, he had chided Aleksandr Nikonov for "not loving corn" and publically shamed Jānis Kalnbērziņs. In November 1961, Khrushchev admonished Leonid Maksimov, director of the Kuban sovkhoz in Krasnodar Krai. Although exemplary for large wheat harvests, the sovkhoz did not utilize the potential of corn. A few days before, a newspaper article called attention to its failure to reject the grassfield system. The sovkhoz devoted only 16 percent of its plowland to corn and made only 1.6 percent of its harvest into silage for use as livestock feed, a contrast to the 29.4 percent of its land devoted to hay.[81]

The capitalists of the United States and, in particular, Khrushchev's American corn guru, provided the benchmarks. "Comrade Maksimov, you know agriculture well," Khrushchev began, "but if you were a farmer in America... could you compete with Roswell Garst? What would your production cost on the market?" Answer at the ready, he continued, "Garst would trample you. He does not plant grasses. He plants corn." Khrushchev demanded to know how the unfortunate sovkhoz director expected to cheaply produce meat and milk without the most productive feed crop. Thus the leader invoked Garst to legitimize claims that only corn offered a means to equal American abundance and achieve communism. "In peaceful competition with capitalism, the victory of communism is not in doubt," he explained. "But our country can and must solve problems of historical significance: to catch up to and surpass the United States, the most developed capitalist country, in per capita output of food."[82]

Describing how the country would do this, he highlighted Maksimov's failures. "We possess colossal potential, but we must utilize it more quickly and fully," he urged, "bringing all reserves into action. We cannot allow things to progress at their own speed. We are demonstrating before all the world the socialist economy's attributes: the criteria that will decide which system is better are the people's material and spiritual rewards." This was the decisive benchmark. Corn's role in meeting it could not have been clearer: "If we speak of developing livestock raising, then the most important part of this job is feed."[83] This ferocity notwithstanding, Khrushchev could also act magnanimously. In 1963, he dispatched a letter praising Maksimov for having learned the lesson and turned around the sovkhoz's operations by planting corn.[84]

During the antigrass campaign, party authorities resolved disputes about technical matters in favor of whoever successfully invoked Khrushchev's preferences. Each oblast, krai, and district had to develop "scientifically grounded" guidelines for crop rotations. Each kolkhoz and sovkhoz needed an individually tailored plan. In all cases, the schemes had to eliminate pastures and expand corn plantings. In February 1961, the journal *Agriculture of the North Caucasus* published an article by V. K. Moroz, an expert at a Stavropol agricultural research institute. Describing a plan he developed for the October kolkhoz, he argued for the "economic effectiveness" of using pasture in crop rotations to ensure sustainable production at low cost and to manage the danger of low average annual rainfall. Claiming the mantle of Central Committee directives requiring that plans encourage efficiency and lower production cost, Moroz's article used the word grassfield, a transgression sufficient to guarantee a challenge.[85] Even though Moroz only proposed using methods derived from the system's principles, rather than the whole orthodoxy, the chief economist of the krai agricultural department, Vladimir Chachin, denounced Moroz and his findings in a letter to the krai party secretary, Fëdor Kulakov. Chachin noted that Moroz's scheme earmarked as little as one-twelfth of the kolkhoz's fields for corn, cutting the area from the 1960 total of 4,600 hectares, or 23 percent of the land. In place of corn, the plan called for more perennial hayfields, which Khrushchev had condemned.[86]

Chachin and Moroz each adduced data, meaning that politics decided the outcome of the conflict. Denouncing Moroz's plan as covert preservation of the grassfield system, Chachin won support for his request that party officials "ensure that research institute directors make production recommendations based on objective analysis of actual data, rather than the subjective, archaic proposals of certain scientists," meaning Viliams.[87] In response, Moroz defended his proposed crop rotations and plowing methods as the only means to combat the dry conditions, wind erosion, and soil salination threatening the kolkhoz's viability.

He dismissed Chachin's argument that corn offered more feed, countering that the potential harvest of a single year was secondary to preserving the soil.[88]

Moroz could not prevail in a struggle in which Khrushchev's pronouncements sat on the other side of the scale. Reporting to Kulakov on the conflict, a committee functionary maintained that Chachin "was correct to oppose the crop rotations of the October kolkhoz." In fairness, the official acknowledged that Moroz had written the article in 1959, when the recommendations accorded with the party line. It had languished in the hands of the journal editor throughout 1960, during which time Moscow's policies changed. In light of this, Moroz escaped serious consequences by acknowledging that his old plans were no longer valid. The party committee called both men in for a consultation and then let the matter drop.[89]

The conflict resolved itself without lasting harm to Moroz, but the tenor of Chachin's attack illustrates the feverish atmosphere fostered by Khrushchev's initiatives. Advocacy for pasture, hay, or any other elements of Viliams's system brought a swift reaction. Few local authorities could resist demands to eliminate those practices. Many did not want to because it seemed to be an undemanding task requiring only a few orders and some badgering of subordinates.[90] Republics and oblasts responded promptly because their secretaries hoped to curry favor with Moscow, or because they were too weak in the post-Riazan atmosphere to coordinate an effort to dodge the orders.

Khrushchev's zeal for replacing grasses with corn furthermore provides the most substantial evidence for the convention that he pushed his corn crusade beyond reasonable boundaries. Not coincidentally, plantings reached their largest extent in 1962, yet he then allowed the total to decline in 1963 and 1964. In Stavropol Krai, party and agricultural officials ordered kolkhozes and sovkhozes to plow up pastures and plant corn. At the beginning of 1961, farms boasted more than 450,000 hectares of perennial pastures, but by fall had plowed up 27 percent of them. Plans determined that only 75,000 hectares were to remain by the end of 1963.[91] Officials planned to cut their proportion of cropland from about 12 percent to less than 2 percent. In March 1963, they spoke of having "disavowed the grassfield system," even as they acknowledged that in practice the elimination of pastures remained incomplete.[92]

Lithuania differed from Stavropol Krai and most other regions. The republic's communist party and government dodged these orders, cooperating to preserve the cultivated pastures that had long formed the foundation of agricultural land use. With a history of independence during the interwar period, Lithuania boasted solidarity founded on nationalism, which helped local leaders chart an alternative course.[93] That nationalism influenced the entire history of the Soviet Union's rule in the republic. As the Red Army drove the Germans out in 1944, the new authorities repeated police and military actions originally used in 1940 to establish power in Lithuania, as well as in neighboring Latvia and Estonia. In Lithuania, resulting partisan resistance continued until 1950. After Stalin died, Lavrenti Beria moved quickly to stabilize western regions by curbing repression.[94] Mutual trust among republic officials rose amid the exceptionally long tenure of Antanas Sniečkus, who led the party in 1940 and again from 1944 until his death in 1974. Headed by an authoritative secretary who negotiated with subordinates rather than dictating to them, the party organization proved itself a stable. Coexisting with formal procedures, informal relationships protected local authorities and their initiatives from what they considered Moscow's meddling.

Beria's action quietly permitted previously inhibited sentiments among the republic's titular nationality to reemerge, promoted by cadres drawn from among the local people. Lithuania differed from its Baltic neighbors, however, because of Sniečkus's longevity and the resulting stability.[95] Reversing policy in place since the 1930s, postwar initiatives looked back to those of the 1920s, which encouraged promoting cadres belonging to local ethnonational groups.[96] In 1953, Beria's moves signaled favor for appointing Lithuanian leaders, rather than to importing ones from Russia. These leaders nonetheless had to toe the line Moscow drew between permissible local interests and deviations termed "bourgeois nationalism."[97] Local leaders had to mind not to overstep that line, which shifted over time. Khrushchev did not devote great attention to the issue of national sentiment in the union republics, an attitude that changed only episodically, when some event called attention to an individual republic.[98]

Historians of the Soviet Union have found that nationality is not an essential and stable category, but instead one created and maintained through historical processes.[99] Peasants in Lithuania expressed their Lithuanian identity through relations with the land. They responded to collectivization at the end of the 1940s not only through the subtle strategies often termed "weapons of the weak," but also in the ways they organized their personal plots. These practices revealed interconnections among land, labor, community, and power, allowing peasants to exercise some autonomy in circumscribed spheres, offsetting their subordinate status. These processes suggest that Soviet central authority was not absolute, a fact ensuring that Soviet citizenship entailed constant efforts negotiate a range of sometimes-incompatible priorities.[100]

Moscow ordered the republic's political leaders to alter patterns of land use, which was tied to notions of national identity. However tenuous the connection of Lithuanian political leaders to actual rural ways of living and working, they adapted and avoided implementing dictates from above, suggesting a link between land use and national identification. In practical terms, this meant that the republic bosses guarded their prerogatives to appoint local officials and to govern agrarian economic activity. They quietly questioned what they interpreted as interference from Moscow. A running contest over the importance of pastures and corn cultivation indicates that Moscow determined formal policy, but could not guarantee subordinates carried it out. Yet Lithuania stood out because its party network could draw not only on built-up trust, but also identification with the titular nationality.

Sniečkus and the republic party responded to Khrushchev's demands for corn with foot-dragging not unlike that in neighboring Latvia. As part of the modest agitation for corn in 1954, Lithuanian farms received enough seed to plant 10,000 hectares, a tiny percentage of the republic's cropland. They planted only 4,000, however, and fed the remainder of the seed to livestock.[101] Hitherto unknown in Lithuania, corn as yet had no place. At the January 1955 Central Committee plenum in Moscow, Sniečkus acknowledged that party leaders had "yet to overcome the stubbornness of certain kolkhoz managers, agricultural specialists, as well as party and government officials who consider it impossible to cultivate corn in the republic." He pledged to increase plantings tenfold to 40,000 hectares.[102] In February, party leaders gathered in Vilnius to endorse the plenum's directives, ensuring that district officials understood the new campaign for corn. The crop featured among the many measures required to augment output of meat, as well as of the dairy products for which the republic was known. Highlighting corn's potential, Sniečkus acknowledged that local farms underappreciated the crop and the party had not effectively promoted it.[103]

That year's agricultural planning reforms tasked the republic's party, council of ministers, and ministry of agriculture with designing a five-year plan for dairy and meat production. Each administrative region across the Soviet Union formulated a document, "Measures for Increasing Output of Grain and of Livestock," which outlined annual production targets for 1960. Party committees then promised to meet them two, three, or even four years sooner.[104] Cautiously describing only "significant expansion of corn and feed crops, as well as boosting their yields," the Lithuanian plan did not define exact proportions of cropland devoted to each crop.[105]

District leaders reported to Vilnius about the dismal results achieved in growing corn. On April 25, 1955, even before spring planting, the republic's premier lamented the attitude of the ministry of agriculture. The republic's machine-tractor stations had ordered 100 corn planters, but received only 40. They had on hand 66 planters adapted to planting corn, enough for only 3,300 hectares, or 2.3 percent of the plan. That figure also indicates that the plan had doubled since February to approximately 80,000 hectares. Farms had yet to designate fields, apply fertilizer to them, or collect seeds from government distribution stations.[106] Little had improved by the end of July, when inspectors reported "unsatisfactory progress." Overall, kolkhozes had assigned laborers to weed only 56.9 percent of corn plantings, but in some districts this figure fell as low as 18 percent.[107] Farms used available machinery ineffectively, a problem compounded by managers' "continued underappreciation" of and "most cautious" commitment to the crop. Many claimed that the cool spring weather had caused slow growth, a problem only aggravated by delays in weeding. The weather did not account for the fact that on one farm only 10 of the 174 kolkhozniks assigned to weeding showed up in the fields.[108]

Lithuanian farms differed little from those of other oblasts and republics. Moreover, the situation changed little by the end of the 1950s, as the republic achieved few corngrowing objectives.[109] Similarly, Lithuania experienced its own small-scale cases of fraudulent livestock purchases by kolkhozes.[110]

Lithuania became exceptional only in the early 1960s, when its leaders responded to Khrushchev's campaign against hay and pastures. The leader demanded that the Baltic republics replace long-standing crop rotations based on pastures with row crops grown using industrial farming methods. Lithuania had a cool, humid climate poorly suited to corn and lacked necessary technologies. Local experts favored a system of managed pastures, a local adaptation of the principles underlying the grassfield system that Khrushchev condemned. The legumes and grasses of these rotations replenished the soil, requiring little labor and few synthetic fertilizers or other inputs.

A November 1961 scandal pulled back the curtain on some of the processes at work. A group of agronomists and researchers at the republic's Institute for Land Management questioned Khrushchev's new dogma. Petras Vasinauskas, director of the institute, and a group of colleagues signed a letter advocating their solution to Lithuania's need for livestock feed. They rejected the dogma that Vasilii Viliams's system had become under Stalin, a position in accord with that of Khrushchev. Instead of corn and other row crops, however, they supported managed perennial pastures, which they argued better suited the Lithuania's cool temperatures and regular, abundant rainfall.[111] The republic's Russian-language newspaper, *Sovetskaia Litva* [Soviet Lithuania], published a version of their arguments for plantings of clover and other legumes, which return nitrogen to the soil.[112] This made their position antithetical to Khrushchev's vision of industrial farming founded on synthetic fertilizers, which obviated the need for fallow.

Khrushchev had created a fervor that forced the republic's party leaders to condemn the letter as "politically harmful." On December 22, 1961, they resolved that the specialists' views "contradicted the party line on the grassfield system of land management, . . . leaving the grassfield system in place and opposing fodder crops such as corn, sugar beet, and other row crops." The document formally reprimanded Vasinauskas, as well as his superior, the minister of agriculture, who failed to denounce the letter and suppress its message. Finally, it threatened both with strict punishment in the event of future similar occurrences.[113] Even before the resolution passed, *Sovetskaia Litva* shifted to denouncing the scientists and asserting that corn was superior. The newspaper did not publicize the Central Committee resolution, but on the day it passed an article appeared that signaled the new party line by arguing for reforms to land management. To do so, it singled out Ignalina District because its kolkhozes had failed to replace grasses with corn and other row crops.[114]

Believing the Lithuanian Central committee's moves inadequate, Khrushchev attacked its members for lacking faith in corn and modern methods. Having learned of the letter, he denounced it, and the "grassfielders" who wrote it, in the presence of Sniečkus and other Lithuanian leaders. In Minsk on January 12, 1962, he seized on an excerpt from the letter arguing that cattle had evolved the rumen, a complex biological system for digesting grasses, meaning that these were the logical solution to the feed shortage, rather than grain, beets, or other feeds. "Why do the Lithuanian researchers call on the kolkhozniks to continue using old-fashioned methods?" Khrushchev questioned. His answer: they had insufficient faith in corn. He insisted that corn was equally vital for farms in Belarus, where he was speaking, as well as in neighboring northern Ukraine. Growing agitated, he resorted to ridicule: "Some might say: 'What's this, Khrushchev has come just to rip us apart, to criticize us?'" He then sarcastically queried, "Did you think I came to read you poems by Pushkin?" Instead, he considered it his duty to call attention to failures, forcing officials and organizations to reform flawed old ways steeped in the grassfield dogma. He taunted the audience, "You can read poetry on your own time."[115]

Throughout early 1962, Khrushchev's rebuke colored Lithuanian officials' actions and the resulting content of *Sovetskaia Litva*. On January 16, 1962, a session of the republic central committee's executive committee addressed several agenda items related to agriculture. The resulting resolution required the party to quickly "achieve the goals established by Comrade Khrushchev [in Minsk], to increase production of grain, meat, milk, and other agricultural products." To achieve this, the party pledged "to adopt row-crop land-management systems . . . and [new] crop rotations, and to replace less bountiful crops with higher-yielding ones, especially corn, sugar beets, and pulses."[116] The following day, the newspaper initiated a long run of articles encouraging managers to replace old crop rotations with new ones.[117] A district party secretary declared, "The grassfield system . . . has harmed the agriculture of our republic. Comrade Khrushchev very justly criticized Comrade Vasinauskas" and the others for their alleged support for that system.[118] Republic authorities then held a series of conferences for district leaders to detail the case for corn. Sniečkus coolly reported to one of them that, in Minsk, "shortcomings in the leadership of kolkhozes and sovkhozes were revealed."[119] In February and March, Sniečkus censured district leaders for "failure to restructure crop rotations and, especially, to adopt high-yielding row crops—corn and sugar beets."[120] Another meeting of the republic's party bureau took up the issue yet again, raising concerns about the teaching in agricultural colleges.[121]

Republic authorities themselves linked the dispute over land management to larger fears about nationalism and loyalty. Sniečkus underscored concerns about manifestations of impermissible nationalist sentiments found in the colleges. Much as the republic's land-management institute had, these schools attracted scrutiny for disseminating so-called dogmatic interpretations of agricultural science—meaning the Viliams system—and for their students' lack of knowledge about corn. Moreover, republic central committee officials voiced unease about lagging enthusiasm for courses in scientific atheism and historical materialism required as part of the general curriculum. Finally, they saw nonconformist tendencies in the alarmingly low attendance at lessons on Russian language and party history.[122]

Khrushchev's attacks on the Lithuanians for loyalty to the grassfield system mounted. On March 7, 1962, he again castigated them in an address to a Central Committee plenum in Moscow. There, Sniečkus had to acknowledge that Khrushchev's apparently accurate criticism of the academics' letter had "aided in a thorough understanding of the grassfield system's fallacious nature. . . . [It helped] our specialists grasp their mistakes."[123] Back in Vilnius in late March, Sniečkus condemned the agricultural bureaucracy for continuing "the extensive grassfield system."[124] He also noted Khrushchev's January tirade in Minsk, calling the criticism "just and deserved." Sniečkus concluded by calling on kolkhozes and sovkhozes to reduce their pasturelands by approximately 20 percent in 1962 alone, planting corn instead.[125] In July, still another meeting issued a report to Moscow pledging to increase the republic's corn plantings by 61,000 hectares, or 51 percent. Sniečkus described how Lithuanian leaders "had drawn practical conclusions" from Khrushchev's censure, embarking on a program to "reconsider crop structures" and "alter them toward fuller and more rational land use by expanding plantings of row crops."[126]

It seemed that Khrushchev had compelled the Lithuanian party to carry out Moscow's policy, furthering his crusade for corn and industrial farming. He overturned Lithuanian experts' conservative, albeit agronomically informed, approaches. However, the evidence points to the conclusion that there was more to this history than meets the eye. Given the secretive nature of the actions involved, the following is only a circumstantial account of how Lithuanian officials evaded Moscow's demands, a strategy that enjoyed the support of the republic's highest officials, including Antanas Sniečkus.

A speech Sniečkus gave shortly after Khrushchev's October 1964 ouster provides compelling evidence for this finding, a starting point from which to work backward. After the plenum in Moscow ratified Khrushchev's forced retirement, Sniečkus spoke frankly to the republic's branch of the KGB. The text of his speech is located not among documents of the republic Central Committee, but in his personal files, suggesting that he ascribed to it particular importance. In the triumphant tone characteristic of the days following Khrushchev's removal, Sniečkus first recounted events leading to the plenum. "For months, even years, there has been no confidence in tomorrow," he recalled, blaming "all of these reorganizations." Khrushchev had even "terrorized" high-ranking officials for perceived mistakes by pinning "labels and nicknames" on them. He had created sham "democratic methods of leadership, when this was actually only for appearances." "A true dictatorship existed within the Presidium," which reinforced his "boundless authority." Sniečkus allowed that the Soviet Union had made some progress under Khrushchev, but only in areas where the now former first secretary had not "violated Leninist methods of leadership," as he had with growing frequency after 1957. That date was critical because Sniečkus naturally applauded the defeat of "the antiparty group" because it had brought to power those protégés who had in recent weeks succeeded Khrushchev.[127]

Sniečkus emphasized crises in agriculture, repeating charges made at the October plenum that nothing had improved since 1958. He boasted that Lithuania had achieved better outcomes. Holding Khrushchev responsible for Riazan and environmental challenges in the Virgin Lands of Kazakhstan, Sniečkus turned to the campaign against grasses. Khrushchev forced unnecessary change on all regions, pressuring party officials to eliminate pastures wholesale, plowing up even the productive ones. Sniečkus reminded his audience about the earlier case of the researchers' letter, which Khrushchev had transformed from a technical issue into a political one by viewing support for pastures as a challenge to his personal authority.[128]

Facing demands to destroy the pastures entrenched in Lithuanian practice, Sniečkus and the republic party leaders had responded with dissimulation and delay. Other regions had bowed to the "enormous pressure" from above, but not Lithuania. Sniečkus described how the republic's leaders had meekly accepted reprimands, pledging each spring to plow up pastures in the fall, and each fall to do it in the spring. "In truth, we sabotaged this business," Sniečkus revealed. "We believed in the practitioners and specialists in our republic, not in Khrushchev." "During these years when they pressured us," he continued, "we reduced our area of perennial grasses by 3 percent, but that was only old clover and the like. We held out."[129]

Sniečkus then implied that formal documents and reports served only to misdirect Moscow, while in private leaders encouraged farms to carry out policies they considered correct. Describing how they relieved the pressure, he stated, "You speak with the kolkhoz chairmen: you'll learn that we sent them directives [to plow up pastures] and then we gathered meetings . . . in Dotnuva [site of the agricultural institute] or some other place." The key was to do so away from Vilnius, because in the capital the meeting might be overheard "by any sort of person."[130] Indeed, *Sovetskaia Litva* had reported on meetings convened in Kaunas and Dotnuva, where Sniečkus, the minister of agriculture, and other officials meet with district leaders and kolkhoz personnel.[131] "There, without the stenographers," Sniečkus revealed, "we told them, 'Comrades, we must do this and do that.' And people understood. . . . They knew that it was truly necessary to do it that way. . . . Many nights, we did not sleep because [Moscow] attacked us from all sides."[132] He claimed that, as a result, Lithuania's output remained high, having grown 24 percent even in 1964, in contrast to falling milk output in regions where farms plowed up pastures and planted corn.[133]

Kolkhozes devised several strategies to circumvent orders. Some planted corn in a strip a few rows deep along roads leading to and from the farm.[134] Any unsympathetic visitors would see only that façade, while hay, clover, and other crops grew on the remainder of the field. In an alternative strategy, kolkhozes planted corn far from the road on the premise that if it remained out of sight, inspectors were unlikely to measure the size of the plantings and the farm's statistics could easily be falsified. As second secretary of the Lithuanian Central Committee and the only Russian in the republic's top leadership, Boris Sharkov was Moscow's representative, meaning that the others likely did not inform him of the schemes.[135] In June 1961, Sharkov denounced one district party committee because it "had not learned from the criticism" of superiors, as party discipline required, or ceased its stubborn resistance to changing crop rotations. "What have you learned from this?" he asked. "You said, 'We must plant the corn further from the road, so that it is not visible.'"[136] Notably, this aside was edited out of the version of the stenographer's transcription distributed to local party organizations.

As early as December 1958, Lithuania's policy on grasses had caused a stir. Speaking to a Central Committee plenum in Moscow, Sniečkus praised small successes in meeting demands for corn and promised future improvements.[137] At the same plenum, he had to respond to the party secretary of neighboring Kaliningrad Oblast, who complained that kolkhozes in his oblast sold their grass seeds to neighbors across the border in Lithuania, where prices were noticeably higher. Red clover that sold to the state for only 20 rubles per kilogram in the

RSFSR brought 32 rubles in Lithuania. The secretary posited that this was the result of "an abnormal situation, some sort of brouhaha." In response, Sniečkus presented the republic's five-year plan to raise output by planting more clover, which constituted 25 percent of cropland in 1959, double the 1953 total.

Although it predated Khrushchev's feverish campaign against the grassfield system, the incident put Sniečkus on shaky ground. He replied that the higher prices were required to ensure a supply of seeds sufficient for that program. When challenged about the source of the extra seeds, Sniečkus coyly replied that they came "from friends," a response that, according to the stenographer's record, drew laughter from the audience. Furthermore, the file in Sniečkus's personal papers containing the October 1964 speech also contained a transcript excerpting this exchange, further evidence of connections between them.[138]

In August 1964, Nikolai Egorychev, secretary of the Moscow city party committee, had met Sniečkus during a vacation on the Baltic Sea coast. As Egorychev recalled later, the usual dinner, drinking, and "comradely socializing" that accompanied such meetings gave him an opening to start a conversation about the cabal forming to oppose Khrushchev. Sniečkus and the other Lithuanian leaders steadfastly avoided being drawn into any such talk. Later, after Khrushchev's fall, Sniečkus called Egorychev to apologize, revealing that he had rejected the overture out of fear it was a "provocation" designed to trap him.[139] The Lithuanian party boss's anxiety seems especially comprehensible in the event that he had something to hide from any emissaries from Moscow.

Such scheming to protect pastures required trust among the republic's officials. A disgruntled party member could easily write to the authorities in Moscow, as those in many oblasts did amid the post-Riazan scandals. In Lithuania, a sense of belonging to the republic's titular nationality combined with built-up trust to make Sniečkus's scheme possible by discouraging officials from breaking ranks and writing to Moscow. In his October 1964 speech, Sniečkus described maintaining unity through personal relationships with local officials developed over long years leading the republic. He recounted a private meeting in which he dissuaded a kolkhoz chairman from writing to Khrushchev to account for the republic's problems and plead the case for pastures. The man "was extraordinarily agitated," Sniečkus related, "but I told him, 'Whatever you write, the repercussions will be visited on me,'" implying that such a letter would only draw Moscow's attention, rather than solve the problems caused by its overbearing authority. Having agreed to wait, the chairman admitted, "'We see how our republic leadership mitigates many of Khrushchev's improper policies.'"[140] Sniečkus used this story to illustrate links between party bosses and those responsible for carrying out back room maneuvers. The leader spoke about these tactics openly before subordinates only after Khrushchev's removal, reflecting the relief amid the post-Khrushchev atmosphere, which soon emphasized stability giving local party organizations relative freedom.

It is unlikely that other regional party organizations carried out similar large-scale subterfuge, or that these schemes could work without the added element of nationalism. The strong trust that Larionov had built up in Riazan was rare, and even it was proven to have limits. National identity set Lithuania apart, aiding local officials in presenting a united front against external authorities seeking to alter historical agricultural practices and relations to the land. This affinity facilitated projects to construct and maintain a norm-bound, or compromise network both efficient at achieving results and presenting a comparatively unified front to agents of Moscow's authority. According to a metaphor formulated by historian Yuri Slezkine, the multinational federation was a communal apartment, in which each national republic was a separate room connected to the whole primarily through the main hall, the RSFSR.[141] To expand on the concept, the walls of each room permitted inhabitants a certain amount of license, so long as they kept activities out of the sight of the apartment building's residential committee, which is to say the central authorities in Moscow.

Under Khrushchev, Lithuanian party leaders and rank-and-file members built solidarity against interference in local economic activity. The period from Stalin's death to the early 1970s was one of weakening control and increasing independence from Moscow. Republic leaders had license to strengthen connections to the local people by appealing to adapted versions of ethnic identities. These allowed them leeway to operate, which they continued to enjoy so long as they stayed within the bounds set by Moscow and ensured continued economic performance.[142]

After 1953, new Lithuanian-but-Soviet (or Soviet-but-Lithuanian) elites arose under the guidance of Sniečkus. Moscow only concerned itself with ensuring that nationalism in republics remained within boundaries. Republic leaders gained room to maneuver so long as assertions of belonging to a nation remained couched in terms of an overarching Soviet identity. Thanks to Mikhail Gorbachev's policy of *glasnost*, national identities aided groups seeking to shed the overarching framework, contributing to the union's dissolution. Nationalist movements' rapid reemergence into the open, with Lithuanians among the first and most passionate, suggests that these sentiments survived all along, coexisting during the Khrushchev era with senses of belonging acceptable to Moscow.

Accumulating authority throughout the decade under his leadership, Khrushchev required subordinates to put industrial farming technologies and corn to work in the fields. Tirelessly promoting those approaches, he hoped to replace crops and land-management practices he dismissed as conservative and, worse still, inadequate to the goal of raising production to American benchmarks.

He attacked local authorities who hesitated or failed to meet his expectations, challenging his authority and obstructing progress toward the abundance necessary for communism. During the first seven years of his leadership, Khrushchev backed strong regional secretaries who promised to carry out his policies, and seemed to do so. These powerful secretaries apparently made his agricultural revolution into reality. In 1960 and 1961, the resulting scandals damaged Khrushchev's legitimacy, harmed the project, and destabilized the foundations that had facilitated considerable successes up to 1958. Having ousted the regional secretaries deemed responsible, Khrushchev still demanded swift action on a new campaign to plow up hayfields and disavow anything that smacked of the old grassfield dogma. Corn, sugar beets, and other row crops growing using industrial technologies took their place, driving corn toward its 1962 peak of 37 million hectares.

Far from complete, Khrushchev's ability to get things done was in fact much more tenuous than scholars have long imagined. Regional responses to policy directives from Moscow shed light on relationships between center and periphery. Local histories bring into focus the two-way relationships in which the center was in part a product of interactions with the periphery, rather than the sole motive force.[143] Appearing from the outside to be overwhelming, Moscow's power was actually relatively weak out in the regions. Khrushchev's authority to determine formal policy did not guarantee compliance by regional networks, which each maintained its own leaders, priorities, internal dynamics, and relationships to Moscow. Unwritten rules and informal practices facilitated the outward appearance of compliance, but did not necessarily correlate with substantive action. In turn, Khrushchev and Moscow had to respond, which they did with sweeping change beginning in 1961. This is not to suggest that these stories of Zhytomyr Oblast, the Red Plowman kolkhoz, or Lithuania encompass all possible forms of subterfuge. The cases exemplify only some of the strategies available to leaders who stove to respond to Khrushchev's demands to achieve miracles. Furthermore, not every farm practiced fraud in reporting about corn or other crops, and not in every year. If this had been the case, affairs would have been even more ungovernable.

CONCLUSION

Even technological societies remain dependent on farming because, as the essayist and critic of industrial agriculture Wendell Berry implores his readers to understand, "eating is an agricultural act."[1] Plagued by chronic agrarian crises, the Soviet Union was a lumbering military-industrial colossus with feet made of clay. After 1953, corn helped provision a rapidly urbanizing population, but fell into disrepute because Nikita Khrushchev only partially made good on his promises of the good life like that enjoyed by Americans. Sensing inconsistencies in official narratives about the United States, citizens created alternative images that reversed the terms of the propaganda contrasting the two societies.[2] In 1961, a voter in Perm Oblast criticized the official braggadocio. "Elect who you'll elect," the anonymous citizen wrote on a ballot, "but there is no meat, no fish." Questioning official narratives about the capitalist world, the citizen concluded, "They say we've caught up to America, but why is it necessary to catch up to them when they live 'in poverty'?"[3]

Ubiquitous in the mass media, corn and Khrushchev's campaign became symbolic of these apparent failures and therefore attracted ridicule. The Soviet Union had embraced industrial agriculture and industrial food, which divorced urban consumers from farming. These consumers experienced food as a commodity, not a product of farmers' labor to harness delicate biological processes powered by the sun. Instead of a tangible object, corn became abstract and therefore capable of absorbing blame for almost anything. Official efforts to equate the crop with abundance boomeranged, making it fodder for jokes. Cultural critics Aleksandr Genis and Pëtr Vail later described the era as one of extremes. If the Soviet Union was to overtake America, it would do so in only three years. Khrushchev's "impulsive dogmatism," the two concluded, caused a mania for planting corn in every climate and corner of the country.[4] Even today, the Russian press quietly maintains memories of Khrushchev and his obsession with corn.[5] Survey respondents rate Khrushchev fifth out of seven twentieth-century leaders.[6] In such polling, agricultural initiatives top the list of his era's defining events. The Virgin Lands campaign and so-called experiments claim places ahead of his mass housing program and revelations about Iosif Stalin's crimes.[7]

Ever since 1964, scholars have echoed Khrushchev's internal critics, who responded to his attacks on the bureaucracies he blamed for slow output growth and scandals. Gathering nearly universal support for ousting him, former comrades derided him and his policies as "adventurist," "irresponsible," "ill-conceived," "unscientific," and more. Still guided by Khrushchev's policies, the Soviet Union brought in a bumper crop in 1964, but his opponents pressed home attacks that blamed him alone for past failures, shielding themselves and their constituencies from responsibility. Speaking to the Central Committee in March 1965, Tambov Oblast party boss Grigorii Zolotoukhin described how:

> An anti-Marxist, subjective, volunteerist approach to agriculture has been allowed in recent years, causing much damage. At plenums, in print, and in directives it was mistakenly reported that our farms had everything, or nearly everything, needed for development. Every problem was the fault of the local officials, who have become scapegoats. At the same time, fundamental questions of agriculture have not been resolved. Force was used. Speeches were pronounced about local initiative, but nothing was done. A scientific approach, an analysis of the actual state of things was supplanted by harebrained scheming. Because of this subjective approach, no one in the planning agencies ever defended the interests of agriculture or paid attention to the needs of this fundamental economic sphere. From year to year, they trimmed finances as well as material and technical aid, while attempting to extract as much wealth as possible.[8]

Zolotoukhin even avoided Khrushchev's name, which remained unprintable until after 1985. But Zolotoukhin also avoided holding anyone else responsible for these failings. Although acknowledging failures in investment and management, he made the case that officialdom deserved no blame.

Alternative interpretations allocated guilt more equitably. In December 1964, two Gosplan economists submitted to the Central Committee an analysis arguing that the corn policy had failed "to properly account for various regions' natural and economic peculiarities, as well as the farms' material and technical capacities." Calculating the cost of livestock feed from corn, the researchers quantified the harm done by officials who, acting "formulaically," had forced farms to plant corn despite inappropriate climate conditions. Preferring political expediency, local bosses had not considered farms' capacity to plant, cultivate, and harvest corn in a timely manner and more profitably than alternatives. Even in locales with conditions hospitable to corn, harvests remained measly because farms harvested up to 50 percent of the crop before it had matured even to milky-wax stage, raising costs and producing inferior feed.[9]

Farms harvested so early because they lacked necessary machinery, the manufacture of which Khrushchev had ordered but the bureaucracy had mismanaged. After 1960, when Khrushchev pleaded for more machines and more corn to be harvested as grain, average yields for silage actually fell. Additionally, farms' feed supplies began to run short in late summer because Khrushchev had insisted that they plant corn in place of the pastures that traditionally had fed animals during that season. Finally, farms in southern regions faced time constraints imposed by their crop rotations. They had to harvest the corn, plow the fields, and plant the next year's winter wheat or barley before the frosts set in, a further incentive to bring in corn early.[10] Despite all of these technical failures, the experts cautioned against rashly abandoning corn, which had potential in warm southern regions and on irrigated lands.[11]

Even though scholars long presumed that a rapid decline in corn plantings began in 1965, Khrushchev's successors did not actually reject the crop. It became a regular feature of the farm economy, albeit one far less prominent in the mass media. After peaking at 37 million in 1962, the number of hectares planted soon settled at an equilibrium much higher than that seen before 1953. In 1963 and 1964, Khrushchev had sanctioned a cumulative 20 percent reduction. Incorrectly concluding that this had presaged a precipitous decline, Roy and Zhores Medvedev wrote, "In 1965, the amount planted fell below the 1940 level. Even those kolkhozes where it had been a success now refused to plant corn!"[12] Plantings in 1940 were all for grain and totaled only 2.4 million hectares. In 1965, the figure for grain was 3.2 million hectares. The dissident scholars may have meant that the *proportion* of cropland had fallen to the 1940 level. Khrushchev had expanded the total area planted, meaning that 3.2 million amounted to 1.4 percent of the total, a percentage equal to that of 1940. After 1965, however, annual totals rose anew, reaching 4.2 million by 1970.

The dissidents also misinterpreted plantings for silage, a much larger matter. Far from falling "at a double rate," as they suggested, these amounted to some 20.2 million hectares in 1965, a decline from the peak reached in 1962, but otherwise only 2.9 million hectares less than the 1960 figure and far larger than the vanishingly small pre-1953 total. By the end of the decade, silage plantings had decreased, but only to a still substantial 18 million hectares.[13] These figures for grain and silage remained typical of the fluctuating totals for each of the remaining years of the Soviet Union's existence. In 1987, the country's farms planted 4.5 million hectares of grain and 18 million of silage and fodder.[14] Today, farmers in Russia, Ukraine, and most other post-Soviet states grow corn. Reports of the crop's demise were greatly exaggerated.

By October 1964, foreign scholars had grown certain in their conception of an agrarian system allowing Khrushchev to compel every farm to plant corn, one largely immune to economic calculation and fundamental change. During the Khrushchev era, Sovietologists investigating economics gradually grew skeptical of efforts to measure indicators with any degree of certainty. Counterintuitively, they lent increasing weight to politics. They began to echo positions common among those who, subscribing to interpretations of the Soviet Union as totalitarian, accepted the party's assertions of monolithic structure and other fundamentals, but transformed them from a positive into a negative.[15] Scholars asserted that Khrushchev exclusively employed mobilizational tactics, ignoring economic calculations altogether.[16] Sovietologists additionally overestimated Khrushchev's authority to make formal policies a reality. Arguing a range of positions on the depth of Khrushchev's reforms, scholars largely agreed that he had futilely attempted to maintain unsustainable principles.[17] Other Sovietologists discounted the possibility that Khrushchev had achieved significant changes in the agrarian economy, disregarding the programs to abolish the machine-tractor stations, transform the kolkhozes, and spread industrial farming methods. Khrushchev had not even attempted necessary basic structural changes, they charged, because of the straitjacket imposed by ideology.[18]

In reality, Khrushchev had altered the principles of the agrarian economy inherited from Stalin by employing differential prices, wage labor, and financial mechanisms in an attempt to inspire cost-effective production. Influenced by events that overshadowed these innovations and limited their effectiveness, scholars lost sight of them. Subsequently, few sustained interest in the bygone Khrushchev era after 1964.[19] Foreign and dissident scholars tended to balance the failings of Khrushchev's final years with a positive appraisal of some of his efforts.[20] Into the 1980s, official histories emphasized continuity between prewar and postwar developments, leaving Khrushchev out entirely. They lionized popular exertions after 1945 to realize goals established by the party in the 1930s, as if the postwar era preserved sovkhozes, kolkhozes, central planning, and state procurement as the unchanging bedrock of a fixed agrarian system.[21]

Since 1985, historians have pored over Khrushchev-era culture and society, but have left largely unexamined economic activity, on which the Communist Party and government spent the majority of their effort. Scholars have especially neglected the agrarian sector tasked with feeding the population a rich and varied diet.[22] After decades of official silence, Mikhail Gorbachev's *glasnost* reforms permitted renewed debate about Khrushchev. His efforts became a proxy for *perestroika*, the contemporary economic reforms.[23] Amid the new possibilities of the 1990s, only a few Russian historians began to inquire into the postwar period and, of those, an even smaller number into agrarian policy.[24] The perception that Gorbachev had hastened the Soviet Union's demise by initiating fundamental reforms prompted pessimism about the prospects for genuine reform in any preceding transformative era.[25]

If not Khrushchev alone, who or what deserves the blame for the dashing of the hopes he invested in industrially farmed corn? He certainly deserves some of it, but other components of the Soviet Union's constantly evolving economic and political formations also contributed. The apparently overwhelming authority he possessed made others powerless to openly resist corn. Ample evidence demonstrates that many local officials had to comply when he ordered more corn plantings, even though this authority was far from complete. Yet Khrushchev also ordered local officials to allow kolkhozes certain freedom in planning, to calculate expenditures on commodity production, and to press for corn only where it would work. Subordinates flouted these orders, harming Khrushchev's corn crusade and broader agrarian reforms. Scholars have somehow considered these failings also his fault. Far from sitting atop a "single bureaucratic pyramid," as one Sovietologist concluded in 1967, Khrushchev wrestled with bureaucracies unable or unwilling to effectively govern agrarian spaces, rural people, and agricultural enterprises. Together, Khrushchev and the Soviet Union's economic and political arrangements were simply incapable of overcoming the effects of climate, poverty, wartime destruction, and Stalin's rapacious extraction of resources to finance industrialization.

Yet Khrushchev also deserves credit for significant achievements. A relatively vibrant society in the 1950s and 1960s, the Soviet Union began to improve the living standards not only of urban consumers, but also of rural ones, all of whom had been subjected by Stalin to a spartan regime that suppressed consumption to fund industrialization and military buildup. The country could achieve Khrushchev's goals only by raising efficiency. Expecting a revolution in agricultural production, he enacted labor reforms that offered kolkhozniks a guaranteed wage. Thus advancing their integration into society as producers and consumers, the policies drew a contrast to the decades in which kolkhozniks' status as second-class citizens had been obvious to all. This decade saw perhaps the largest and most sustained improvement in peasants' position of any period between 1917 and 1991. Nonetheless, in 1964 kolkhozniks on average still had poorer access to manufactured goods and store-bought food than unskilled urban workers. However, Khrushchev had raised farmers' expectations. Urban dwellers had come to expect more meat, milk, and eggs amid a growing variety of goods accessible in state shops. Rural counterparts similarly expected their lot to improve, even though it rose from a lower baseline.

From 1953 to 1964 and beyond, the Soviet Union was experiencing the resurgence of the historical trend of rural modernization, a process that in many countries saw not only the application of machines and chemicals to agriculture, but also the integration of producers into the country's economic formation.[26] In return for producing more, farmers in many countries around the world earned money permitting them to become consumers. This strengthened links to the urbanizing national polity and granted peasants a status approximating the citizenship of their town-dwelling compatriots. Earlier analyses have considered the postwar Soviet Union in isolation from contemporary societies, missing such parallels in the country's experiences of rapid urbanization, rural modernization, and technological change.

Khrushchev's reform program created conditions facilitating these processes of rural modernization, which had a history dating to before 1917 but had long been stemmed and reshaped by Stalin's exploitative policies. Khrushchev expanded into wider areas of production a commitment to industrial farming established in the 1920s, but hitherto limited in practical application. Designed to make corn a success, his efforts are comprehensible as ambitious adaptations of this industrial ideal, which he designed to respond to the entrenched problems that hindered his initiatives. Khrushchev's gamble on corn was more calculated than critics have imagined, forming part of a history of asymmetrical transnational interactions in the sphere of farming technology. As a result of Khrushchev's efforts, the agrarian arrangements the leader left to successors bore at most a family resemblance to that of 1953. Farms had grown much larger, the machine-tractor stations had disappeared, wage labor paid in cash had become common, bureaucrats' power had waned (albeit in principle more than in practice), financial calculations had entered into everyday affairs, and modern technology had spread into a wider range of production activities.

The Soviet Union's experiences of rural modernization and the advent of industrial agriculture call into question the perceived separation of the country from the rest of the industrialized world. That separation has been integral to its portrayal as an experiment whose leaders strived to create a separate modernity, antithetical to the one defined by liberal capitalism. Although embedded in the larger world, the Soviet Union's leaders did employ distinctive mechanisms. Collectivization was the most chaotic, brutal form of the frequently violent processes of rural modernization that, in other contexts, culminated in peasant insurrections and counterinsurgency campaigns to repress them. The movement into and out of the Soviet Union of technologies, ideas, and practices related to industrial agriculture from the 1920s to the 1960s and beyond, however, suggests deeper commonalities unseen because of Cold War–inspired discourses about an impenetrable Iron Curtain. These parallel processes resulted in part from a shared set of high-modernist, Promethean conceptions of human societies and the natural world.

The postwar decades saw a fundamental transformation in the way farmers across the world worked and grew food, and therefore how others consumed it. In industrial societies especially, agriculture ceased to be powered by the solar energy accumulated in a given year. Instead, synthetic fertilizers, agricultural chemicals, fuel for tractors, and other innovations permitted farmers to apply to fields not only energy harvested from the sun that year, but an extra quantity in the form of hydrocarbons, a supply of energy stored tens of millions of years ago. This has fundamentally altered societies' relationship to food and farming.[27]

The Soviet Union produced these petrochemicals but never achieved their comparatively efficient transformation into food, which marked rival industrial societies of Western Europe, North America, and beyond. There, these technologies produced short-term benefits in the form of cheap food, but they have begun to summon increasing skepticism as their social, health, ecological, economic, and cultural consequences have become clear. This is not a call to rehabilitate Khrushchev's vision of industrial agriculture based on corn. Even if his policies had increased farm output in the short term and matched the results achieved in the United States, the pressing economic and ecological consequences of industrial farming highlighted by critics over recent decades would also have given rise to complications in the Soviet Union. Americans' apparent postwar successes only postponed the need to address the repercussions of industrial farming.

Discarding the idea that the Soviet Union remained a polar opposite of its opponents, a reconsideration of the Cold War divide becomes possible. The Cold War was a geopolitical conflict and a contest between totalizing ideologies, a division that encouraged a binary worldview shared by both sides. Between 1950 and 1970, the Soviet Union measured itself against contemporary capitalist societies experiencing uncharacteristic stability and rising opportunities for working people. The Soviet Union's propagandistic claims about class-based oppression rang hollow during decades of improving income, wealth, and diet for professional and working classes of Europe, North America, and beyond. Despite this, the postwar era did not ring in a new phase of capitalist stability. As economist Thomas Piketty has shown, those years were exceptional because the world was recovering from the destruction wrought between 1914 and 1945, and because governments consequently were pursuing the political project that constructed welfare states.[28]

Looking back through an era of resurgent instability that began in the 1970s, we see that farmers everywhere—including in the United States—experienced that harsh reality earlier than most, much as they had the precursors to the Great

Depression before 1929. The apparent productivity and prosperity of capitalist farming that stood in contrast to the struggles of farms in the Soviet Union was never guaranteed, universal, or stable. Already in the 1990s, historian Eric Hobsbawm reflected on those years as a "golden age." "We have all been marked," he wrote, by the Russian Revolution, "inasmuch as we got used to thinking of the modern industrial economy in terms of binary opposites, 'capitalism' and 'socialism' as alternatives mutually excluding each other." "This was an arbitrary and to some extent artificial construction," he reiterated, "which can only be understood as part of a particular historical context."[29]

Khrushchev's agrarian initiatives illustrate one form of the degree of commonality among industrial societies. His corn crusade was not harebrained, self-evidently unfit for the Soviet Union, or subject to any single countervailing force. Khrushchev's impetuous nature, practices inherited from Stalin, local officials, farm managers, exploitative labor relations, the farmworkers themselves, and—although scholars have overemphasized them—technological and climatic constraints all combined to dash Khrushchev's sky-high expectations. Dissatisfying outcomes notwithstanding, his crusade coincided with larger fundamental changes in basic practices of production, wages, labor, and economic calculation to the farms, processes that continued to define subsequent agrarian development. These two facts together cemented Khrushchev's ambiguous legacy and ensured that food and agriculture remained weak links in the chain of the economic life.

Despite the efforts of Khrushchev's successors, falling returns on investment ensured that Mikhail Gorbachev confronted similar weaknesses twenty years later. The later reformer first had entered politics as a Komsomol official in Stavropol Krai during the corn crusade. Hailing from that agricultural region, he was, when summoned to Moscow at the end of the 1970s, viewed as a good candidate for holding the portfolio of Central Committee secretary overseeing agriculture. Launching a first round of cautious reform in 1985, he introduced the need to activate "the human factor" even before launching "acceleration" and "perestroika," his more famous later efforts to modernize productive capacity and restructure the economy. He announced aims to bind social and economic relations more closely to the needs of the people, to reduce command and control in favor of new incentives that would reinvigorate all elements of society.[30]

Although these first reforms proved insufficient to solving the problems at hand, they reflected the formative period in the lives of Gorbachev and his advisers, members of the cohort he named the Children of Twentieth Party Congress. The concept of human capital had been present in the writings of Soviet social scientists of the 1960s. The lessons of the Khrushchev era taught that the Soviet Union and its leaders needed to value the people and their abilities as a national resource.[31] Nowhere was the challenge more profound or evident than in rural areas. Indeed, agrarian relations had given the impetus to developing the Novosibirsk Report, an effort by a group of sociologists led by Tatiana Zaslavskaia to diagnose the deep social malaise of the early 1980s. These inherited challenges ensured that Gorbachev was unable to simply tinker with a stable system. He had to seek to transform one beset by deep-seated problems. The previous era when a new leader challenged ingrained practices to overcome a rising crisis of modernization, the Khrushchev years subtly influenced this final effort to reform the flawed economy, which unleashed the ultimate end of the Soviet Union.

NOTES

PROLOGUE

1. Corn refers throughout to the crop known outside North America as maize and in binomial nomenclature as *Zea mays*.
2. M. E. Saltykov-Shchedrin, *The History of a Town*, trans. I. P. Foote (Oxford, UK: W. A. Meeuws, 1980), 83–86.
3. Ibid., 100. For more on satire in Saltykov-Shchedrin's writings, see Emil Draitser, *Techniques of Satire: The Case of Saltykov-Ščedrin* (New York: Mouton de Gruyter, 1994).
4. Irina Mak, "Professor Sergei Khrushchev: Esli by ottsa ne sniali, v kontse 1960-kh v SSSR byla by rynochnaia ekonomika," *Izvestiia*, April 9, 2010, 17. This and all subsequent translations are mine, unless otherwise noted.

INTRODUCTION

1. Stephen Rosenfeld, "Corn, the Crop Khrushchev Pushed, Appears to Be Sharing His Disgrace. Butt of Soviet Jokes," *Washington Post*, December 14, 1964.
2. Dora Shturman and Sergei Tiktin, *Sovetskii soiuz v zerkale politicheskogo anekdota* (London: Overseas Publications Interchange, 1985), 200.
3. N. S. Khrushchev, *Stroitel'stvo kommunizma v SSSR i sel'skoe khoziaistvo*, 8 vols. (Moscow: Gosudarstvennoe izdatel'stvo politicheskoi literatury, 1962–1964).
4. Naum Jasny, *Khrushchev's Crop Policy* (Glasgow, UK: Outram, 1965), 142.
5. RGANI, f. 2, op. 1, d. 780, l. 105.
6. I. V. Rusinov, "Agrarnaia politika KPSS v 50-e–pervoi polovine 60-kh gg.: Opyt i uroki," *Voprosy istorii KPSS* no. 9 (1988): 41–43.
7. Jasny, *Khrushchev's Crop Policy*, 21.
8. Elena Zubkova, "The Rivalry with Malenkov," in *Nikita Khrushchev*, ed. Sergei Khrushchev, William Taubman, and Abbott Gleason (New Haven, CT: Yale

University Press, 2000), 83–84; I. E. Zelenin, *Agrarnaia politika N. S. Khrushcheva i sel'skoe khoziaistvo* (Moscow: Institut istorii Rossiiskoi Akademii Nauk, 2001), 275; and Roy Medvedev and Zhores Medvedev, *Khrushchev: The Years in Power*, trans. Andrew Durkin (New York: Columbia University Press, 1975), 65.

9. William Taubman, *Khrushchev: The Man and His Era* (New York: W. W. Norton, 2003), 373.

10. Jerzy Karcz, "Khrushchev's Impact on Soviet Agriculture," *Agricultural History* 40, no. 1 (1966): 26.

11. Taubman, *Khrushchev*, xi.

12. On trade relations and the former colonies, as well as economic relations with the rest of the world, see Oscar Sanchez-Sibony, *Red Globalization: The Political Economy of the Soviet Cold War from Stalin to Khrushchev* (New York: Cambridge University Press, 2014). On Stalin's legacy of repression, see Miriam Dobson, *Khrushchev's Cold Summer: Gulag Returnees, Crime, and the Fate of Reform after Stalin* (Ithaca, NY: Cornell University Press, 2009). On the housing program, see Mark Smith, *Property of Communists: The Urban Housing Program from Stalin to Khrushchev* (DeKalb: Northern Illinois University Press, 2013); and Steven Harris, *Communism on Tomorrow Street: Mass Housing and Everyday Life after Stalin* (Baltimore: Johns Hopkins University Press, 2013). On relations with China, see Lorenz Lüthi, *The Sino-Soviet Split: Cold War in the Communist World* (Princeton, NJ: Princeton University Press, 2008).

13. For the few recent works in English on agriculture, see Jenny Leigh Smith, *Works in Progress: Plans and Realities on Soviet Farms, 1930–1963* (New Haven, CT: Yale University Press, 2014); Maya Haber, "The Soviet Ethnographer as Social Engineer: Socialist Realism and the Study of Rural Life, 1945–1958," *Soviet and Post-Soviet Review* 41, no. 2 (2014): 193–219; Haber, "Socialist Realist Science: Constructing Knowledge about Rural Life in the Soviet Union, 1943–1958" (PhD diss., University of California at Los Angeles, 2013); and Auri Berg, "Reform in the Time of Stalin: Khrushchev and the Fate of the Russian Peasantry" (PhD diss., University of Toronto, 2012).

14. James Scott, *Seeing Like a State: How Certain Schemes to Improve the Human Condition Have Failed* (New Haven, CT: Yale University Press, 1998), 201.

15. Rooted in Soviet studies, the idea that the country was an industrial one comparable to others began with the Harvard Interview Project. Surveying wartime expatriates remaining in Western Europe and the United States in 1949 and 1950, scholars used responses to construct a model consisting of three elements. First, the state pursued ambitions of totalitarian control over political, social, and cultural life. Second, components of society specific to the Soviet Union hindered those aspirations. Third, the Soviet Union shared features with other industrial societies, making it comparable to economically advanced countries in Western Europe, North America, and beyond. Mark Edele, "Soviet Society, Social Structure,

and Everyday Life: Major Frameworks Reconsidered," *Kritika: Explorations in Russian and Eurasian History* 8, no. 2 (2007): 353–54. See also Raymond Bauer, Alex Inkeles, and Clyde Kluckhohn, *How the Soviet System Works: Cultural, Psychological, and Social Themes* (Cambridge, MA: Harvard University Press, 1956); and Raymond Bauer and Alex Inkeles, *The Soviet Citizen: Daily Life in a Totalitarian Society* (Cambridge, MA: Harvard University Press, 1959).

16. Historian Adeeb Khalid writes that the Soviet Union's authorities, "[by] leading their populations on a forced march to progress and development," sculpted citizens and society in ways comparable to the Republic of Turkey founded by Mustafa Kemal Atatürk, "Backwardness and the Quest for Civilization: Early Soviet Central Asia in Comparative Perspective," *Slavic Review* 65, no. 2 (2006): 244. Peter Holquist argues that practices developed by belligerents during World War I to manage people, production, and information defined the nascent Bolshevik state, where they grew more radical due to the strains of the Russian Civil War. By contrast, France, Great Britain, and other societies reverted—at least temporarily—to prewar norms. *Making War, Forging Revolution: Russia's Continuum of Crisis, 1914–1921* (Cambridge, MA: Harvard University Press, 2002).

17. On this process and the crop's spread, see Arturo Warman, *Corn and Capitalism: How a Botanical Bastard Grew to Global Dominance*, trans. Nancy Westrate (Chapel Hill: University of North Carolina Press, 2003).

18. I do not mean the model, explicit in modernization theory, in which societies move from less advanced to more advanced, i.e., market-oriented and capitalist. Cold War–inspired fears about the Soviet Union's appeal pushed American social scientists to conceive of liberal societies as normal and the Soviet Union as a deviation. Michael Latham, "Modernization, International History, and the Cold War World," in *Staging Growth: Modernization, Development, and the Global Cold War*, ed. David Engerman et al. (Amherst: University of Massachusetts Press, 2003), 4–6. For a contemporary example, see Cyril Black, "The Modernization of Russian Society," in *The Transformation of Russian Society: Aspects of Social Change since 1861*, ed. Cyril Black (Cambridge, MA: Harvard University Press, 1960), 661–62.

19. On consumption, see Jenny Leigh Smith, "Empire of Ice Cream: How Life Became Sweeter in the Postwar Soviet Union," in *Food Chains: From Farmyard to Shopping Cart*, ed. Warren Belasco and Roger Horowitz (Philadelphia: University of Pennsylvania Press, 2009), 142–57.

20. On the larger point, see Eric Hobsbawm, *Age of Extremes: The Short Twentieth Century, 1914–1991* (New York: Vintage, 1996), 415. In France, the rural population fell from 30 percent to 9.2 percent, and in West Germany from 23 to 6.8. Tony Judt, *Postwar: A History of Europe since 1945* (New York: Penguin, 2005), 327. On a global scale, this process is a thread running through Jeffry Frieden's *Global Capitalism: Its Rise and Fall in the Twentieth Century* (New York: W. W. Norton, 2006).

21. Noël Kingsbury, *Hybrid: The History and Science of Plant Breeding* (Chicago: University of Chicago Press, 2009), 289.
22. Deborah Fitzgerald, *Every Farm a Factory: The Industrial Ideal in American Agriculture* (New Haven, CT: Yale University Press, 2003), 3–5.
23. Opponents denounce the ecological, social, and cultural costs of industrial agriculture that economic analysis ignores. Several of them have particularly influenced my approach. Fitzgerald's skepticism is evident in her work. Journalist Michael Pollan has brought problems of corn monoculture to popular attention. *The Omnivore's Dilemma: A Natural History of Four Meals* (New York: Penguin, 2006). Farmer, poet, essayist, and activist Wendell Berry condemns industrial farming for sundering societies, while wasting environmental and cultural patrimonies. *Citizenship Papers: Essays* (Berkeley, CA: Counterpoint, 2004); and *Bringing It to the Table: On Farming and Food* (Berkeley, CA: Counterpoint, 2009).
24. Fitzgerald, *Every Farm a Factory*, 187.
25. Anthropologist Katherine Verdery cautioned that the Cold War had been not only a geopolitical struggle, but also a way of understanding the world. *What Was Socialism and What Comes Next?* (Princeton, NJ: Princeton University Press, 1996), 4–5.
26. Sanchez-Sibony, *Red Globalization*, 1–10.
27. Michael Werner and Bénédicte Zimmermann, "Beyond Comparison: *Histoire Croisée* and the Challenge of Reflexivity," *History and Theory* 45, no. 1 (2006): 30–50. For an example of how historians can use this toolkit in a Russian context, see Stephen Bittner, "American Roots, French Varietals, Russian Science: A Transnational History of the Great Wine Blight in Late-Tsarist Bessarabia," *Past and Present* no. 227 (2015): 151–77. Michael David-Fox has considered how the Soviet Union engaged with the world even while partially isolating itself, providing an understanding of how the field of Soviet history has developed in the direction of transnational history. He argues for a conception of "entangled history" that emphasizes knowledge, mutual influences, and perceptions of the foreign. The idea of "transsystemic" history, however, implies the existence of two distinct systems which ideas and people must cross borders to bridge. "The Implications of Transnationalism," *Kritika: Explorations in Russian and Eurasian History* 12, no. 4 (2011): 887–88. See also David-Fox, "Entangled Histories in the Age of Extremes," in *Fascination and Enmity: Russia and Germany as Entangled Histories, 1914–1945*, ed. Michael David-Fox, Peter Holquist, and Alexander M. Martin (Pittsburgh: University of Pittsburgh Press, 2012), 4–7.
28. This argument that the Soviet Union was subject to developments related to global trends is distinct from the theories of convergence of the 1960s and 1970s, which challenged the presumed binary between communist and capitalist. They suggested that bureaucratically organized industrial societies, whether capitalist

or state-socialist, were gradually taking on features of the other. This trend led, advocates contended, toward a midpoint representing an optimum use of resources, which was somewhere between the polar opposites. For a critical review of convergence theories, see Michael Ellman, "Against Convergence," *Cambridge Journal of Economics* 4, no. 3 (1980): 199–210.

29. Recently, historians have begun to emphasize continuities in politics and doctrines across the 1953 divide. Thomas Bohn, Rayk Einax, and Michel Abesser, "From Stalinist Terror to Collective Constraints: 'Homo Sovieticus' and the 'Soviet People' after Stalin," in *De-Stalinization Reconsidered: Persistence and Change in the Soviet Union*, ed. Thomas Bohn, Rayk Einax, and Michel Abesser (New York: Campus, 2014), 14–15; and Denis Kozlov and Eleonory Gilburd, eds., *The Thaw: Soviet Society and Culture during the 1950s and 1960s* (Toronto: University of Toronto Press, 2013). Recent books have called attention to complexities and continuities belying a simple separation of Stalin from post-Stalin periods. Dobson, *Khrushchev's Cold Summer*; Smith, *Property of Communists*; Benjamin Tromly, *Making the Soviet Intelligentsia: Universities and Intellectual Life under Stalin and Khrushchev* (New York: Cambridge University Press, 2014); Yoram Gorlizki and Oleg Khlevniuk, *Cold Peace: Stalin and the Soviet Ruling Circle, 1945–1953* (New York: Oxford University Press, 2004); and more.

30. Juliane Fürst, Polly Jones, and Susan Morrissey, "The Relaunch of the Soviet Project, 1945–1964," *Slavonic and East European Review* 86, no. 2 (2008): 201. The scholars confirm "a growing sense that the era holds the key to both the Soviet Union's eventual demise, but also to its longevity."

31. I do not want these examples to suggest excessive optimism. They often rested—sometimes literally—on foundations laid at gigantic cost using crude methods. Housed in a pit that Gulag prisoners dug by hand, the Soviet Union's first nuclear reactor, for instance, lacked shielding to protect personnel from radiation when it was constructed in the late 1940s. Kate Brown, *Plutopia: Nuclear Families, Atomic Cities, and the Great Soviet and American Plutonium Disasters* (New York: Oxford University Press, 2013), 87–108.

32. In 1989, political scientist Don van Atta wrote that party militants dispatched by Stalin to the countryside during collectivization encountered a "state of siege" that had survived essentially unchanged into the Gorbachev era. "The USSR as a 'Weak State': Agrarian Origins of Resistance to Perestroika," *World Politics* 42, no. 1 (1989): 129–49. Even while analyzing Gorbachev's reforms to a system transformed by decades of evolution unbroken by war or violence since 1953, van Atta presumed continuities in administrative mobilization from above and peasant resistance from below. Influential historian Stephen Kotkin subsequently drew on this analysis, making it particularly significant. Even as he argued that the Soviet Union's interwar mass culture and nascent welfare state were comparable to those of

counterparts abroad, Kotkin concluded that the postwar Soviet Union succumbed to "time-bound structural weaknesses" that hampered efforts to reallocate resources toward consumer goods and food production after the war. "Modern Times: The Soviet Union and the Interwar Conjuncture," *Kritika: Explorations in Russian and Eurasian History* 2, no. 1 (2001): 162–63. For a view of the Soviet Union as essentially ideological and static, see Amir Weiner, "Robust Revolution to Retiring Revolution: The Life Cycle of the Soviet Revolution, 1945–1968," *Slavonic and East European Review* 86, no. 2 (2008): 208–31.

33. For more on this critique, see Andrew Sloin and Oscar Sanchez-Sibony, "Economy and Power in the Soviet Union, 1917–1939," *Kritika: Explorations in Russian and Eurasian History* 15, no. 1 (2014): 17–18.

34. Anna Krylova writes that historians of this period have "an amount of work in store . . . comparable to the decades of work that have been devoted" to the prewar era under Stalin. "Soviet Modernity: Stephen Kotkin and the Bolshevik Predicament," *Central European History* 23, no. 2 (2014): 172.

35. In this inquiry, the concept of "the historical study of economic life," which Andrew Sloin and Oscar Sanchez-Sibony have borrowed from historian William Sewell, proves useful. "Economy and Power in the Soviet Union," 8.

36. It therefore looks to social histories of a country in the throes of rural modernization. In 1980s, Moshe Lewin highlighted profound changes wrought by urbanization and industrialization, even as he noted the limits of these processes. *The Making of the Soviet System: Essays in the Social History of Interwar Russia* (New York: Pantheon Books, 1985). Although he unconvincingly characterized the Soviet Union under Stalin as backward and failing, particularly after the war, Lewin did concede that, after Stalin, it became comparatively energetic amid a tide of social change. Moshe Lewin, *The Soviet Century*, ed. Gregory Elliott (New York: Verso, 2005).

37. In Laura Engelstein's terms, this inquiry into their "habits of mind" puts the emphasis on culture "under the sign of practice." "Culture, Culture Everywhere: Interpretations of Modern Russia, across the 1991 Divide," *Kritika: Explorations in Russian and Eurasian History* 2, no. 2 (2001): 377–83 and 393.

38. For scholarship on these, see Sanchez-Sibony, *Red Globalization*; Kristen Roth-Ey, *Moscow Prime Time: How the Soviet Union Built the Media Empire That Lost the Cultural Cold War* (Ithaca, NY: Cornell University Press, 2011); and many of the contributions to Kozlov and Gilburd, *Thaw*.

39. Donald Raleigh, *Soviet Baby Boomers: An Oral History of Russia's Cold War Generation* (New York: Oxford University Press, 2011).

40. Smith, *Property of Communists*; Lewis Seigelbaum, *Cars for Comrades: The Life of the Soviet Automobile* (Ithaca, NY: Cornell University Press, 2008); and Stephen Lovell, *Summerfolk: A History of the Dacha, 1710–2000* (Ithaca, NY: Cornell University Press, 2003).

CHAPTER 1

1. Alexander Solzhenitsyn, *The Solzhenitsyn Reader: New and Essential Writings, 1947–2005*, ed. Edward Ericson, Jr., and Daniel Mahoney (Wilmington, DE: ISI Books, 2006), 28.
2. As with any source, these reports on popular mood, or *svodki*, reflected the preconceptions of the police officials who compiled them. On their methodological pitfalls, see Jeffrey Jones, *Everyday Life and the "Reconstruction" of Soviet Russia during and after the Great Patriotic War, 1943–1948* (Bloomington, IN: Slavica, 2008), 13.
3. RGANI, f. 5, op. 24, d. 536, ll. 2–9.
4. Moshe Lewin, *The Soviet Century*, ed. Gregory Elliott (New York: Verso, 2005), 52.
5. Carol Leonard, *Agrarian Reform in Russia: The Road from Serfdom* (New York: Cambridge University Press, 2010), 25–59. Naturally, this brief account cannot capture complex local conditions. Moreover, recent research has emphasized reasons for optimism in contrast to decades of scholarship that, echoing contemporaries, viewed the empire's agrarian problems as intractable. David Kerans, "Toward a Wider View of the Agrarian Problem in Russia, 1861–1930," *Kritika: Explorations in Russian and Eurasian History* 1, no. 4 (2000): 657–78.
6. Lynne Viola et al., eds., *The War against the Peasantry, 1927–1930: The Tragedy of the Soviet Countryside*, trans. Steven Shabad (New Haven, CT: Yale University Press, 2005), 11–12.
7. On these events' relationship to global economic circumstances, see Oscar Sanchez-Sibony, *Red Globalization: The Political Economy of the Soviet Cold War from Stalin to Khrushchev* (New York: Cambridge University Press, 2014), especially chapter 1, "Depression Stalinism," 25–56.
8. On the procurements crisis, see Viola et al., eds., *War against the Peasantry*, 17. For the intellectual origins in Lenin's writings of Stalin's class analysis of rural society, including the term *kulak*, see Orlando Figes, Introduction to *Rural Russia under the New Regime*, by V. P. Danilov, trans. and ed. Orlando Figes (Bloomington: Indiana University Press, 1988), 14–16. On the Ural-Siberian method and the multisided conflict of collectivization, see James Hughes, *Stalinism in a Russian Province: A Study of Collectivization and Dekulakization in Siberia* (New York: St. Martin's Press, 1996); as well as Lynne Viola, *The Best Sons of the Fatherland: Workers in the Vanguard of Soviet Collectivization* (New York: Oxford University Press, 1987); *Peasant Rebels under Stalin: Collectivization and the Culture of Peasant Resistance* (New York: Oxford University Press, 1996); and *The Unknown Gulag: The Lost World of Stalin's Special Settlements* (New York: Oxford University Press, 2007). On outcomes of collectivization, see R. W. Davies and Stephen Wheatcroft, *The Years of Hunger: Soviet Agriculture, 1931–1933* (New York: Palgrave Macmillan, 2004), 436.
9. I. V. Stalin, "God velikogo pereloma," *Pravda*, November 7, 1929, 2. On Stalin's vision, see Sheila Fitzpatrick, *Stalin's Peasants: Resistance and Survival in the*

Russian Village after Collectivization (New York: Oxford University Press, 1994), 39. Even moderates and noncommunist experts drawn in the 1920s to the People's Commissariat of Agriculture shared these goals. Opposed to Stalin's methods, they favored peaceful means for introducing modern farming. James Heinzen, *Inventing a Soviet Countryside: Soviet Power and the Transformation of Rural Russia, 1917–1928* (Pittsburgh: University of Pittsburgh Press, 2008), 189.

10. R. W. Davies, Mark Harrison, and Stephen Wheatcroft, eds., *The Economic Transformation of the Soviet Union, 1913–1945* (New York: Cambridge University Press, 1993), 121.
11. Viola et al., eds., *War against the Peasantry*, 325–26.
12. On plans to expand crops and raise yields, see Davies and Wheatcroft, *Years of Hunger*, 436–39. Numbers of horses, cattle, sheep, and hogs fell by between 40 and 60 percent each. Those animals, especially workhorses, given to the kolkhozes were low in weight and fitness. Davies, Harrison, and Wheatcroft, eds., *Economic Transformation of the Soviet Union*, 114–15. On tractors, see Dana Dalrymple, "American Technology and Soviet Agricultural Development," *Agricultural History* 40, no. 3 (1966): 193.
13. On peasant resistance, see Fitzpatrick, *Stalin's Peasants*. On the famine and its effects, see Davies and Wheatcroft, *Years of Hunger*. The Kazakh famine is less known in the English-language scholarship. Isabelle Ohayon, *La sédentarisation des Kazakhs dans l'URSS de Staline: Collectivisation et changement social, 1928–1945* (Paris: Maisonneuve et Larose, 2006).
14. Robert C. Allen, *Farm to Factory: A Reinterpretation of the Soviet Industrial Revolution* (Princeton, NJ: Princeton University Press, 2003), 107.
15. Davies, Harrison, and Wheatcroft, eds., *Economic Transformation of the Soviet Union*, 123.
16. Heinzen, *Inventing the Soviet Countryside*, 5.
17. On limits on the state's influence on rural mentalities, see Mark Edele, *Stalinist Society, 1928–1953* (New York: Oxford University Press, 2011), 142–43.
18. Allen, *Farm to Factory*, 101.
19. Davies, Harrison, and Wheatcroft, eds., *Economic Transformation of the Soviet Union*, 127.
20. RGAE, f. 7486, op. 7, d. 1391, ll. 23, 32–33, and 36–37.
21. This account of Khrushchev's early career relies primarily on William Taubman's biography, *Khrushchev: The Man and His Era* (New York: W. W. Norton, 2003).
22. As Iurii Shapoval writes, Sergei and Ksenia Khrushchev "struggled to make ends meet." "The Ukrainian Years, 1894–1949," in *Nikita Khrushchev*, ed. William Taubman, Sergei Khrushchev, and Abbott Gleason (New Haven, CT: Yale University Press, 2000), 9.
23. Taubman, *Khrushchev*, 37.
24. Ibid., 67.

25. Ibid., 122.
26. Alec Nove, "Soviet Peasantry in World War II," in *The Impact of World War II on the Soviet Union*, ed. Susan Linz (Totowa, NJ: Rowman and Allenheld, 1985), 78–86.
27. Iu. V. Arutiunian, *Sovetskoe krest'ianstvo v gody velikoi otechestvennoi voiny*, 2nd ed. (Moscow: Nauka, 1970), 89–90.
28. O. M. Verbitskaia, *Rossiiskoe krest'ianstvo: Ot Stalina k Khrushchevu, sredina 40-kh–nachalo 60-kh g.* (Moscow: Nauka, 1992), 42–43.
29. E. Iu. Zubkova, *Poslevoennoe sovetskoe obshchestvo: Politika i povsednevnost', 1945–1953* (Moscow: ROSSPEN, 2000), 61.
30. Verbitskaia, *Rossiiskoe krest'ianstvo*, 42–43.
31. Arutiunian, *Sovetskoe krest'ianstvo v gody velikoi*, 79.
32. Wendy Goldman, "Not by Bread Alone: Food, Workers, and the State," in *Hunger and War: Food Provisioning in the Soviet Union during World War II*, ed. Goldman and Donald Filtzer (Bloomington: Indiana University Press, 2015), 44–97.
33. "On the Material Damage Inflicted by the German-Fascist Invaders on Government Enterprises and Institutions, Kolkhozes, Social Organizations, and Citizens of the USSR" (Moscow: Gospolitizdat, 1945), 2–3 and 17, quoted in Verbitskaia, *Rossiiskoe krest'ianstvo*, 22. At the unreliable official exchange rate, this sum equaled $128.1 billion, or $1.693 trillion adjusted for inflation to 2014.
34. For a succinct evaluation of postwar life in the literature, see Juliane Fürst, "Late Stalinist Society: History, Policies, and People," in *Late Stalinist Russia: Society between Reconstruction and Reinvention*, ed. Fürst (New York: Routledge, 2006), 1–20.
35. M. L. Bogdenko et al., *Istoriia Krest'ianstva SSSR*, vol. 4, *Krest'ianstvo v gody uprocheniia i razvitiia sovetskogo obshchestva, 1945–konets 50-kh gg.* (Moscow: Nauka, 1988), 5.
36. Jones, *Everyday Life and the "Reconstruction" of Soviet Russia*, 1.
37. Elena Zubkova, *Russia after the War: Hopes, Illusions, and Disappointments, 1945–1957*, trans. Hugh Ragsdale (Armonk, NY: M. E. Sharpe, 1998), 61.
38. Zubkova, *Poslevoennoe sovetskoe obshchestvo*, 64–66.
39. Vera Dunham, *In Stalin's Time: Middleclass Values in Soviet Fiction* (New York: Cambridge University Press, 1976).
40. Anatolii Strelianyi, "The Last Romantic," in *Memoirs of Nikita Khrushchev*, vol. 2, *Reformer (1945–1964)*, ed. Sergei Khrushchev, trans. Stephen Shenfield and George Shriver (University Park: Pennsylvania State University, 2006), 83.
41. Taubman, *Khrushchev*, 211.
42. Nikolai Dronin and Edward Bellinger, *Climate Dependence and Food Problems in Russia, 1900–1990: The Interaction of Climate and Agricultural Policy and Their Effect on Food Problems* (New York: Central European University Press,

2005), 161. On the famine itself, see Nicholas Ganson, *The Soviet Famine of 1946–1947 in Global and Historical Perspective* (New York: Palgrave Macmillan, 2009).
43. RGANI, f. 89, op. 57, d. 20, ll. 9–12, quoted in Zubkova, *Poslevoennoe sovetskoe obshchestvo*, 77–78.
44. RGANI, f. 5, op. 30, d. 64, l. 2.
45. Taubman, *Khrushchev*, 199–205.
46. Donald Filtzer, *The Hazards of Urban Life in Late Stalinist Russia: Health, Hygiene, and Living Standards, 1943–1953* (New York: Cambridge University Press, 2010), 163–67.
47. RGANI, f. 5, pp. 30, d. 64, ll. 2–3.
48. Filtzer, *Hazards of Urban Life*, 240–43.
49. Dronin and Bellinger, *Climate Dependence and Food Problems*, 166.
50. *Memoirs of Nikita Khrushchev*, 2:398–99.
51. *Memoirs of Nikita Khrushchev*, vol. 3, *Statesman (1953–1964)*, ed. Sergei Khrushchev, trans. Stephen Shenfield and George Shriver (University Park: Pennsylvania State University, 2007), 141.
52. On Mikoyan, see *Memoirs of Nikita Khrushchev*, 2:573. See also "K novym uspekham sotsialisticheskogo sel'skogo khoziaistva Ukrainy: Doklad tov. Khrushcheva N. S. na soveshchanii partiinogo, sovetskogo i kolkhoznogo aktiva Kievskoi oblasti, 28 ianvaria 1941 g.," *Pravda*, February 10, 1941, 4.
53. Khrushchev spoke about the 1949 harvest on several occasions. A. A. Fursenko, ed., *Prezidium TsK KPSS, 1954–1964*, vol. 1, *Chernovye protokol'nye zapisi zasedanii: Stenogrammy* (Moscow: ROSSPEN, 2004), 466; and N. G. Tomilina et al., eds., *Nikita Sergeevich Khrushchev: Dva tsveta vremeni; Dokumenty iz lichnogo fonda N. S. Khrushcheva*, vol. 2 (Moscow: Mezhdunarodnyi fond "Demokratiia," 2009), 94.
54. L. V. Sazanova, *Istoriia rasprostraneniia kukuruzy v nashe strane* (Minsk: Urozhai, 1964); and Arturo Warman, *Corn and Capitalism: How a Botanical Bastard Grew to Global Dominance*, trans. Nancy Westrate (Chapel Hill: University of North Carolina Press, 2003), 108–09.
55. Naum Jasny, *Khrushchev's Crop Policy* (Glasgow, UK: Outram, 1965), 140.
56. Zhores Medvedev, *The Rise and Fall of T. D. Lysenko*, ed. Lucy Lawrence, trans. I. Michael Lerner (New York: Columbia University Press, 1969); and David Joravsky, *The Lysenko Affair* (Cambridge, MA: Harvard University Press, 1970). For more on Khrushchev, Lysenko, and corn, see ch. 7.
57. Taubman, *Khrushchev*, 130–31.
58. RGANI, f. 5, op. 46, d. 394, ll. 30. In the only full-scale study of the process, Auri Berg argues that this campaign demonstrated the influence of high modernist ideals, but also their coexistence with traditional state goals. "Reform in the Time

of Stalin: Khrushchev and the Fate of the Russian Peasantry" (PhD diss., University of Toronto, 2012), 5–7.
59. Taubman, *Khrushchev*, 299.
60. Verbitskaia, *Rossiiskoe krest'ianstvo*, 37–41; and Fitzpatrick, "Postwar Soviet Society," 146–47.
61. Mark Edele, "Veterans and the Village: The Impact of Red Army Demobilization on Soviet Urbanization, 1945–1955," *Russian History* 36 (2009): 171–76.
62. Verbitskaia, *Rossiiskoe krest'ianstvo*, 81–83.
63. Michael Ryan and Richard Prentice, *Social Trends in the Soviet Union from 1950* (London: Macmillan, 1987), 17.
64. As late as in 1970, the overall population consisted of 855 men for every 1,000 women. Fitzpatrick, "Postwar Soviet Society," 144; and Ryan and Prentice, *Social Trends in the Soviet Union*, 15.
65. As Zubkova writes, they had "difficulty making ends meet." *Poslevoennoe sovetskoe obshchestvo*, 67. For more on these effects, see Verbitskaia, *Rossiiskoe krest'ianstvo*, 32; and Jean Lévesque, "'Part-Time Peasants': Labour Discipline, Collective Farm Life, and the Fate of Soviet Socialized Agriculture after the Second World War, 1945–1953" (PhD diss., University of Toronto, 2003), ii–iii.
66. RGANI, f. 5, op. 24, d. 536, l. 2.
67. M. E. Beznin and T. M. Dimoni, "Krest'ianstvo i vlast'' v Rossii v kontse 1930- kh–1950-e gg.," in *Mentalitet i agrarnoe razvitie Rossiii (XIX–XX vv.): Materialy mezhdunarodnoi konferentsii, 14–15 iiunia 1994 g.*, ed. V. P. Danilov and L. V. Milov (Moscow: ROSSPEN, 1996), 163. For more on the commission, called the Council for Kolkhoz Affairs, see Lévesque, "'Part-Time Peasants.'"
68. Zubkova, *Poslevoennoe obshchestvo*, 66.
69. Lévesque, "'Part-Time Peasants,'" 43.
70. J. Arch Getty, Gábor Rittersporn, and Viktor Zemskov, "Victims of the Soviet Penal System in the Pre-War Years: A First Approach on the Basis of Archival Evidence," *American Historical Review* 98, no. 4 (1993): 1017–49.
71. Documents relating to the policy and its eventual abandonment are in RGANI, f. 3, op. 30, d. 200. See also Lévesque, "'Part-Time Peasants,'" 137–83.
72. RGANI, f. 5, op. 45, d. 111, l. 91.
73. RGANI, f. 5, op. 45, d. 111, l. 61.
74. RGANI, f. 5, op. 45, d. 111, l. 14.
75. RGANI, f. 5, op. 24, d. 589, ll. 82–121.
76. RGANI, f. 5, op. 45, d. 111, l. 16.
77. GANISK, f. 1, op. 2, d. 6537, l. 7.
78. Tomilina, at al., eds., *Dva tsveta vremeni*, 2:25.
79. N. S. Khrushchev, *Stroitel'stvo kommunizma v SSSR i sel'skoe khoziaistvo*, vol. 1 (Moscow: Gosudarstvennoe izdatel'stvo politicheskoi literatury, 1962), 11–12.

80. Richard Taylor, "Singing on the Steppes for Stalin: Ivan Pyr'ev and the Kolkhoz Musical in Soviet Cinema," *Slavic Review* 58, no. 1 (1999): 143–59. As Taylor explains, the passage in Khrushchev's February 1956 Secret Speech to the Twentieth Party Congress has been regarded as a reference to Pyrev's award-winning *Kuban Cossacks* (1949).
81. N. S. Khrushchev, *Stroitel'stvo kommunizma v SSSR i sel'skoe khoziaistvo*, vol. 2 (Moscow: Gosudarstvennoe izdatel'stvo politicheskoi literatury, 1962), 237 and 462.
82. Strelianyi, "Last Romantic," 569–70.

CHAPTER 2

1. *Memoirs of Nikita Khrushchev*, vol. 3, *Statesman (1953–1964)*, ed. Sergei Khrushchev, trans. Stephen Shenfield and George Shriver (University Park: Pennsylvania State University, 2007), 141–45.
2. The automobile, which was loaned by the local chamber of commerce, and other details appeared in the contemporary American press accounts summarized in Peter Carlson, *K Blows Top: A Cold War Comic Interlude Starring Nikita Khrushchev, America's Most Unlikely Tourist* (New York: Public Affairs, 2009), 208.
3. Michael Werner and Bénédicte Zimmermann, "Beyond Comparison: *Histoire Croisée* and the Challenge of Reflexivity," *History and Theory* 45, no. 1 (2006): 30–50.
4. Not recognizing the ideals guiding Khrushchev, agricultural historians have viewed their subject in strictly domestic terms. Russian historian Ilia Zelenin argues that policy embraced "full and complete intensification" only in 1958, evidence of the "contradictory character" of the reforms. *Agrarnaia politika N. S. Khrushcheva i sel'skoe khoziaistvo* (Moscow: Institut istorii Rossiiskoi Akademii Nauk, 2001), 16.
5. Elena Zubkova, "The Rivalry with Malenkov," in *Nikita Khrushchev*, ed. Sergei Khrushchev, William Taubman, and Abbott Gleason (New Haven, CT: Yale University Press, 2000), 76–81.
6. Edward Geist, "Cooking Bolshevik: Anastas Mikoian and the Making of the *Book about Delicious and Healthy Food*," *Russian Review* 71, no. 1 (2012): 3; and Jukka Gronow, *Caviar with Champagne: Common Luxury and the Ideals of the Good Life in Stalin's Russia* (New York: Berg, 2003).
7. N. S. Khrushchev, *Stroitel'stvo kommunizma v SSSR i razvitie sel'skogo khoziaistva*, vol. 5 (Moscow: Gosudarstvennoe izdatel'stvo politicheskoi literatury, 1963), 258.
8. N. S. Khrushchev, *Stroitel'stvo kommunizma v SSSR i razvitie sel'skogo khoziaistva*, vol. 1 (Moscow: Gosudarstvennoe izdatel'stvo politicheskoi literatury, 1962), 342.
9. RGASPI, f. 556, op. 22, d. 64, l. 32.

Notes to pages 30–32 • 247

10. *Memoirs of Nikita Khrushchev*, vol. 2, *Reformer (1945–1964)*, ed. Sergei Khrushchev, trans. Stephen Shenfield and George Shriver (University Park: Pennsylvania State University, 2006), 558.
11. György Péteri, "The Oblique Coordinate Systems of Modern Identity," in *Imagining the West in Eastern Europe and the Soviet Union*, ed. György Péteri (Pittsburgh: University of Pittsburgh Press, 2010), 8.
12. Robert Hornsby, *Protest, Reform, and Repression in Khrushchev's Soviet Union* (New York: Cambridge University Press, 2013). Historian Vladimir Kozlov termed this a "crisis of modernization." *Mass Uprisings in the USSR: Protest and Rebellion in the Post-Stalin Years*, trans. and ed. Elaine McClarnand MacKinnon (Armonk, NY: M. E. Sharpe, 2002), 11–12.
13. Samuel Baron, *Bloody Saturday in the Soviet Union: Novocherkassk, 1962* (Stanford, CA: Stanford University Press, 2001).
14. Favoring ideology over economic and practical considerations, Vladislav Zubok argues that Stalin's successors analyzed the world using a "revolutionary-imperial paradigm," which melded traditional great-power concerns with idealistic internationalism. *A Failed Empire: The Soviet Union in the Cold War from Stalin to Gorbachev* (Chapel Hill: University of North Carolina Press, 2007), 104.
15. Khrushchev, *Stroitel'stvo kommunizma v SSSR*, 1:331.
16. Many have argued that visions of material abundance reaching citizens from the West helped delegitimize and bring down the Soviet Union. Zubok, *Failed Empire*, 176; and Yale Richmond, *Cultural Exchange and the Cold War: Raising the Iron Curtain* (University Park: Pennsylvania State University Press, 2003), xiii. Susan Reid counters that although citizens became aware of differences in terms of consumption, they long judged the contest in terms of official ideals and promises of future abundance. Differences in consumption did not preordain the Cold War's result or the dissolution of the Soviet Union. "Who Will Beat Whom? Soviet Popular Reception of the American National Exhibition in Moscow, 1959," *Kritika: Explorations in Russian and Eurasian History* 9, no. 4 (2008): 855–904.
17. Michael David-Fox, "The Iron Curtain as Semipermeable Membrane: Origins and Demise of the Stalinist Superiority Complex," in *Cold War Crossings: International Travel and Exchange across the Soviet Bloc, 1940s–1960s*, ed. Patryk Babiracki and Kenyon Zimmer (College Station: Texas A&M University Press, 2014), 20–23.
18. Nikolai Krementsov, *The Cure: A Story of Cancer and Politics from the Annals of the Cold War* (Chicago: University of Chicago Press, 2002).
19. György Péteri draws attention to how societies articulate identities in spatial terms: not only "East" and "West," but "ahead" and "behind," "within" and "without." These divisions permitted party propaganda to concentrate on social ills in the United States. "Oblique Coordinate Systems of Modern Identity," 2. Anthropologist Alexi Yurchak has explored how codes developed under Stalin later switched to endow the Western "other" with positive associations and the

Soviet "ours" with negative ones. *Everything Was Forever until It Was No More: The Last Soviet Generation* (Princeton, NJ: Princeton University Press, 2006).

20. György Péteri, "Nylon Curtain: Transnational and Transsystemic Tendencies in the Cultural Life of State-Socialist Russia and East-Central Europe," *Slavonica* 10, no. 2 (2003): 113–23. Michael David-Fox has suggested that people experienced these barriers as a "semipermeable membrane." David-Fox, "Iron Curtain as Semipermeable Membrane," 18.

21. On Hemingway and Brynner, see Petr Vail and Alexander Genis, *Shestidesiatye: Mir sovetskogo cheloveka* (Ann Arbor, MI: Ardis, 1986), 53. On youth culture, see Donald Raleigh, *Soviet Baby Boomers: An Oral History of Russia's Cold War Generation* (New York: Oxford University Press, 2011), especially ch. 3, "Unconscious Agents of Change," 120–67. On specialists, see Eleonory Gilburd, "The Revival of Soviet Internationalism in the Mid to Late 1950s," 362–401; Larissa Zakharova, "Soviet Fashion in the 1950s–1960s: Regimentation, Western Influences, and Consumption Strategies," 402–435; and Oksana Bulgakowa, "Cine-Weathers: Soviet Thaw Cinema in the International Context," in *The Thaw: Soviet Society and Culture during the 1950s and 1960s*, ed. Denis Kozlov and Eleonory Gilburd (Toronto: University of Toronto Press, 2013), 436–81.

22. N. S. Khrushchev, *Stroitel'stvo kommunizma v SSSR i sel'skoe khoziaistvo*, vol. 2 (Moscow: Gosudarstvennoe izdatel'stvo politicheskoi literatury, 1962), 27.

23. Authorities noted the surpluses and the US government's responses. RGAE, f. 7486, op. 22, d. 88, l. 59. See also Arturo Warman, *Corn and Capitalism: How a Botanical Bastard Grew to Global Dominance*, trans. Nancy Westrate (Chapel Hill: University of North Carolina Press, 2003), 188–89.

24. David Kerans, "Toward a Wider View of the Agrarian Problem in Russia, 1861–1930," *Kritika: Explorations in Russian and Eurasian History* 1, no. 4 (2000): 657–78. On Americans' travels, see Thomas Isern, "Wheat Explorer the World Over: Mark Carleton of Kansas," *Kansas History* 23, no. 1–2 (2000): 12–25. On Russian subjects' interest in America, see Robert V. Allen, *Russia Looks at America: The View to 1917* (Washington, DC: Library of Congress, 1988).

25. Fred Carstensen, *American Enterprise in Foreign Markets: Studies of Singer and International Harvester in Imperial Russia* (Chapel Hill: University of North Carolina Press, 1984).

26. *Letters from an American Farmer: The Eastern European and Russian Correspondence of Roswell Garst*, ed. Richard Lowitt and Harold Lee (DeKalb: Northern Illinois University Press, 1987), 105; and Louis Guy Michael, *More Corn for Bessarabia: Russian Experience, 1910–1917* (East Lansing: Michigan State University Press, 1983), 1. Attracting some attention from the Free Economic Society, the crop played a small role in nineteenth-century agricultural improvement in the empire's south. David Moon, *The Plough That Broke the Steppes: Agriculture and*

Environment on Russia's Grasslands, 1700–1914 (Oxford, UK: Oxford University Press, 2013), 18, 254, and 256.

27. Richard Stites, *Revolutionary Dreams: Utopian Vision and Experimental Life in the Russian Revolution* (New York: Oxford University Press, 1989), 146–49.

28. Kendall Bailes, "The American Connection: Ideology and the Transfer of American Technology to the Soviet Union, 1917–1941," *Comparative Studies in Society and History* 23, no. 3 (1981): 427; Paul Josephson, *Would Trotsky Wear a Bluetooth? Technological Utopianism under Socialism, 1917–1989* (Baltimore: Johns Hopkins University Press, 2010), 19–63; and Alan Ball, *Imagining America: Influence and Images in Twentieth-Century Russia* (Lanham, MD: Rowman & Littlefield, 2003), 145–76. On the critics of Americanism, see Loren Graham, *The Ghost of the Executed Engineer: Technology and the Fall of the Soviet Union* (Cambridge, MA: Harvard University Press, 1993), 39.

29. Josephson, *Would Trotsky Wear a Bluetooth?*, 26. On Americans' contributions to Magnitogorsk, see Stephen Kotkin, *Magnetic Mountain: Stalinism as a Civilization* (Berkeley: University of California Press, 1995), 37–71.

30. Dana Dalrymple, "The American Tractor Comes to Soviet Agriculture: The Transfer of Technology," *Technology and Culture* 5, no. 2 (1964): 193.

31. Stites, *Revolutionary Dreams*, 146–49; and Ball, *Imagining America*, 124–25.

32. David Joravsky, "The Vavilov Brothers," *Slavic Review* 24, no. 3 (1965): 383–85; and R. W. Davies, Mark Harrison, and S. G. Wheatcroft, eds., *The Economic Transformation of the Soviet Union, 1913–1945* (New York: Cambridge University Press, 1994), 124.

33. Jenny Leigh Smith, *Works in Progress: Plans and Realities on Soviet Farms, 1930–1963* (New Haven, CT: Yale University Press, 2014), 21–62.

34. Deborah Fitzgerald, *Every Farm a Factory: The Industrial Ideal in American Agriculture* (New Haven, CT: Yale University Press, 2003), 187 and 157.

35. Douglas Weiner, *Models of Nature: Ecology, Conservation, and Cultural Revolution in Soviet Russia* (Bloomington: Indiana University Press, 1998), 2 and 169.

36. William Cronon, *Nature's Metropolis: Chicago and the Great West* (New York: W. W. Norton, 1991).

37. Paul Josephson et al., *An Environmental History of Russia* (New York: Cambridge University Press, 2013), 119.

38. Christopher Ward, *Brezhnev's Folly: The Building of BAM and Late Soviet Socialism* (Pittsburgh: University of Pittsburgh Press, 2009), 4.

39. Michaela Pohl, "From White Grave to Tselinograd to Astana: The Virgin Lands Opening, Khrushchev's Forgotten First Reform," in *Thaw*, ed. Kozlov and Gilburd, 269–307.

40. John Perkins, *Geopolitics and the Green Revolution: Wheat, Genes, and the Cold War* (New York: Oxford University Press, 1997), 5.

41. Historian Edward Melillo uses these terms to describe the development of industrial farming, a historical instance of changes in what Karl Marx termed humans' metabolic relationship to the natural world. "The First Green Revolution: Debt Peonage and the Making of the Nitrogen Fertilizer Trade, 1840–1930," *American Historical Review* 117, no. 4 (2012): 1035.
42. Thomas Hager, *The Alchemy of Air: A Jewish Genius, a Doomed Tycoon, and the Scientific Discovery That Fed the World but Fueled the Rise of Hitler* (New York: Harmony Books, 2008).
43. Khrushchev, *Stroitel'stvo kommunizma v SSSR*, 1:220.
44. N. S. Khrushchev, *Stroitel'stvo kommunizma v SSSR i sel'skoe khoziaistvo*, vol. 6 (Moscow: Gosudarstvennoe izdatel'stvo politicheskoi literatury, 1963), 60–61.
45. RGANI, f. 5, op. 45, d. 81, ll. 1–3 and 6–9.
46. Warman, *Corn and Capitalism*, 186–91.
47. Tony Judt, *Postwar: A History of Europe since 1945* (New York: Penguin, 2005), 305; and Jeremy Anderson, "The Soviet Corn Program: A Study in Crop Geography" (PhD diss., University of Washington, Seattle, 1964), 26–28.
48. Chaia Heller, *Food, Farms, and Solidarity: French Farmers Challenge Industrial Agriculture and Genetically Modified Crops* (Durham, NC: Duke University Press, 2013), 42–43.
49. Deborah Fitzgerald, *The Business of Breeding: Hybrid Corn in Illinois, 1890–1940* (Ithaca, NY: Cornell University Press, 1990).
50. Warman, *Corn and Capitalism*, 26.
51. Yixin Chen, "Cold War Competition and Food Production in China, 1957–1962," *Agricultural History* 83, no. 1 (2009): 51–78.
52. Lorenz Lüthi, *The Sino-Soviet Split: Cold War in the Communist World* (Princeton, NJ: Princeton University Press, 2008).
53. "Why Khrushchev Fell," *Hongqi* (November 21, 1964), in *The Polemic on the General Line of the International Communist Movement* (Beijing: Foreign Languages Press, 1965), 481–92, https://www.marxists.org/history/international/comintern/sino-soviet-split/cpc/nkfall.htm.
54. RGANI, f. 52, op. 1, d. 247, ll. 71–119
55. On the concept of "Bolshevik tempo," see Jutta Scherrer, "'To Catch Up and Overtake' the West: Soviet Discourse on Socialist Competition," in *Competition in Socialist Society*, ed. Katalin Miklóssy and Melanie Ilic (New York: Routledge, 2014), 11–12.
56. "Po povodu vystupleniia amerikanskoi gazety 'De-Moin Redzhister,'" *Pravda*, March 2, 1955, 3.
57. RGANI, f. 5, op. 45, d. 76, ll. 67–68 and 71–75.
58. RGAE, f. 7486, op. 22, d. 86, l. 1.
59. William Taubman, *Khrushchev: The Man and His Era* (New York: W. W. Norton, 2003), 372.

60. Harold Lee, *Roswell Garst: A Biography* (Ames: Iowa State University Press, 1984), 189.
61. Walter Hixson, *Parting the Curtain: Propaganda, Culture, and the Cold War, 1945–1961* (New York: St. Martin's Press, 1997).
62. RGAE, f. 7486, op. 22, d. 88, l. 58.
63. RGAE, f. 7486, op. 22, d. 88, l. 26.
64. V. V. Matskevich, "Sovetskaia sel'skokhoziaistvennaia delegatsiia v SShA i Kanade," *Sel'skoe khoziaistvo*, January 11, 1956, 3.
65. Khrushchev, *Stroitel'stvo kommunizma v SSSR*, 1:111.
66. M. A. Kharlamov and O. Vadeev, eds., *Litsom k litsu s Amerikoi: Rasskaz o poezdke N. S. Khrushcheva v SShA, 15–27 sentiabria 1959 g.* (Moscow: Gosudarstvennoe izdatel'stvo politicheskoi literatury, 1960), 327–57.
67. A. S. Shevchenko, *Kukuruza: Dlia obmena opyta dveri shiroko otkryty* (Moscow: Izdatel'stvo sel'skokhoziaistvennoi literatury, zhurnalov, i plakatov, 1961), 346–47.
68. N. S. Khrushchev, *Stroitel'stvo kommunizma v SSSR i sel'skoe khoziaistvo*, vol. 8 (Moscow: Gosudarstvennoe izdatel'stvo politicheskoi literatury, 1964), 399.
69. RGAE, f. 7486, op. 22, d. 88, l. 116.
70. RGAE, f. 7486, op. 22, d. 88, l. 27.
71. RGAE, f. 7486, op. 22, d. 88, ll. 40–50 and RGANI, f. 5, op. 30, d. 107, l. 22.
72. RGAE, f. 7486, op. 22, d. 89, l. 154. On the climate, see Nikolai Dronin and Edward Bellinger, *Climate Dependence and Food Problems in Russia, 1900–1990: The Interaction of Climate and Agricultural Policy and Their Effect on Food Problems* (New York: Central European University Press, 2005), 7.
73. RGANI, f. 5, op. 30, d. 361, ll. 196–202; and Khrushchev, *Stroitel'stvo kommunizma v SSSR*, 6:31.
74. RGANI, f. 5, op. 30, d. 107, l. 180.
75. Khrushchev, *Stroitel'stvo kommunizma v SSSR*, 1:331.
76. RGAE, f. 7486, op 22, d. 88, l. 10.
77. Garst, *Letters from an American Farmer*, 124–25. Adjusted to inflation to 2015, this amounts to more than $8 million.
78. Roswell Garst, "Experiences in Eastern Europe," *North American Review* 249, no. 1 (1964): 28–30.
79. "Missionary of Food: Roswell Garst," *New York Times*, September 23, 1959, 28.
80. Kharlamov and Vadeev, *Litsom k litsu s Amerikoi*, 340.
81. Lee, *Roswell Garst*, 186–90.
82. RGASPI, f. 556, op. 22, d. 64, l. 32. In the area of farming technology, the Soviet Union had shed its role of "scavenger" seeking out industrial equipment as reparations after World War II. Austin Jersild, "The Soviet State as Imperial Scavenger: 'Catch Up and Surpass' in the Transnational Socialist Bloc, 1950–1960," *American Historical Review* 116, no. 1 (2011): 109–32.

83. RGANI, f. 5, op. 45, d. 131, ll. 150–152.
84. RGAE, f. 7486, op. 22, d. 129, ll. 36–32.
85. RGANI, f. 5, op. 45, d. 75, ll. 49–52; and d. 78, ll. 97–103. Numerous examples of Khrushchev's advocacy can be found in the eight-volume collection of his speeches.
86. RGANI, f. 5 op., 45, d. 130, l. 31 and GARF, f. R-5446, op. 93, d. 656, ll. 14–23; and 47–52. On the 1959 order, see RGANI, f. 5, op. 30, d. 306, ll. 122–123.
87. RGANI, f. 5, op. 45, d. 131, ll. 32–34.
88. RGANI, f. 5, op. 45, d. 131, ll. 1–2; and RGAE, f. 7486, op. 22, d. 104, ll. 32, 33, and 34.
89. RGAE, f. 7486, op. 22, dd. 157, 158, and 177. See also, RGANI, f. 5, op. 45, d. 75.
90. RGANI, f. 5, op. 45, d. 166, and d. 199; and RGANI, f. 5, op., 46, d. 119.
91. RGAE, f. 7486, op. 22, d. 78, ll. 13–21.
92. RGAE, f. 7486, op. 22, d. 78, l. 27. See also Perkins, *Geopolitics and the Green Revolution*, 187–209.
93. Some of their reports can be found in RGANI, f. 5, op. 46, d. 388.
94. RGAE, f. 7486, op. 22, d. 119, ll. 11–12 and ll. 65–67.
95. "Soviets Eye Iowa's Ears," *Life* (August 1, 1955): 30–32.
96. D. Gale Johnson, "Corn, Commissars, and Collectives: An American Expert Reports on What a US Delegation Learned about Soviet Agriculture during a Conducted Tour of Russian Farmlands," *New York Times Sunday Magazine*, September 4, 1955, 12 and 32. See also Welles Hangen, "Farming in Soviet in Race with Time: Must Keep Pace with Birth Rate, Says Reporter Who Made Agricultural Tour," *New York Times*, August 28, 1955, 1.
97. In 1955, Khrushchev boasted that Garst had praised the Soviet Union's Odessa-10 hybrid during his visit. Khrushchev, *Stroitel'stvo kommunizma v SSSR*, 2:167. Similarly, KGB chief Vladimir Semichastnyi's documents indicated a similar interest among the members of a 1963 delegation. RGANI, f. 5, op. 45, d. 336, l. 114.
98. RGANI, f. 5, op. 45, d. 336, ll. 108–110. A sense of the state of American knowledge about agriculture in the Soviet Union at the time, and Americans' desire to unlock its secrets, can be found in W. A. Douglas Jackson, *The Nature and Structure of Soviet Agriculture: A Report for the Use of Specialists in the Field of Agriculture Planning to Visit the Soviet Union* (New York: Institute of International Education, 1963).
99. N. S. Khrushchev, *Stroitel'stvo kommunizma v SSSR i sel'skoe khoziaistvo*, vol. 3 (Moscow: Gosudarstvennoe izdatel'stvo politicheskoi literatury, 1962), 439–42.
100. Khrushchev, *Stroitel'stvo kommunizma v SSSR*, 3:421; and GARF, f. A-259, op. 7, d. 8048, ll. 9–13.
101. N. S. Khrushchev, *Stroitel'stvo kommunizma v SSSR i sel'skoe khoziaistvo*, vol. 4 (Moscow: Gosudarstvennoe izdatel'stvo politicheskoi literatury, 1963), 96.
102. *Programma kommunisticheskoi partii Sovetskogo soiuza: Priniata XXII s"ezdom KPSS* (Moscow: Izdatel'stvo politicheskoi literatury, 1964), 133.
103. Khrushchev, *Stroitel'stvo kommunizma v SSSR*, 8:27.

104. Ibid., 8:399.
105. Ibid., 8:411. See also Taubman, *Khrushchev*, 607.
106. USSR Council of Ministers Central Statistical Department, *Sel'skoe khoziaistvo SSSR: Statisticheskii sbornik* (Moscow: Statistika, 1971), 359, 373, and 378.
107. Robert Deutsch, *The Food Revolution in the Soviet Union and Eastern Europe* (Boulder, CO: Westview Press, 1985), 40. Philip Hansen places these trends within the larger arc of postwar economic development and the structural problems that slowed growth in subsequent decades. *The Rise and Fall of the Soviet Economy: An Economic History of the USSR from 1945* (New York: Longman, 2006).
108. Sanchez-Sibony, *Red Globalization*, 128.
109. Noël Kingsbury, *Hybrid: The History and Science of Plant Breeding* (Chicago: University of Chicago Press, 2009), 278–79; and Perkins, *Geopolitics and the Green Revolution*, 115.
110. Kingsbury, *Hybrid*, 300
111. For a full explanation, see Nick Cullather, *The Hungry World: America's Cold War Battle against Poverty in Asia* (Cambridge, MA: Harvard University Press, 2010), especially the chapter "A Parable of Seeds," 159–79.
112. Ibid., 48–50.
113. Ibid., 64.
114. Ibid., 44–45.
115. Ibid., 267.
116. Ibid., 7–8.
117. Anderson, "Soviet Corn Program."
118. Cullather, *Hungry World*, 5.
119. Oscar Sanchez-Sibony, *Red Globalization: The Political Economy of the Soviet Cold War from Stalin to Khrushchev* (New York: Cambridge University Press, 2014), 128.
120. Ibid., 138–40. The Soviet Union made loans at 2.5 percent interest for 12 years. Its aid to all countries between 1945 and 1991 totaled $68 billion, the same amount of US aid to Israel, which, although a major recipient, was only one of many.
121. For a summary of these relations, see RGANI, f. 5, op. 30, d. 306, ll. 127–136; as well as Shri Ram Sharma, *India-USSR Relations, 1947–1971*, Part I, *From Ambivalence to Steadfastness* (New Delhi: Discover Publishing House, 1999), 26–27.
122. On planning practices, see Nirmal Kumar Chandra, "Relevance of Soviet Economic Model for Non-Socialist Countries," *Economic and Political Weekly* 39, no. 22 (2004): 2287–305. In fact, Indian development most resembled import-substitution practices developed in South America. Sanchez-Sibony, *Red Globalization*, 155–57. On the influence of India on Soviet thinking, see David Engerman, "Learning from the East: Soviet Experts and India in the Era of Competitive Coexistence," *Comparative Studies of South Asia, Africa, and the Middle East* 33, no. 2 (2013): 227–38.

123. RGANI, f. 5, op. 30, d. 272, l. 165; and f. 5, op. 45, d. 130, l. 4. Augmented by later grants of machines and of equipment and expertise to repair the machines, the total for the farm in India grew over several years to 10 million rubles of the 440 million granted in aid to all countries by 1959. RGANI, f. 5, op. 30, d. 304, l. 124.
124. RGANI, f. 5, op. 30, d. 303, ll. 12–13.
125. RGANI, f. 5, op. 45, d. 266, l. 47.
126. RGANI, f. 5, op. 45, d. 361, ll. 1–4.
127. Sanchez-Sibony, *Red Globalization*, 163–64.
128. RGANI, f. 5, op. 30, d. 302, ll. 2–9.
129. RGANI, f. 5, op. 45, d. 266, l. 18; and GARF, f. R-5446, op. 93, d. 656, l. 63.
130. RGANI, f. 5, op. 45, d. 266, ll. 23–25.
131. RGANI, f. 5, op. 45, d. 266, l. 51.
132. RGANI, f. 5, op. 45, d. 334, ll. 8–16.
133. RGAE, f. 7486, op. 22, d. 119, ll. 70–76.
134. RGANI, f. 5, op. 45, d. 236, ll. 2–31.
135. RGANI, f. 5, op. 45, d. 236, ll. 85–86.
136. Ball, *Imagining America*, 122–23; and Graham, *Ghost of the Executed Engineer*, 50–55.
137. On the TVA as a model, see James Scott, *Seeing Like a State: How Some Schemes to Improve the Human Condition Have Failed* (New Haven, CT: Yale University Press, 1998), 224. On American expertise and the Helmand Valley Authority, see Cullather, *Hungry World*, 299.
138. They reached their epitome in the program to remake the Hungry Steppe. As Christine Bischel has argued, this project exemplified the vast irrigation schemes, testifying to the scale and attempt to control nature described by Scott's concept of "high modernism." "'The Drought Does Not Cause Fear': Irrigation History in Central Asia through James C. Scott's Eyes," *Revue d'études comparatives Est-Ouest* 43, no. 1–2 (2012): 73–108.
139. Khrushchev, *Stroitel'stvo kommunizma v SSSR*, 2:38–39.
140. RGANI, f. 5, op. 45, d. 266, l. 51.
141. Khrushchev, *Stroitel'stvo kommunizma v SSSR*, 4:296.
142. RGANI, f. 5, op. 30, d. 272, l. 158.
143. RGANI, f. 5, op. 30, d. 272, l. 164.
144. RGANI, f. 5, op. 30, d. 272, ll. 166–169.
145. RGANI, f. 5, op. 30, d. 272, l. 181.
146. RGANI, f. 5, op. 46, d. 387, ll. 1–3.
147. Sanchez-Sibony, *Red Globalization*, 125; Michael Latham, "Modernization, International History, and the Cold War World," in *Staging Growth: Modernization, Development, and the Global Cold War*, ed. David Engerman et al. (Boston: University of Massachusetts Press, 2003), 3; and David

Engerman, "The Second World's Third World," *Kritika: Explorations in Russian and Eurasian History* 12, no. 1 (2011): 183–211.
148. Warman, *Corn and Capitalism*, 193–96.

CHAPTER 3

1. N. G. Tomilina et al., eds., *Nikita Sergeevich Khrushchev: Dva tsveta vremeni; Dokumenty iz lichnogo fonda N. S. Khrushcheva*, vol. 2 (Moscow: Mezhdunarodnyi fond "Demokratiia," 2009), 31 and 41.
2. Richard Lowenthal, "The Logic of One-Party Rule," 27–45; and Merle Fainsod, "What Happened to 'Collective Leadership'?," in *Russia under Khrushchev: An Anthology from* Problems of Communism, ed. Abraham Brumberg (New York: Praeger, 1962), 91–113.
3. Carl Linden, *Khrushchev and the Soviet Leadership, 1957–1964* (Baltimore: Johns Hopkins University Press, 1966). For a critique of this approach, see Jerry Hough, *How the Soviet Union Is Governed* (Cambridge, MA: Harvard University Press, 1979), 232–33.
4. Werner Hahn, *The Politics of Soviet Agriculture, 1960–1970* (Baltimore: Johns Hopkins University Press, 1972); and Sidney Ploss, *Conflict and Decision-Making in Soviet Russia: A Case Study of Agricultural Policy, 1953–1963* (Princeton, NJ: Princeton University Press, 1965).
5. George Breslauer, *Khrushchev and Brezhnev as Leaders: Building Authority in Soviet Politics* (Boston: Allen & Unwin, 1982), 3; and Stephen Cohen, "The Friends and Foes of Change: Reformism and Conservatism in the Soviet Union," *Slavic Review* 38, no. 2 (1979): 187–202.
6. Miriam Dobson has explored how their efforts to address Stalin's legacy of repression strengthened and weakened according to this process. *Khrushchev's Cold Summer: Gulag Returnees, Crime, and the Fate of Reform after Stalin* (Ithaca, NY: Cornell University Press, 2009), 157–68.
7. By contrast, historian Ilia Zelenin argues that Khrushchev did not secure full authority until 1957. Zelenin suggests that although this period saw the "totalitarian regime" weaken, Khrushchev possessed dictatorial powers granted by "authoritarian control over the party and the state." *Agrarnaia politika N. S. Khrushcheva i sel'skoe khoziaistvo* (Moscow: Institut istorii Rossiiskoi Akademii Nauk, 2001), 4 and 107.
8. Lazar Volin, "Khrushchev and the Soviet Agricultural Scene," in *Soviet and East European Agriculture*, ed. Jerzy Karcz (Berkeley: University of California Press, 1967), 1.
9. Ibid., 14.
10. Alec Nove, *An Economic History of the USSR, 1917–1991*, 3rd ed. (New York: Penguin, 1993), 376.

11. Alec Nove, "Peasants and Officials," in *Soviet and East European Agriculture*, ed. Karcz, 57.
12. Yoram Gorlizki and Oleg Khlevniuk, *Cold Peace: Stalin and the Soviet Ruling Circle, 1945–1953* (New York: Oxford University Press, 2004), 10.
13. The conflict school particularly emphasized this. Linden, *Khrushchev and the Soviet Leadership*, 29; and Martin McCauley, *Khrushchev and the Development of Soviet Agriculture: The Virgin Land Programme, 1953–1964* (New York: Holmes & Meier, 1976), 41. Iurii Aksiutin and Aleksandr Pyzhikov later linked the power struggle to a contest between competing visions of the relative importance of party and government. *Poststalinskoe obshchestvo: Problema liderstva i transformatsiia vlasti* (Moscow: Nauchnaia kniga, 1999).
14. Yoram Gorlizki, "Party Revivalism and the Death of Stalin," *Slavic Review* 54, no. 1 (1995): 1–22.
15. William Taubman, *Khrushchev: The Man and His Era* (New York: W. W. Norton, 2003), 229.
16. A. I. Mikoian, *Tak bylo: Razmyshleniia o minuvshem* (Moscow: Vagrius, 1999), 261.
17. A. A. Nikonov, *Spiral' mnogovekovoi dramy: Agrarnaia nauka i politika Rossii, XVIII–XX vv.* (Moscow: Entsiklopediia rossiiskikh dereven', 1995), 299. This was still the practice as late as July 1964. A. N. Artizov et al., eds., *Nikita Khrushchev, 1964: Stenogrammy Plenuma TsK KPSS i drugie dokumenty* (Moscow: Mezhdunarodnyi fond "Demokratiia," 2007), 50–51.
18. Anatolii Strelianyi, "The Last Romantic," in *Memoirs of Nikita Khrushchev*, vol. 2, *Reformer, 1945–1964*, ed. Sergei Khrushchev, trans. Stephen Shenfield and George Shriver (University Park: Pennsylvania State University, 2006), 569.
19. On Khrushchev's personal aides, see Hough, *How the Soviet Union Is Governed*, 419. Exercising considerable influence over access to his boss, Shevchenko later aroused resentment. Roy Medvedev and Zhores Medvedev, *Khrushchev: The Years in Power*, trans. Andrew Durkin (New York: Columbia University Press, 1975), 137.
20. Dmitrii Shepilov, *The Kremlin's Scholar: A Memoir of Soviet Politics under Stalin and Khrushchev*, ed. Stephen Bittner, trans. Anthony Austin (New Haven, CT: Yale University Press, 2007), 281.
21. Tomilina et al., eds. *Dva tsveta vremeni*, 2:31; and Nikonov, *Spiral' mnogovekovoi dramy*, 299.
22. N. S. Khrushchev, *Stroitel'stvo kommunizma v SSSR i sel'skoe khoziaistvo*, vol. 1 (Moscow: Gosudarstvennoe izdatel'stvo politicheskoi literatury, 1962), 28.
23. RGANI, f. 5, op. 45, d. 14, ll. 27–33.
24. RGANI, f. 5, op. 45, d. 49, ll. 1–8.
25. RGANI, f. 5, op. 30, d. 64, ll. 241–242.
26. Taubman, *Khrushchev*, 265.
27. Breslauer, *Khrushchev and Brezhnev as Leaders*, 24–33.
28. Khrushchev, *Stroitel'stvo kommunizma v SSSR*, 1:429–432.

29. Ibid., 1:27.
30. Ibid., 1:165.
31. Even historians have mistakenly written that he initially expected grain there. Zelenin, *Agrarnaia politika N. S. Khrushcheva*, 108.
32. Khrushchev, *Stroitel'stvo kommunizma v SSSR*, 1:432. Although moisture content, rate of spoilage, maturity of the grain, production cost, and other variables make comparisons between grain and silage difficult, this equivalency is at least sometimes legitimate.
33. RGAE, f. 7486, op. 22, d. 89, ll. 135–36.
34. Jeremy Anderson, "The Soviet Corn Program: A Study in Crop Geography" (PhD diss., University of Washington, Seattle, 1964), 17 and 25.
35. Khrushchev, *Stroitel'stvo kommunizma v SSSR*, 1:423.
36. RGANI, f. 2, op. 1, d. 126, l. 108.
37. N. S. Khrushchev, *Stroitel'stvo kommunizma v SSSR i sel'skoe khoziaistvo*, vol. 2 (Moscow: Gosudarstvennoe izdatel'stvo politicheskoi literatury, 1962), 38. In post–Soviet Russia, the region is known as the Sakha Republic.
38. For example, see I. V. Rusinov, "Agrarnaia politika KPSS v 50-e–pervoi polovine 60-kh gg.: Opyt i uroki," *Voprosy istorii KPSS* no. 9 (1988): 41.
39. "Ob uvelichenii proizvodstva produktov zhivotnovodstva: Doklad tovarishcha N. S. Khrushcheva na Plenume Tsentral'nogo Komiteta KPSS, 25 ianvaria 1955 g.," *Pravda*, February 3, 1955, 1–4. Later the butt of jokes about their length, such speeches were also carefully edited to remove sensitive content and ad-libbed remarks. Shepilov, *Kremlin's Scholar*, 286. Shevchenko exercised wide power over speeches on agriculture. Medvedev and Medvedev, *Khrushchev*, 136–37.
40. See, respectively, "Resheniia Plenuma TsK KPSS—boevaia programma deistvii," *Pravda*, February 4, 1955, 1; and "Sovetskii narod goriacho odobriaet resheniia Plenuma TsK KPSS i razvertyvaet bor'bu za ikh osushchestvlenie," *Pravda*, February 5, 1955, 1.
41. RGASPI, f. 556, op. 22, d. 64, ll. 31–32.
42. "Uvelichenie posevov kukuruzy—krupneishii rezerv uvelicheniia proizvodstva zerna," *Pravda*, March 1, 1955, 1.
43. RGASPI, f. 556, op. 22, d. 64, l. 31.
44. The rubric first appeared on February 22 and repeated February 27 and 26, as well as on March 2, 3, 4, 8, and 10. Articles on substantive technical matters then began on March 9 and continued at regular intervals until planting. See, for instance, "Massovoe obuchenie agrotekhniki i mekhanizatsii vozdelyvanii kukuruzy," *Sel'skoe khoziaistvo*, April 1, 1955, 1.
45. RGASPI, f. 556, op. 22, d. 64, l. 32.
46. Sovietologist Leopold Labedz concluded that Khrushchev "must realize the utter incongruity of reducing the problem of achieving the Realm of Freedom [communism] to a question of per-capita production of meat and milk." "Ideology: The

Fourth Stage," in *Russia under Khrushchev*, ed. Brumberg, 46–48. Ridiculing Khrushchev's convictions, Shepilov purported that the leader had asserted, "Communism is pancakes with melted butter *and* sour cream [emphasis added]." *Kremlin's Scholar*, 292.
47. Khrushchev, *Stroitel'stvo kommunizma v SSSR*, 2:126–127.
48. Ibid., 2:27
49. Nikonov, *Spiral' mnogovekovoi dramy*, 300.
50. RGANI, f. 2, op. 1, d. 126, ll. 110–12.
51. RGANI, f. 2, op. 1, d. 126, ll. 115–19.
52. TsKhDOPIM, f. 3, op. 159, d. 6, l. 8.
53. *Direktivy KPSS i Sovetskogo pravitel'stva po khoziaistvennym voprosam, 1917–1957*, vol. 4, *1953–1957 gg.* (Moscow: Gosudarstvennoe izdatel'stvo politicheskoi literatury, 1958), 365–71; and GARF, f. R-5446, op. 89, d. 1149, l. 289.
54. RGASPI, f. 556, op. 22, d. 70, l.1.
55. TsDAHOU, f. 1, op. 31, d. 411, l. 66.
56. RGASPI, f. 556, op. 22, d. 70, l. 3.
57. RGANI, f. 5, op. 45, d. 84, l. 166.
58. RGASPI, f. 556, op. 22, d. 70, l. 3.
59. RGASPI, f. 556, op. 22, d. 70, l. 22.
60. RGANI, f. 5, op. 30, d. 157, l. 17.
61. GARF, f. R-8300, op. 24, d. 809a, l. 62.
62. RGANI, f. 5, op. 45, d. 84, ll. 168–69.
63. For example, RGANI, f. 5, op. 45, d. 84, ll. 168 and 177.
64. RGASPI, f. 556, op. 22, d. 70, l. 5.
65. Khrushchev, *Stroitel'stvo kommunizma v SSSR*, 2:196.
66. RGANI, f. 5, op. 30, d. 107, ll. 159–60.
67. "Doveli kukuruzu do dela!" *Sel'skoe khoziaistvo*, September 18, 1955, 3.
68. RGANI, f. 5, op. 30, d. 107, l. 162.
69. F. Shamet'ko, "Ob ochkovtirateliakh i uborke kukuruzy," *Sel'skoe khoziaistvo*, October 13, 1955, 2.
70. Respectively, see M. Usov, "Na Stavropol'e zatiagivaiut uborku kukuruzy," *Sel'skoe khoziaistvo*, October 6, 1955, 2; on Kustanai (in Kazakh-Qostanay) Oblast, Z. Gorbunov, "Uborka kukuruzy pushchen na samotek," *Sel'skoe khoziaistvo*, October 11, 1955; on Kherson Oblast, D. Meksin, "Na kukuruznykh poliakh net poriadka," *Sel'skoe khoziaistvo*, October 14, 1955; and on Frunze Oblast (post-Soviet Bishkek), I. Dziubenko, "Narushaiut pravila silosovaniia," *Sel'skoe khoziaistvo*, October 19, 1955: 2.
71. We can infer this from the systematic notes about Presidium sessions made at Khrushchev's behest by Vladimir Malin, head of the Central Committee's General Department. A. A. Fursenko, ed., *Prezidium TsK KPSS, 1954–1964*, vol. 1, *Chernovye protokol'nye zapisi zasedanii: Stenogrammy* (Moscow: ROSSPEN, 2004).

72. Taubman, *Khrushchev*, 266–67. Khrushchev did so in private, as well, for instance at a Presidium session in 1960. RGANI, f. 52, op. 1, d. 247, l. 77.
73. Shepilov, *Kremlin's Scholar*, 297.
74. RGANI, f. 2, op. 1, d. 118, l. 17.
75. RGANI, f. 5, op. 31, d. 23, l. 22.
76. RGANI, f. 2, op. 1, d. 118, l. 17.
77. Nikonov, *Spiral' mnogovekovoi dramy*, 302.
78. Khrushchev, *Stroitel'stvo kommunizma v SSSR*, 2:29.
79. Nikonov, *Spiral' mnogovekovoi dramy*, 302.
80. This remained true in the early 1960s, as shown below, but also into the 1970s. Despite repeated directives affirming kolkhozes' freedom of action, the local authorities continued to dictate. Caroline Humphrey, *Marx Went Away—But Karl Stayed Behind*, rev. ed. (Ann Arbor: University of Michigan Press, 1998), 161.
81. GARF, f. R-8300, op. 24, d. 809a, ll. 44–46.
82. GARF, f. R-8300, op. 24, d. 809a, l. 88.
83. GARF, f. R-8300, op. 24, d. 809a, l. 96.
84. GARF, f. R-8300, op. 24, d. 809a, l. 123.
85. RGANI, f. 5, op. 31, d. 23, l. 22. See also RGANI, f. 5, op. 30, d. 106, ll. 82–97.
86. Khrushchev, *Stroitel'stvo kommunizma v SSSR*, 2:368.
87. Scholarly consensus suggests that before June 1957, Khrushchev's policies were more judicious and successful because of the constraints imposed by the rivalry. Afterward, Khrushchev's ever optimistic predictions became excessively sanguine, leaving him "defenseless against his own weaknesses and against entrenched bureaucratic resistance." Taubman, *Khrushchev*, 366; and Elena Zubkova, "The Rivalry with Malenkov," in *Nikita Khrushchev*, ed. Sergei Khrushchev, William Taubman, and Abbott Gleason (New Haven, CT: Yale University Press, 2000), 84. The policies that achieved success, however, show little sign that anyone attempted to temper Khrushchev's vision.
88. O. V. Khlevniuk, "Regional'naia vlast' v SSSR v 1954–kontse 1950-kh gg.: Ustoichivost' i konflikty," *Otechestvennaia istoriia* no. 3 (2007): 31.
89. N. Kovaleva et al., eds., *Molotov, Malenkov, Kaganovich, 1957: Stennogramma iiun'skogo plenuma TsK KPSS i drugie dokumenty* (Moscow: Mezhdunarodnyi fond "Demokratiia," 1998), 26.
90. For example, Zelenin, *Agrarnaia politika N. S. Khrushcheva*, 110.
91. Taubman, *Khrushchev*, 305–6 and 727n22.
92. Breslauer, *Khrushchev and Brezhnev as Leaders*, 12.
93. Kovaleva et al, eds., *Malenkov, Molotov, Kaganovich*, 26.
94. Khrushchev, *Stroitel'stvo kommunizma v SSSR*, 2:69.
95. Ibid., 2:101.
96. In fact, the idea had been around since at least 1951, when Stalin adamantly rejected a proposal by agrarian economists. O. M Verbitskaia, *Rossiiskoe krest'ianstvo: Ot*

Stalina k Khrushchevu, sredina 40-kh–nachalo 60-kh gg. (Moscow: Nauka, 1992), 33. For more on the economists and the reform, see Maya Haber, "Socialist Realist Science: Constructing Knowledge about Rural Life in the Soviet Union, 1943–1958" (PhD diss., University of California–Los Angeles, 2013).

97. Not given to undue praise of Khrushchev, Zelenin termed this "[one] of the most progressive, antitotalitarian of Khrushchev's socioeconomic reforms" on the basis of its aims, if not its results. *Agrarnaia politika N. S. Khrushcheva*, 276.
98. GARF, f. A-340, op. 1, d. 116, ll. 1–14.
99. GARF, f. A-340, op. 1, d. 116, ll. 10–11.
100. Naum Jasny, *Khrushchev's Crop Policy* (Glasgow, UK: Outram, 1965), 167–79, offered the more pessimistic evaluation, while Anderson's was the more favorable account. "Soviet Corn Program," 93, 101, and 106.
101. USSR Council of Ministers Central Statistical Department, *Sel'skoe khoziaistvo SSSR: Statisticheskii sbornik* (Moscow: Statistika, 1971), 178 and 220. Contemporary American grain yields, by contrast, were 3.6 tons per hectare. The figure of 1.93 was closer to the American average of the 1930s, when hybrid seeds and other technologies were only emerging. Similarly, the silage yield was lower than the American average of 20 tons per hectare. Anderson, "Soviet Corn Program," 15 and 17.
102. GARF, f. A-340, op. 1, d. 116, l. 16.
103. GARF, f. A-340, op. 1, d. 116, ll. 29–42, ll. 95–103, and ll. 104–10
104. GARF, f. A-340, op. 1, d. 116, ll. 62–63.
105. GARF, f. R-9477, op. 1, d. 107, ll. 8–14.
106. GARF, f. R-9477, op. 1, d. 107, ll. 68–78 and d. 108, ll. 1–14.
107. GARF, f. R-9477, op. 1, d. 107, l. 71.
108. N. S. Khrushchev, *Stroitel'stvo kommunizma v SSSR i sel'skoe khoziaistvo*, vol. 3 (Moscow: Gosudarstvennoe izdatel'stvo politicheskoi literatury, 1962), 383–84.
109. Ibid., 3:283–84.
110. RGANI, f. 2, op. 1, d. 341, l. 106. In a play on words, Khrushchev used a single instance of the Russian noun *vzgliad* in the quote above to signify both the act of seeing, which I have translated as "gaze," as well as a person's "point of view" on an issue.
111. Khrushchev, *Stroitel'stvo kommunizma v SSSR*, 3:385.
112. RGANI, f. 2, op. 1, d. 348, l. 30. Khrushchev returned to the theme of personnel in his closing remarks to the plenum. RGANI, f. 2, op. 1, d. 348, l. 107.
113. RGANI, f. 2, op. 1, d. 348, ll. 122–23. The ratios were in fact much closer to Canadian practice, where in 1960 nearly 42 percent of corn went for use as fodder and silage. Anderson, "Soviet Corn Program," 25.
114. Khrushchev, *Stroitel'stvo kommunizma v SSSR*, 3:472–74. This equaled $250 million in 1959 terms, or $2.047 billion when adjusted for inflation to 2015.
115. McCauley, *Khrushchev and the Development of Soviet Agriculture*, 137.

116. "Kontrol'nye tsifry plana razvitiia narodnogo khoziaistva SSSR na 1959–1965 gg.," *Pravda*, February 9, 1959, 4.
117. RGANI, f. 5, op. 45, d. 231, ll. 2–4. On the program, see Viacheslav Nekrasov, "Neftekhimicheskii proekt N. S. Khrushcheva (vtoraia polovine 1950-kh—perviaia polovina 1960-kh gg.): Strategiia modernizatsii sovetskoi ekonomiki, eksport nefti i raspredelenie resursnoi renty," *Istoriia* 6, no. 11 (2015), http://history.jes.su/s207987840001371-7-1.
118. In 1954, Stavropol Krai fulfilled only 18 percent of its plan, a measure that dipped as low as 5 percent in some districts. Of 2 million tons of organic fertilizer planned, only 372,300 tons were actually applied. The vast krai had a supply of synthetic fertilizer of only 6,000 tons. GANISK, f. 1, op. 2, d. 6539, l. 5. These problems persisted as late as 1961, meaning that one district applied only 238.5 of 1,546 tons on hand. GANISK, f. 1, op. 2, d. 8594, l. 101. A similar story applies to the krai's application of its small supply of pesticides and herbicides. GANISK, f. 1, op. 2, d. 8597, ll. 6–7.
119. RGANI, f. 2, op. 1, d. 345, l. 129.
120. RGANI, f. 5, op. 30, d. 298, ll. 91–92.
121. RGASPI, f. 556, op. 22, d. 256, ll. 4–5.
122. Anatolii Strelianyi, "Khrushchev and the Countryside," in *Khrushchev*, ed. Khrushchev, Taubman, and Gleason, 121.
123. McCauley, *Khrushchev and the Development of Soviet Agriculture*, 137.
124. RGANI, f. 5, op. 30, d. 298, l. 191.
125. RGASPI, f. 566, op. 22, d. 256, l. 22 and l. 187.
126. For an example, see Hahn, *Politics of Soviet Agriculture*, 4.
127. William Thompson, "Industrial Management and Economic Reform," in *Khrushchev*, ed. Khrushchev, Taubman, and Gleason, 154.
128. RGANI, f. 52, op. 1, d. 247, ll. 71–74.
129. Breslauer, *Khrushchev and Brezhnev as Leaders*, 120–21.
130. Medvedev and Medvedev, *Khrushchev*, 111–12.
131. N. S. Khrushchev, *Stroitel'stvo kommunizma v SSSR i sel'skoe khoziaistvo*, vol. 8 (Moscow: Gosudarstvennoe izdatel'stvo politicheskoi istorii, 1964), 478.
132. RGANI, f. 5, op. 30, d. 107, l. 199.
133. Medvedev and Medvedev, *Khrushchev*, 115.
134. Roy Laird, "The Dilemma of Soviet Agricultural Administration: The Short and Unhappy Life of the TPA," *Agricultural History* 40, no. 1 (1966): 12–13.
135. Medvedev and Medvedev, *Khrushchev*, 155.
136. GARF, f. R-5446, op. 98, d. 1045, ll. 44–53.
137. GARF, f. R-5446, op. 98, d. 1045, ll. 2–8.
138. GARF, f. R-5446, op. 98, d. 1045, l. 22.
139. GARF, f. R-5446, op. 98, d. 1045, ll. 42–43.
140. GARF, f. R-5446, op. 98, d. 1045, ll. 54–56.
141. Taubman, *Khrushchev*, 598.

142. Taubman cites the 1997 edition of the memoir by Nikolai Leonov, a translator and intelligence agent. I have used my own translation of a more detailed version of the story in the 1995 edition. N. S. Leonov, *Likholet'e* (Moscow: Mezhdunarodnoe otnosheniia, 1995), 90.
143. Anderson, "Soviet Corn Program," 116.
144. RGANI, f. 5, op. 30, d. 419, ll. 28–30.
145. Khrushchev, *Stroitel'stvo kommunizma v SSSR*, 8:261.
146. RGANI, f. 5, op. 45, d. 368, l. 101.
147. A. V. Sushkov, *Prezidium TsK KPSS v 1957–1964 gg.: Lichnost' i vlast'* (Ekaterinburg: Ural'skii tsentr akademicheskogo obsluzhivaniia, 2009), 226–28.
148. See the biographical note in *Memoirs of Nikita Khrushchev*, 2:839–40.
149. Sushkov, *Prezidium TsK KPSS*, 248.
150. P. E. Shelest, *Da ne sudimye budete: Dnevnikovye zapisi, vospominaniia chlena Politburo TsK KPSS* (Moscow: Edition Q, 1995), 173, 181–87, and 200–2.
151. RGANI, f. 2, op. 1, d. 711, l. 26.
152. Khrushchev, *Stroitel'stvo kommunizma v SSSR*, 8:105–13.
153. Artizov, ed., *Nikita Khrushchev*, 14–42 and 44–51.
154. Sovietologists typically presupposed that "serious" reform required concessions to existing but circumscribed private farming at the expense of collective and state-owned enterprise. Jasny, *Khrushchev's Crop Policy*, 21.

CHAPTER 4

1. N. S. Khrushchev, *Stroitel'stvo kommunizma v SSSR i sel'skoe khoziaistvo*, vol. 4 (Moscow: Gosudarstvennoe izdatel'stvo politicheskoi literatury, 1963), 39.
2. N. S. Khrushchev, *Stroitel'stvo kommunizma v SSSR i sel'skoe khoziaistvo*, vol. 3 (Moscow: Gosudarstvennoe izdatel'stvo politicheskoi literatury, 1962), 445.
3. Ibid., 3:498.
4. Tatiana Chumachenko, *Church and State in Soviet Russia: Russian Orthodoxy from World War II to the Khrushchev Years*, ed. and trans. Edward Roslof (Armonk, NY: M. E. Sharpe, 2002), 7. Note that Chumachenko finds that Stalin had many other justifications for reviving the religious hierarchies.
5. For a succinct summary of this literature, see Emily Baran, *Dissent on the Margins: How Soviet Jehovah's Witnesses Defied Communism and Lived to Preach about It* (New York: Oxford University Press, 2014), 4–5.
6. On the role of socialist realism in constructing the ideal of the subject operative under Stalin, see Haber, "The Soviet Ethnographer as Social Engineer: Socialist Realism and the Study of Rural Life, 1945–1958," *Soviet and Post-Soviet Review* 41, no. 2 (2014): 193–219. On the transformation under Khrushchev, see Haber, "Socialist Realist Science: Constructing Knowledge about Rural Life in the Soviet Union, 1943–1958" (PhD diss, University of California at Los Angeles, 2013), 14.

7. Thomas Wolfe, *Governing Soviet Journalism: The Press and the Socialist Person after Stalin* (Bloomington: Indiana University Press, 2005), 13–17.
8. Ibid., 2.
9. Raymond Bauer and Alex Inkeles, *The Soviet Citizen: Daily Life in a Totalitarian Society* (Cambridge, MA: Harvard University Press, 1959), especially "Keeping Up with the News," 159–88.
10. See Donald Raleigh, *Experiencing Russia's Civil War: Politics, Society, and Revolutionary Culture in Saratov, 1917–1922* (Princeton, NJ: Princeton University Press, 2002), 53. Culture, as Laura Engelstein has argued, constitutes a stable yet historical "system of values, signs, and conventions" permitting members of a society to understand both individual and collective experiences. Laura Engelstein, "Culture, Culture Everywhere: Interpretations of Modern Russia, across the 1991 Divide," *Kritika: Explorations in Russian and Eurasian History* 2, no. 2 (2001): 363.
11. "Pervy nomer zhurnala 'Kukuruzy,'" *Sel'skoe khoziaistvo*, May 15, 1956, 3.
12. I. E. Emel'ianov, *Kukuruza: Bibliograficheskii ukazatel' otechestvennoi literatury za 1794–1959 gg.* (Moscow: Izdatel'stvo Ministerstva sel'skogo khoziaistva SSSR, 1961).
13. RGANI, f. 5, op. 45, d. 368, l. 149.
14. *Letopis' zhurnal'nykh statei* (Moscow: Izdatel'stvo Vsesoiuznoi knizhnoi palaty, 1926–); and *Letopis' gazetnykh statei* (Moscow: Izdatel'stvo Vsesoiuznoi knizhnoi palaty, 1936–).
15. *Letopis' izobrazitel'nogo isskustva* (Moscow: Vsesoiuznaia knizhnaia palata, 1944–).
16. "A conversation with the head of the Department of Feed Crops of the USSR Ministry of Agriculture, B. F. Solov'ev, 'Corn in Every Region!'" February 28, 1955, 20:00–20:29, GARF, f. R-6903, op. 12, d. 296, ll. 370–71. "Corn in the Fields of Smolensk," March 12, 1955, 6:45–6:59, GARF, f. R-6903, op. 12, d. 296, ll. 427–36.
17. GANISK, f. 1, op. 2, d. 6396, ll. 107–8.
18. RGANI, f. 5, op. 16, d. 707, ll. 39–41, 53–56, 57–59, and 75–78. Part of Saratov Oblast for most of this period, from 1954 to 1957, the town of Balashov was the center of its own oblast, which was formed from districts separated from neighboring Stalingrad (today Volgograd), Voronezh, and Tambov oblasts.
19. Mary Buckley, *Mobilizing Soviet Peasants: Heroines and Heroes of Stalin's Fields* (New York: Rowman and Littlefield, 2006), 6–8. On the industrial Stakhanovites, see Lewis Siegelbaum, *Stakhanovism and the Politics of Productivity in the USSR, 1935–1941* (New York: Cambridge University Press, 1988).
20. LYA, f. 1771, op. 191, d. 423, l. 16.
21. Donald Filtzer, *Soviet Workers and De-Stalinization: The Consolidation of the Modern System of Soviet Production Relations, 1953–1964* (New York: Cambridge University Press, 1992), 128 and 137.

22. N. S. Khrushchev, *Stroitel'stvo kommunizma v SSSR i sel'skoe khoziaistvo*, vol. 1 (Moscow: Gosudarstvennoe izdatel'stvo politicheskoi literatury, 1962), 28. See also "K novym uspekham sotsialisticheskogo sel'skogo khoziaistva Ukrainy: Doklad N. S. Khrushcheva na soveshchanii partiinogo, sovetskogo, i kolkhoznogo aktiva Kievskoi oblasti, 28 ianvaria 1941 g.," *Pravda*, February 10, 1941, 4.
23. M. E. Ozërnyi, "Nasha partizanka," *Moskovskaia pravda*, February 25, 1954; "Tysicha pudov kukuruzy s gektara," *Komsomol'skaia pravda*, June 22, 1954, 2; "Moi opyt vyrashchivaniia vysokikh urozhaev kukuruzy," *Sel'skoe khoziaistvo*, April 7, 1955, 3; *Kak ia vyrashchivaiu kukuruzu* (Moscow: Ministerstvo sel'skogo khoziaistva SSSR, 1955); *Kukuruzu—vo vse raiony* (Moscow: Sel'khozizdat, 1955); and *Sovety vyrashchivanie kukuruzy: Otvety M. E. Ozërnogo na voprosy kolkhoznikov* (Moscow: Moskovskii rabochii, 1955).
24. "Uvelichenie proizvodstva zerna—reshchaiushchee uslovie pod"ema zhivotnovodstva," *Pravda*, February 5, 1955, 1; and "Rasshiriaiutsia posevy kukuruzy," *Pravda*, February 5, 1955, 1.
25. "Posevy kukuruzy—istochnik kolkhoznogo bogatstva," *Pravda*, February 6, 1955, 1.
26. Harrison Salisbury, "Farm Fair in Moscow," *New York Times*, September 5, 1954, 19.
27. RGANI, f. 5, op. 45, d. 36, l. 79 and l. 102; and GARF, f. R-5446, op. 89, d. 1167, l. 127.
28. *Direktivy KPSS i Sovetskogo pravitel'stvo po khoziaistvennym voprosam, 1917–1957*, vol. 3, *1946–1952 gg.* (Moscow: Gosudarstvennoe izdatel'stvo politicheskoi literatury, 1958), 625–31.
29. GARF, f. R-5446, op. 89, d. 1012, ll. 140–62.
30. GARF, f. R-5446, op. 89, d. 1167, l. 127.
31. RGANI, f. 5, op. 45, d. 182, ll. 27–36.
32. RGANI, f. 5, op. 45, d. 129, ll. 42–47.
33. O. Kretova, "Kukuruznyi universitet," *Literaturnaia gazeta*, September 15, 1955, 1. For some of the many other examples of corn propaganda at the Moscow exhibition, see P. Makrushenko, "Bezymianaia kukuruza," *Literaturnaia gazeta*, August 27, 1955, 2; and "Smotr dostizheniia peredovykh kukuruzovodov," *Sel'skoe khoziaistvo*, October 19, 1957, 1.
34. GARF, f. R-5446, op. 89, d. 1012, l. 124.
35. I. Sokolov, "V gorode chudes," *Sel'skaia zhizn'*, May 19, 1961, 1. See also B. Medvedev, "Otkrylsia vsesoiuznyi smotr kukuruzy na zerno," *Sel'skoe khoziaistvo*, October 11, 1959, 2.
36. Sonja Schmid, "Celebrating Tomorrow Today: The Peaceful Atom on Display in the Soviet Union," *Social Studies of Science* 36, no. 3 (2006): 331–65.
37. "Na Plenume Tsentral'nogo Komiteta KPSS: Vystuplenie tovarishcha A. V. Gitalova," *Sel'skaia zhizn'*, January 18, 1961, 2; "Na Plenume Tsentral'nogo Komiteta KPSS: Vystuplenie tovarishcha N. F. Manukovksii," *Sel'skaia zhizn'*, January 18,

1961, 3; and "Vruchenie ordena Lenina Ukrainskoi SSR: Rech' tovarishcha A. V. Gitalova," *Sel'skoe khoziaistvo*, May 12, 1959, 2.

38. "Zachinateli kompleksnoi mekhanizatsii vozdelyvanii kukuruzy vstupili v sorevnovanie," *Sel'skoe khoziaistvo*, March 22, 1959, 1. See also "V kolkhoznoi trakternoi brigade A. V. Gitalova," *Sel'skoe khoziaistvo*, June 18, 1958, 2; and D. V. Meksin, "U posledovatelei Aleksandra Gitalova," *Kukuruza* no. 9 (1963): 29–30.

39. A. Gitalov, "U nas i v Amerike," *Sel'skoe khoziaistvo*, January 3, 1959, 3.

40. *K izobiliiu* (1958). RGAKFD, ed. khr. 15682.

41. On the expansiveness of film viewing, see Kristin Roth-Ey, *Moscow Prime Time: How the Soviet Union Built the Media Empire That Lost the Cultural Cold War* (Ithaca, NY: Cornell University Press, 2011).

42. Victoria Bonnell, "The Peasant Woman in Stalinist Political Art of the 1930s," *American Historical Review* 98, no. 1 (1993): 79–80.

43. "Za vysokii urozhai kukuruzy: Rasskaz E. A. Doliniuk" (Moscow: Izdatel'stvo Ministerstva sel'skogo khoziaistva, 1955).

44. See, for instance, a sketch accompanying one profile. Ia. Makarenko, "Vsegda idti vperëd!" *Pravda*, March 29, 1961, 3.

45. V. Govorkov, poem by A. Zharov, "Plakaty rasskazyvaiut," *Sel'skoe khoziaistvo*, December 5, 1959, 3.

46. "Na Plenume Tsentral'nogo Komiteta KPSS: Vystuplenie tovarishcha E. A. Doliniuk," *Sel'skaia zhizn'*, January 14, 1961, 2.

47. RGANI, f. 2, op. 1, d. 423, l. 23.

48. V. Bol'shak, "Ocherk: Shkola Evgenii Doliniuk," *Pravda*, December 10, 1959, 2.

49. RGANI, f. 5, op. 31, d. 168, ll. 121–23.

50. RGASPI, f. M-1, op. 31, d. 598, l. 66.

51. Wolfe, *Governing Soviet Journalism*, xiv.

52. For example, see "Vot ona, 'koroleva polei'!" *Sel'skaia zhizn'*, July 14, 1960, 5; P. Savchuk, "Est' 1,000 tsentnerov kukuruzy na gektare!" *Sel'skaia zhizn'*, June 1, 1961, 2; and "Bogatyrskaia kukuruza," *Sel'skaia zhizn'*, July 14, 1961, 1.

53. V. Kliuev, "A zavtra novyi trudovoi den'," *Kukuruza* no. 1 (1963): 4–5; and Kliuev, "Shkola Liubi Li," *Kukuruza* no. 5 (1963): 7–9.

54. Kliuev, "A zavtra novyi trudovoi den'," 5.

55. Chumachenko, *Church and State in Soviet Russia*, 143–88.

56. Sally West, *I Shop in Moscow: Advertising and the Creation of Consumer Culture in Late Tsarist Russia* (DeKalb: Northern Illinois University Press, 2011), 193.

57. Iu. Feodorov, "Korolevskii vyezd," *Krokodil*, October 30, 1961, 5.

58. "Untitled," *Krokodil*, June 30, 1957, 3.

59. Iu. Grigor'ev, "Kino: 'Chudesnitsa,'" *Sel'skoe khoziaistvo*, November 14, 1957, 4.

60. N. S. Khrushchev, *Stroitel'stvo kommunizma v SSSR i sel'skoe khoziaistvo*, vol. 2 (Moscow: Gosudarstvennoe izdatel'stvo politicheskoi literatury, 1962), 27.

61. "Pomozhem vyrastit' kukuruzu: Pis'mo studentov Voronezhskogo sel'skokhoziaistvennogo instituta k studentam sel'skokhoziaistvennykh vysshikh i srednykh uchebnykh zavedenii," *Komsomol'skaia pravda*, May 5, 1955, 1.
62. Khrushchev, *Stroitel'stvo kommunizma v SSSR*, 3:184.
63. See, for example "Moguchii potok," *Pravda*, February 2, 1960, 1; "Gimn kukuruze," *Pravda*, November 14, 1960, 2; "Tak vyrashchivat' kukuruzu," *Izvestiia*, December 2, 1960, 1; and "Bogatyrskoe zerno," *Sel'skaia zhizn'*, September 26, 1962, 1.
64. On the one in Leningrad, see N. B. Lebina and A. N. Chistikov, *Obyvatel' i reformy: Kartiny povsednevnoi zhizni gorozhan v gody NEPa i khrushchevskogo desiatiletiia* (St. Petersburg: Dmitrii Bulanin, 2003), 239. On that in Moscow, see GARF, f. A-259, op. 45, d. 852, l. 115.
65. "Za steklianye dvery magazina," *Kukuruza* 9, no. 1 (1963): 52–53.
66. Susan Reid, "Cold War in the Kitchen: Gender and the Destalinization of Consumer Taste in the Soviet Union under Khrushchev," *Slavic Review* 61, no. 2 (2002): 214.
67. Edward Geist, "Cooking Bolshevik: Anastas Mikoian and the Making of the *Book about Delicious and Healthy Food*," *Russian Review* 71, no. 1 (2012): 11–14.
68. GARF, f. A-259, op. 45, d. 852, ll. 1–17 and l. 58.
69. Aaron Hale-Dorrell, "Industrial Farming, Industrial Food: Transnational Influences on Soviet Convenience Food in the Khrushchev Era," *Soviet and Post-Soviet Review* 42, no. 2 (2015): 174–96.
70. N. S. Khrushchev, *Stroitel'stvo kommunizma v SSSR i sel'skoe khoziaistvo*, vol. 7 (Moscow: Gosudarstvennoe izdatel'stvo politicheskoi literatury, 1963), 85–86.
71. Pëtr Vail' and Aleksandr Genis, *Shestidesiatye: Mir sovetskogo cheloveka* (Ann Arbor, MI: Ardis, 1986), 209.
72. Hale-Dorrell, "Industrial Farming, Industrial Food," 193. On that system, see Jenny Leigh Smith, *Works in Progress: Plans and Realities on Soviet Farms, 1930–1963* (New Haven, CT: Yale University Press, 2014), 151–87.
73. V. I. Lenin, *Polnoe sobranie sochenenie*, 55 vols. (Moscow: Gosudarstvennoe izdatel'stvo politicheskoi literatury, 1958–1965).
74. Nina Tumarkin, *Lenin Lives! The Lenin Cult in Soviet Russia* (Cambridge, MA: Harvard University Press, 1983), 255–61.
75. "Kul'tura neischerpaemykh vozmozhnostei: V. I. Lenin o dostoinstvakh kukuruzy," *Kukuruza* 7, no. 5 (1962): 4–5.
76. RGASPI, f. 556, op. 22, d. 286, l. 74.
77. Mark Dubovskii, *Istoriia SSSR v anekdotakh, 1917–1992* (Minsk: Smiadyn', 1991), 90.
78. M. E. Beznin and T. M. Dimoni, "Krest'ianstvo i vlast' v Rossii v kontse 1930-kh–1950-e gg.," in *Mentalitet i agrarnoe razvitie Rossiii (XIX–XX vv.): Materialy mezhdunarodnoi konferentsii, 14–15 iiunia 1994 g.*, ed. V. P. Danilov and L. V. Milov (Moscow: ROSSPEN, 1996), 159.

CHAPTER 5

1. RGASPI, f. M-1, op. 9, d. 483, l. 30. This chapter draws on research in the records of the Komsomol Central Committee in Moscow, especially its Department for Rural Youth.
2. Importantly, "youth" is a historical rather than a universal category. After the war, authorities used the term to define people up to the age of 27 years, including those entering early adulthood as soldiers, students, and workers. Juliane Fürst, *Stalin's Last Generation: Soviet Post-War Youth and the Emergence of Mature Socialism* (New York: Oxford University Press, 2010), 7–8.
3. Donald Raleigh, *Experiencing Russia's Civil War: Politics, Society, and Revolutionary Culture in Saratov, 1917–1922* (Princeton, NJ: Princeton University Press, 2002), 129.
4. Stephen Kotkin, *Magnetic Mountain: Stalinism as a Civilization* (Berkeley: University of California Press, 1995), 76.
5. Michaela Pohl, "Women and Girls in the Virgin Lands," in *Women in the Khrushchev Era*, ed. Melanie Ilič, Susan Reid, and Lynne Attwood (New York: Palgrave Macmillan, 2004), 52–74.
6. On the codex, see Pëtr Vail' and Aleksandr Genis, *Shestidesiatye: Mir sovetskogo cheloveka* (Ann Arbor, MI: Ardis, 1988), 7.
7. Iu. V. Aksiutin and A. V. Pyzhikov, *Poststalinskoe obshchestvo: Problema liderstva i transformatsii vlasti* (Moscow: Nauchnaia kniga, 1999), 200–45; and George Breslauer, *Khrushchev and Brezhnev as Leaders: Building Authority in Soviet Politics* (Boston: Allen & Unwin, 1982), 104–5.
8. RGASPI, f. M-1, op. 9, d. 352, ll. 1–8.
9. Allen Kassof, *The Soviet Youth Program: Regimentation and Rebellion* (Cambridge, MA: Harvard University Press, 1965), 63. For recent research on the organization's work with urban elites, see Benjamin Tromly, *Making the Soviet Intelligentsia: Universities and Intellectual Life under Stalin and Khrushchev* (New York: Cambridge University Press, 2014).
10. RGANI, f. 5, op. 31, d. 133, ll. 39–40.
11. Studying an urban cohort of the Soviet Union's baby boomers, Donald Raleigh finds that they were shaped into optimistic and enthusiastic young adults by formative experiences far less turbulent than those of their parents, who had experienced fear and tribulation. *Soviet Baby Boomers: An Oral History of Russia's Cold War Generation* (New York: Oxford University Press, 2011), 162. On the role of formative experiences in defining generations, see ibid., 381n32.
12. Donald Filtzer, *Soviet Workers and De-Stalinization: The Consolidation of the Modern System of Soviet Production Relations, 1953–1964* (New York: Cambridge University Press, 1992), 41.
13. RGASPI, f. M-1, op. 9, d. 315, ll. 7–18.
14. RGASPI, f. M-1, op. 9, d. 316, ll. 64–70.
15. RGANI, f. 5, op. 30, d. 106, ll. 82–97.

16. RGASPI, f. M-1, op. 9, d. 316, ll. 64–70.
17. RGASPI, f. M-1, op. 2, d. 343, ll. 60–68.
18. RGASPI, f. M-1, op. 3, d. 858, ll. 15–19.
19. "V pokhod za vyrashchivanie kukuruzy!" *Komsomol'skaia pravda*, February 25, 1955, 1.
20. "V pokhod za vyrashchivanie kukuruzy!" *Moskovskii komsomolets*, February 26, 1955, 1; and "Vse sily molodëzhi—na vypolnenie reshenii ianvarskogo Plenuma TsK KPSS!: Plenum Tsentral'nogo Komiteta VLKSM," *Stavropol'skaia pravda*, February 26, 1955, 1. In another instance, the newspaper of the vanguard Stalin kolkhoz of the Chuvash ASSR catalogued the expectations for Komsomol members and youth. "Boevye zadachi sel'skoi molodëzhi," *Stalinets*, March 4, 1955, 1.
21. RGASPI, f. M-1, op. 9, d. 315, l. 57.
22. RGASPI, f. M-1, op. 9, d. 315, l. 10.
23. GARF, f. R-6903, op. 12, d. 296, l. 362.
24. *XIII s"ezd Vsesoiuznogo leninskogo kommunisticheskogo Soiuza molodëzhi: Stenograficheskii otchet, 14–18 aprelia 1958 g.* (Moscow: Izdatel'stvo TsK VLKSM "Molodaia gvardiia," 1959), 269.
25. Michaela Pohl, "From White Grave to Tselinograd," in *The Thaw: Soviet Society and Culture during the 1950s and 1960s*, ed. Denis Kozlov and Eleonory Gilburd (Toronto: University of Toronto Press, 2013), 276 and 296–97.
26. RGANI, f. 2, op. 1, d. 120, l. 17.
27. RGASPI, f. M-1, op. 2, d. 388, l. 86.
28. RGASPI, f. M-1, op. 2, d. 343, l. 64.
29. TsKhDOPIM, f. 634, op. 10, d. 44, ll. 17–18.
30. TsKhDOPIM, f. 634, op. 10, d. 45, ll. 27–29.
31. LYA, f. 1771, op. 191, d. 423, ll. 2–4.
32. RGASPI, f. M-1, op. 9, d. 315, ll. 36–41; and *Komsomol'skaia pravda*, May 5, 1955, 1.
33. A. Zanina, "'Gost'ia' stanovitsia khoziaikoi," *Komsomol'skaia pravda*, May 7, 1955, 2.
34. RGASPI, f. M-1, op. 9, d. 352, ll. 34–36.
35. RGASPI, f. M-1, op. 9, d. 352, l. 73.
36. RGASPI, f. M-1, op. 9, d. 316, l. 37. At other times part of Gorkii oblast—today's Nizhnyi Novgorod—the city of Arzamas served as center of its own oblast from 1954 to 1957.
37. RGASPI, f. M-1, op. 9, d. 415, l. 195.
38. RGASPI, f. M-1, op. 9, d. 352, l. 64.
39. RGASPI, f. M-1, op. 9, d. 352, l. 124.
40. RGASPI, f. M-1, op. 9, d. 316, ll. 71–73.
41. RGASPI, f. M-1, op. 9, d. 316, ll. 64–70.
42. RGASPI, f. M-1, op. 9, d. 316, ll. 38–41.

43. RGASPI, f. M-1, op. 9, d. 316, l. 93.
44. RGASPI, f. M-1, op. 9, d. 315, ll. 47–56.
45. RGASPI, f. M-1, op. 9, d. 316, ll. 76–80.
46. RGASPI, f. M-1, op. 9, d. 316, ll. 81–84.
47. RGASPI, f. M-1, op. 9, d. 316, ll. 38–41.
48. RGASPI, f. M-1, op. 9, d. 315, ll. 20–22
49. RGASPI, f. M-1, op. 9, d. 386, ll. 94–101.
50. RGASPI, f. M-1, op. 9, d. 456, ll. 7–14.
51. RGASPI, f. M-1, op. 9, d. 352, l. 64.
52. RGASPI, f. M-1, op. 9, d. 352, l. 45.
53. RGASPI, f. M-1, op. 9, d. 382, ll. 2–3 and d. 414, ll. 4–7.
54. RGASPI, f. M-1, op. 9, d. 382, l. 26 and d. 414, l. 31. The sum is equal to about $2.7 million when adjusted for inflation to 2015. The documents list costs for each prize: an automobile (12,800 rubles); a motorcycle (5,500); two models of radio (1,000 and 405); sporting equipment (1,000); musical instruments (1,000); two models of wristwatch (500 and 400); trips to the exhibition (1,000); and trips to the Pioneer summer camp "Artek" (600). This total amounted to 0.0009 percent of an annual official USSR government budget of 112 billion rubles, and even less if the 127 billion ruble combined budget of the republic government budgets is included. "Zakon o Gosudarstvennom biudzhete Soiuza Sovetskikh Sotsialisticheskikh respublik na 1955 god," *Pravda*, February 11, 1955, 2.
55. RGANI, f. 5, op. 45, d. 168, ll. 78–80.
56. TsDAHOU, f. 1, op. 24, d. 4168; and TsKhDOPIM, f. 634, op. 10, d. 407, ll. 12–13.
57. A 1956 list of decorated work-team leaders counts 92 top corngrowers across dozens of oblasts and republics. Their names identify 18 men and 70 women, leaving 4 not identifiable. RGASPI, f. M-1, op. 9, d. 354, ll. 23–32.
58. *XIII s'ezd Vsesoiuznogo leninskogo kommunisticheskogo Soiuza molodëzhi*, 269–70.
59. RGASPI, f. M-1, op. 9, d. 398, l. 11. This meant that her work team had grown no fewer than 9 metric tons of grain per hectare or 65 metric tons of silage. GARF, f. A-310, op. 1, d. 1207, ll. 31 and 86–87.
60. *XIII s'ezd Vsesoiuznogo leninskogo kommunisticheskogo Soiuza molodëzhi*, 273.
61. RGASPI, f. M-1, op. 9, d. 307, l. 32.
62. LYA, f. 1771, op. 191, d. 423, l. 8.
63. RGASPI, f. M-1, op. 9, d. 307, ll. 3–4.
64. "Vsenarodnoe sorevnovanie," *Pravda*, September 15, 1958, 1. A later story featured efforts to complete the harvest of grain corn in Rostov Oblast. "Massovoyi voskresnik po uborke kukuruzy," *Pravda*, October 13, 1958, 1.
65. RGASPI, f. M-1, op. 9, d. 445, l. 35.
66. RGASPI, f. M-1, op. 9, d. 445, l. 104.
67. RGASPI, f. M-1, op. 9, d. 505, l. 98.
68. RGASPI, f. M-1, op. 9, d. 505, l. 154.

69. Thus a 1962 report on Penza Oblast listed 19 men as the winners. GARF, f. A-259, op. 45, d. 267, ll. 2–4 and ll. 8–12.
70. RGASPI, f. 556, op. 14, d. 185, l. 49.
71. RGASPI, f. M-1, op. 9, d. 598, l. 2.
72. RGASPI, f. M-1, op. 32, d. 823, l. 6. See also, "'Lëgkaia kavaleriia'—shkola gosudarstvennogo obucheniia: Iz doklada sekretaria TsK VLKSM tov. Kosareva," *Pravda*, July 3, 1934, 3.
73. "Vsem raikomam i gorkomam VLKSM, vsem shtabam 'Lëgkoi kavelerii,'" *Molodoi leninets*, May 30, 1958, 1.
74. GANISK, f. 63, op. 2, d. 1215, l. 318 and d. 1237, l. 4.
75. RGASPI, f. M-1, op. 2, d. 416, ll. 128–29.
76. RGASPI, f. M-1, op. 2, d. 416, l. 99.
77. RGASPI, f. M-1, op. 2, d. 414, l. 136.
78. The most extensive record in the archive is for 1961 in the RSFSR Central Black-Earth Zone, encompassing Voronezh, Orël, Belgorod, Tambov, Kursk, and Lipetsk oblasts. RGASPI, f. M-1, op. 9, d. 547.
79. RGASPI, f. M-1, op. 9, d. 547, l. 2.
80. RGASPI, f. M-1, op. 9, d. 546, l. 171.
81. RGASPI, f. M-1, op. 9, d. 546, l. 7.
82. RGASPI, f. M-1, op. 9, d. 587, l. 28.
83. RGASPI, f. M-1, op. 9, d. 619, l. 3.
84. RGASPI, f. M-1, op. 9, d. 598, l. 1.
85. Examining the relationship of collective and individual, political scientist Oleg Kharkhordin has suggested that the Khrushchev era de-emphasized the repression of the Stalin era in favor of social enforcement, a phenomenon the scholar suggests made these years themselves repressive. *The Collective and the Individual in Russia: A Study of Practices* (Berkeley: University of California Press, 1999), 279.
86. Deborah Field, *Private Life and Communist Morality in Khrushchev's Russia* (New York: Peter Lang, 2007), 5.
87. RGASPI, f. M-1, op. 9, d. 555, l. 4.
88. RGASPI, f. M-1, op. 9, d. 587, l. 123.
89. Fürst, *Stalin's Last Generation*, 3–5.
90. Tromly, *Making the Soviet Intelligentsia*, 27.
91. RGASPI, f. M-1, op. 9, d. 456, l. 23.
92. For examples of the former, see I. I. Mar'iakhina, *Shkol'nikam o kukuruza i kormovykh bobakh: Posobia dlia uchashchikhsia sel'skoi shkoly* (Moscow: Gosudarstvennoe uchebno-pedagogicheskoe izdatel'stvo Ministerstva posveshcheniia RSFSR, 1963); *Molodëzhi o kukuruze: Populiarnyi ocherk* (Vologda: Oblastnaia knizhnaia redaktsiia, 1955); and D. E. Gavrilin, *Kukuruza i eë izuchenie v shkolakh i detskikh domakh: Posobie dlia uchitelei* (Moscow: Ministerstvo posveshcheniia RSFSR, 1955). In the latter case, see N. Deveki, "V bor'be za vysokii urozhai

kukuruzy: Nekotorye itogi raboty kolektivov shkol Riazanskoi oblasti," *Narodnoe obrazovanie* no. 3 (1956): 34–41; and L. Imshenetskaia, "Shkola v bor'be za vysokii urozhai kukuruzy," *Narodnoe obrazovanie* no. 4 (1956): 39–40.

93. On the context, origins, and purposes of this reform, see Laurent Coumel, "The Scientist, the Pedagogue, and the Party Official: Interest Groups, Public Opinion, and Decision-making in the 1958 Education Reform," in *Soviet State and Society under Nikita Khrushchev*, ed. Melanie Ilič and Jeremy Smith (New York: Routledge, 2009), 66–85.
94. RGANI, f. 5, op. 30, d. 107, ll. 139–40.
95. Raleigh, *Soviet Baby Boomers*, 183. The practice became widely known in later years when agricultural labor shortages grew more acute, but directives permitting the dispatch of students to the fields can be found as early as the fall of 1955. GARF, f. A-259, op. 7, d. 5286, l. 1.
96. The terms *izhdivenets* and the related adjective *izhdivencheskii* are difficult to translate. They mean a legal or material dependent, be it a child or a person with disability. In the Khrushchev era, officials used it to connote someone who abused that aid, denouncing those who would supposedly lead an easy life scrounging from government aid; that is, someone with a so-called welfare mentality. For example, Khrushchev used the word to describe sovkhoz directors who, when they failed to harvest enough corn to feed the farm's livestock, expected to purchase feed from the state at artificially low prices to make up the shortfall.
97. A. Zviagintsev, "Izhdivenets," *Molodoi leninets*, November 14, 1958, 3.
98. L. Epaneshnikov, "Fel'eton: Barchuk," *Stavropol'skaia pravda*, November 12, 1958, 3.
99. GANISK, f. 1, op. 2, d. 7917, ll. 70–71.
100. GANISK, f. 1, op. 2, d. 7917, ll. 68–69.
101. Raleigh, *Soviet Baby Boomers*, 183–84.
102. This excerpt from a speech Khrushchev gave in October 1958 appeared in "Shkola, trud, kommunizm," *Molodoi leninets*, November 15, 1958, 3.
103. GANISK, f. 63, op. 2, d. 1132, ll. 8–9.
104. GANISK, f. 63, op. 2, d. 1132, l. 3.
105. Raleigh, *Soviet Baby Boomers*, 101–6.
106. GANISK, f. 63, op. 2, d. 1132, l. 12.
107. GANISK, f. 63, op. 2, d. 1134, l. 152 and d. 1135, l. 8.
108. GANISK, f. 63, op. 2, d. 1132, l. 124.
109. GANISK, f. 63, op. 2, d. 1132, l. 144.
110. Thomas Wolfe, *Governing Soviet Journalism: The Press and the Socialist Person after Stalin* (Bloomington: Indiana University Press, 2005), 1.
111. GANISK, f. 63, op. 2, d. 1134, ll. 106–12.
112. GANISK, f. 63, op. 2, d. 1134, ll. 106–12.
113. Raleigh, *Soviet Baby Boomers*, 102–4.

114. GANISK, f. 63, op. 2, d. 1135, l. 240.
115. GANISK, f. 63, o. 2, d. 1135, l. 266.
116. "Rech' T. Dobrovol'skoi, uchashcheisia 10 klassa srednei shkoly no. 20, stanitsy Goriachevodskoi, Stavropol'skogo kraia, na XIII s"ezde VLKSM," *Molodoi leninets*, April 19, 1958, 5.
117. RGASPI, f. 556, o. 14, d. 185, ll. 31–36.
118. Wolfe, *Governing Soviet Journalism*, 13–17.

CHAPTER 6

1. RGANI, f. 5, op. 30, d. 107, l. 262.
2. N. S. Khrushchev, *Stroitel'stvo kommunizma v SSSR i sel'skoe khoziaistvo*, vol. 1 (Moscow: Gosudarstvennoe izdatel'stvo politicheskoi literatury, 1962), 170.
3. Kolkhozes planted the majority of crops in both 1953 and 1964, even though their relative contributions fell from 88.4 percent to 53.6 percent. In 1950, sovkhozes cultivated 15.9 million hectares and kolkhozes 121 million. By 1964, those numbers had shifted to 95.7 million and 110.8 million, respectively. USSR Council of Ministers, Central Statistical Department, *Sel'skoe khoziaistvo SSSR: Statisticheskii sbornik* (Moscow: Statistika, 1971), 112. Sovkhozes increased in number and size, but their methods for determining their workers' wages changed little in this period. In 1953, sovkhoz workers earned guaranteed wages for shifts worked, leaving only annual bonuses dependent on outcomes. By contrast, kolkhozniks' pay depended entirely on output, a situation Khrushchev-era policies remedied by reforming their pay to resemble that of sovkhoz workers. M. E. Beznin and T. M. Dimoni, "Krest'ianstvo i vlast' v Rossii v kontse 1930-kh–1950-e gody," in *Mentalitet i agrarnoe razvitie Rossiii (XIX–XX vv.): Materialy mezhdunarodnoi konferentsii, 14–15 iiunia 1994 g.*, ed. V. P. Danilov and L. V. Milov (Moscow: ROSSPEN, 1996), 156.
4. Compare this to established arguments that Khrushchev relied principally on the mobilizational tactics inherited from Stalin. Jerzy Karcz, "Khrushchev's Impact on Soviet Agriculture," *Agricultural History* 40, no. 1 (1966): 31.
5. E. P. Thompson, "The Moral Economy of the English Crowd in the Eighteenth Century," *Past and Present* no. 50 (1971): 77–78. The theme is also present in his best-known work, *The Making of the English Working Class* (New York: Vintage, 1966), 62–67. Studies of collectivization have employed the concept, including James Hughes, *Stalinism in a Russian Province: A Study of Collectivization and Dekulakization in Siberia* (New York: St. Martin's Press, 1996), 3–4. Jean Lévesque applied Thompson's approach to the postwar period. He emphasized that kolkhozniks avoided working on the authorities' terms, "habits" and "dispositions" developed in response to exactions under Stalin. " 'Part-Time Peasants': Labour Discipline, Collective Farm Life, and the Fate of Soviet Socialized Agriculture after the Second World War, 1945–1953" (PhD diss., University of Toronto, 2003), 10 and 39–40.

6. Beznin and Dimoni, "Krest'ianstvo i vlast' v Rossii," 156.
7. Peter Fritzsche, "On the Subjects of Resistance," *Kritika: Explorations in Russian and Eurasian History* 1, no. 1 (2000): 147–50.
8. Andrew Sloin and Oscar Sanchez-Sibony, "Economy and Power in the Soviet Union," *Kritika: Explorations in Russian and Eurasian History* 15, no. 1 (2014): 7–22.
9. Khrushchev, *Stroitel'stvo kommunizma v SSSR*, 1:11–12.
10. LYA, f. 1771, op. 191, d. 423, ll. 21–22.
11. RGANI, f. 5, op. 45, d. 111, ll. 72–74.
12. RGANI, f. 5, op. 45, d. 111, l. 89.
13. TsGAMO, f. 191, op. 1, d. 2161, ll. 2–3 and ll. 8–9.
14. RGAE, f. 7486, op. 7, d. 1392, l. 32.
15. RGANI, f. 5, op. 45, d. 111, l. 92.
16. GARF, f. R-5446, op. 89, d. 1174, ll. 17–19. For examples of the institute's correspondence with the Central Committee and Council of Ministers, see RGAE, f. 260, op. 2, dd. 70, 72–73, 125–31, et al. This institute contributed to a larger revival of the discipline of economics under Khrushchev, which Moshe Lewin identified in the 1970s. *Political Undercurrents in Soviet Economic Debates: From Bukharin to the Modern Reformers* (Princeton, NJ: Princeton University Press, 1974).
17. Maya Haber, "The Soviet Ethnographer as Social Engineer: Socialist Realism and the Study of Rural Life, 1945–1958," *Soviet and Post-Soviet Review* 41, no. 2 (2014): 194–219.
18. Maya Haber, "Socialist Realist Science: Constructing Knowledge about Rural Life in the Soviet Union, 1943–1958" (PhD diss., University of California at Los Angeles, 2013), 14.
19. Ibid., 290.
20. For more on the reform, see chapter 3, as well as V. N. Tomilin, *Nasha krepost': Mashinno-traktornye stantsii Chernozenmnogo Tsentra Rossii v poslevoennyi period, 1946–1958 gg.* (Moscow: AIRO-XX, 2009), 339–40.
21. RGANI, f. 5, op. 30, d. 108, l. 90.
22. RGANI, f. 5, op. 45, d. 111, l. 79.
23. RGANI, f. 5, op. 30, d. 108, l. 90.
24. RGANI, f. 5, op. 31, d. 168, ll. 79–84.
25. A. A. Nikonov, *Spiral' mnogovekovoi dramy: Agrarnaia nauka i politika Rossii, XVIII–XX vv.* (Moscow: Entsiklopediia rossiiskikh dereven', 1995), 305.
26. P. Mezhniakova, "Kogda budet vydavat' kukuruzu na trudodni?" *Stavropol'skaia pravda*, October 14, 1955, 3.
27. L. Pankratov, "Nekotorye voprosy oplaty truda v kolkhozakh," *Stavropol'skaia pravda*, April 2, 1955, 3.
28. RGAE, f. 7486, op. 7, d. 1348, l. 94.

29. RGANI, f. 5, op. 45, d. 111, l. 111; and RGAE, f. 7486, op. 7, d. 1348, l. 5.
30. RGAE, f. 7486, op. 7, d. 1348, ll. 14–15.
31. GANISK, f. 1, op. 2, d. 6537, l. 10.
32. On the gendered norms of work within the home and outside of it, see Lynne Attwood, "Celebrating the 'Frail-Figured Welder': Gender Confusion in Women's Magazines of the Khrushchev Era," *Slavonica* 8, no. 2 (2002): 159–77.
33. RGANI, f. 5, op. 45, d. 111, ll. 73–74.
34. RGANI, f. 5, op. 45, d. 111, l. 78.
35. RGANI, f. 5, op. 45, d. 184, ll. 108–12.
36. RGANI, f. 5, op. 46, d. 394, ll. 30.
37. RGANI, f. 5, op. 30, d. 220, ll. 220–25.
38. RGANI, f. 5, op. 45, d. 209, l. 60.
39. RGAE, f. 260, op. 2, d. 70, ll. 5–15.
40. RGAE, f. 260, op. 2, d. 70, ll. 5–15.
41. RGAE, f. 260, op. 2, d. 70, ll. 36–46.
42. GARF, f. A-340, op. 1, d. 116, ll. 8–9.
43. GARF, f. A-340, op. 1, d. 116, l. 38.
44. GARF, f. A-340, op. 1, d. 107, l. 17.
45. In 1955, Voronezh, Kursk, Lipetsk, and Belgorod party committees reported concern about kolkhozes' capacity to bring in the good expected harvests of sugar beets because of the demands of corn. RGANI, f. 5, op. 45, d. 84, ll. 94–95 and 103. Satisfied with the contributions of Soviet Army soldiers, authorities requested permission to extend their assignments from twenty to thirty-five days to aid in harvesting corn, sugar beets, and potatoes. RGANI, f. 5, op. 45, d. 84, l. 104.
46. RGANI, f. 5, op. 45, d. 209, ll. 59–69.
47. RGANI, f. 5, op. 45, d. 209, ll. 59–69.
48. RGANI, f. 5, op. 45, d. 209, ll. 120–21.
49. RGAE, f. 7486, op. 7, d. 1541, ll. 13–16.
50. RGANI, f. 5, op. 24, d. 581, l. 75.
51. GARF, f. R-8300, op. 5, d. 92, l. 210.
52. GARF, f. R-8300, op. 5, d. 92, ll. 213–14.
53. RGANI, f. 5, op. 45, d. 184, ll. 2–3.
54. RGANI, f. 5, op. 45, d. 184, l. 6.
55. RGANI, f. 5, op. 45, d. 184, l. 3.
56. RGANI, f. 5, op. 45, d. 84, ll. 136–39. In October 1964, an inspection of ten oblasts catalogued 2,225 cases against 3,138 kolkhozniks, sovkhoz workers, and officials. RGASPI, f. 556, op. 22, d. 490, ll. 100–4.
57. USSR Council of Ministers Central Statistical Department, *Sel'skoe khoziaistvo SSSR: Statisticheskii sbornik* (Moscow: Statistika, 1971), 152.
58. TsDAHOU, f. 1, op. 24, d. 4182, ll. 46–48.
59. TsDAHOU, f. 1, op. 24, d. 4182, ll. 70–72 and l. 74.

60. RGANI, f. 5, op. 45, d. 184, ll. 59–60.
61. On Ukraine, see TsDAHOU, f. 1, op. 24, d. 4182, ll. 143–45. On Belarus, see RGANI, f. 5, op. 45, d. 184, l. 165.
62. RGANI, f. 5, op. 45, d. 184, l. 62.
63. TsDAHOU, f. 1, op. 24, d. 4182, l. 80.
64. TsDAHOU, f. 1, op. 24, d. 4182, ll. 97–98.
65. TsDAHOU, f. 1, op. 24, d. 4182, ll. 86–87.
66. TsDAHOU, f. 1, op. 24, d. 4182, l. 93 and l. 99.
67. TsDAHOU, f. 1, op. 24, d. 4182, ll. 89–90.
68. TsDAHOU, f. 1, op. 24, d. 4182, l. 70.
69. For more, see Iu. V. Aksiutin and A. V. Pyzhikov, *Poststalinskoe obshchestvo: Problema liderstva i transformatsiia vlasti* (Moscow: Nauchnaia kniga, 1999), 199; and Oleg Kharkhordin, *The Collective and the Individual in Russia: A Study of Practices* (Berkeley: University of California Press, 1999), 279–328.
70. RGANI, f. 2, op. 1, d. 409, ll. 40–41.
71. RGANI, f. 2, op. 1, d. 409, l. 30.
72. RGASPI, f. 556, op. 14, d. 107, l. 22.
73. RGANI, f. 5, op. 45, d. 184, l. 68.
74. Under the tsars, the *stanitsa* had been the basic unit of social, political, and economic organization in a Cossack host. In the USSR, the word denoted a large settlement bridging town and rural areas primarily in the North Caucasus, a legacy of the historical Cossack presence in the area. The Lenin kolkhoz had been formed in 1955, when the amalgamation of the 22 farms once served by its MTS was completed, leaving just 2 large kolkhozes. It comprised 2,000 households and 4,700 people, of whom 62 percent qualified as active members. The farm controlled 13,000 hectares of cropland and produced 12,000–15,000 metric tons of grain annually, as well as fruits, vegetables, milk, meat, and eggs. GANISK, f. 1, op. 2, d. 8830, ll. 17–18.
75. "[Untitled]," *Kolkhoznaia zhizn'*, May 5, 1957, 1. The newspaper also stands out because it survives in a nearly complete print run for these years in the Russian State Library in Moscow.
76. Kh. I. Agnaev, "[Untitled]," *Kolkhoznaia zhizn'*, May 5, 1957, 1; and "[Untitled]," *Kolkhoznaia zhizn'*, June 5, 1957, 1.
77. L. Egorova, "Po-nastoiashchemu borot'sia za vysokii urozhai," *Kolkhoznaia zhizn'*, May 1, 1958, 2; and "Zabotlivo ukhazhivat' za kukuruzoi, ne narushat' pravil agrotekhniki," *Kolkhoznaia zhizn'*, June 10, 1958, 1.
78. I. Sokolov, "Ne otstavat' ot peredovikov," *Kolkhoznaia zhizn'*, June 5, 1957, 1.
79. "Na pravlenii kolkhoza," *Kolkhoznaia zhizn'*, May 25, 1957, 2.
80. "Na pravlenii kolkhoza," *Kolkhoznaia zhizn'*, September 25, 1957, 2.
81. A. Dranov, "Za potravu posevov—k otvetu," *Kolkhoznaia zhizn'*, August 5, 1957, 1.
82. "Na pravlenii kolkhoza," *Kolkhoznaia zhizn'*, September 25, 1957, 2.

83. P. Kovtun, "Kommunist dolzhen byt' obraztsom vo vsekh otnoshcheniiakh," *Kolkhoznaia zhizn'*, February 15, 1958, 2.
84. "Posledovat' primery zvenevykh E. Ul'ianik i N. Kapluna," *Kolkhoznaia zhizn'*, May 25, 1957, 1; "Na puti k izobiliiu," *Kolkhoznaia zhizn'*, November 7, 1957, 1; and "N. I. Kaplun—nastoiashchii cheloveka!" *Kolkhoznaia zhizn'*, November 7, 1958, 1.
85. "Doska pochëta," *Kolkhoznaia zhizn'*, June 15, 1957, 1.
86. "Sorevnovanie kukuruzovodov," *Kolkhoznaia zhizn'*, June 25, 1957, 1.
87. "Zadaniia semiletki—za odin-tri goda!" *Kolkhoznaia zhizn'*, January 15, 1959, 1.
88. "[Untitled]," *Kolkhoznaia zhzin'*, August 5, 1957, 2.
89. For another, later denunciation of the offending official, see "Narushiteli i ikh pokroviteli," *Kolkhoznaia zhizn'*, November 19, 1959, 2.
90. P. Fabrova, "Vsem, kak odin, borot'sia s nedostatkami," *Kolkhoznaia zhizn'*, August 15, 1957, 1.
91. GANISK, f. 1, op. 2, d. 8863, ll. 54–55.
92. "V pokhod za urozhai kukuruzy," *Kolkhoznaia zhzin'*, February 19, 1959, 1.
93. Sibirtsev, "Uchest' proshlegodnye oshibki," *Kolkhoznaia zhizn'*, February 7, 1959, 2. On the wider campaign, see GANISK, f. 1, op. 2, d. 7923, l. 117.
94. A. Ponomarenko, "Chto pokazala proverka," *Kolkhoznaia zhizn'*, June 9, 1959, 1.
95. "Tovarishchi kolkhozniki!" *Kolkhoznaia zhizn'*, June 9, 1959, 1.
96. GANISK, f. 1, op. 2, d. 8594, ll. 18–19.
97. GANISK, f. 1, op. 2, d. 8594, l. 23 and l. 26.
98. GANISK, f. 1, op. 2, d. 7911, ll. 1–2. For a systematic review of the ideas and principles involved, see: Peter Maggs, "The Law of Farm-Farmer Relations," in *The Soviet Rural Community*, ed. James Millar (Urbana: University of Illinois Press, 1971), 147.
99. GANISK, f. 1, op. 2, d. 7576, l. 39.
100. For definitions of net income, or *"chistyi dokhod,"* and related accounting terms, see Caroline Humphrey, *Marx Went Away—But Karl Stayed Behind*, rev. ed. (Ann Arbor: University of Michigan Press, 1998), 77–85.
101. GANISK, f. 1, op. 2, d. 7911, ll. 1–2. For the list of farms, see GANISK, f. 1, op. 2, d. 7459, l. 23.
102. On this earlier era, see Diane Koenker, *Republic of Labor: Russian Printers and Soviet Socialism, 1917–1930* (Ithaca: Cornell University Press, 2005), 34 and 112.
103. GANISK, f. 1, op. 2, d. 7576, l. 3.
104. GANISK, f. 1, op. 2, d. 7576, l. 41.
105. GANISK, f. 1, op. 2, d. 7887, l. 3.
106. GANISK, f. 1, op. 2, d. 7576, l. 74.
107. GANISK, f. 1, op. 2, d. 7576, l. 55.
108. GANISK, f. 1, op. 2, d. 7902, l. 1.
109. GANISK, f. 1, op. 2, d. 7902, l. 10.

110. GANISK, f. 1, op. 2, d. 7576, ll. 75–76. Other kolkhozes paid outright various figures ranging from 83 up to 100 percent. GANISK, f. 1, op. 2, d. 8262, l. 95.
111. GANISK, f. 1, op. 2, d. 8262, l. 101.
112. GANISK, f. 5351, op. 1, d. 309, l. 2; and f. 1, op. 2, d. 7576, l. 3.
113. O. M. Verbitskaia, *Rossiiskoe krest'ianstvo: Ot Stalina k Khrushchevu, sredina 40-kh–nachalo 60-kh gg.* (Moscow: Nauka, 1992), 41.
114. GANISK, f. 1, op. 2, d. 7887, ll. 5–6. For other examples, see GANISK, f. 1, op. 2, d. 7902, l. 11.
115. GANISK, f. 1, op. 2, d. 8262, ll. 93–94.
116. GANISK, f. 1, op. 2, d. 8262, l. 94.
117. GANISK, f. 1, op. 2, d, 8262, l. 102.
118. "Novaia sistema oplaty truda—put' dal'neishego pod''ëma," *Kolkhoznaia zhizn'*, December 23, 1958, 1.
119. "Pooshchreniia za vyrashchivaniia vysokogo urozhaia kukuruzy," *Kolkhoznaia zhizn'*, June 21, 1959, 1.
120. "[Untitled]," *Kolkhoznaia zhizn'*, January 1, 1959, 2.
121. "O tekh, kto govorit: 'Zarplata mala,'" *Kolkhoznaia zhizn'*, June 20, 1959, 2.
122. GANISK, f. 1, op. 2, d. 8260, l. 16.
123. GANISK, f. 1, op. 2, d. 8262, l. 95.
124. GANISK, f. 1, op. 2, d. 8262, ll. 119–20 and l. 121.
125. GANISK, f. 1, op. 2, d. 8262, ll. 98–99.
126. GANISK, f. 1, op. 2, d. 8260, l. 84.
127. GANISK, f. 1, op. 2, d. 8260, l. 102.
128. Kh. I. Agnaev, "Prichinu otstavaniia?—Plokhaia distsiplina!" *Kolkhoznaia zhizn'*, July 23, 1959, 2.
129. GANISK, f. 5351, op. 1, d. 309, l. 2.
130. GANISK, f. 1, op. 2, d. 8260, l. 93. For more on these practices, see Humphrey, *Marx Went Away*, 161.
131. Humphrey, *Marx Went Away*, 316.
132. GANISK, f. 1, op. 2, d. 8262, l. 100.
133. GANISK, f. 1, op. 2, d. 8260, ll. 83–95.
134. GANISK, f. 1, op. 2, d. 7902, ll. 13–14.
135. GANISK, f. 1, op. 2, d. 8830, ll. 17–18.
136. GANISK, f. 1, op. 2, d. 8260, ll. 47–50.
137. GANISK, f. 1, op. 2, d. 8262, l. 101.
138. GANISK, f. 1, op. 2, d. 8260, ll. 56–61. Proposals, supposedly from below, had arisen already at the end of 1958. GANISK, f. 1, op. 2, d. 7576, l. 55.
139. GANISK, f. 1, op. 2, d. 8262, l. 101.
140. RGANI, f. 5, op. 46, d. 394, l. 30.
141. RGANI, f. 5, op. 46, d. 394, ll. 1–2.
142. GARF, f. R-9477, op. 1, d. 360, ll. 336–41.

143. GARF, f. R-9477, op. 1, d. 360, l. 22.
144. GARF, f. R-5446, op. 93, d. 647, ll. 58–64.
145. GANISK, f. 5351, op. 1, d. 307, ll. 1–6.
146. GANISK, f. 5351, op. 1, d. 307, l. 6.
147. GANISK, f. 5351, op. 1, d. 307, ll. 41–43 and ll. 54–62.
148. A. N. Artizov et al., eds., *Nikita Khrushchev, 1964: Stenogrammy Plenuma TsK KPSS i drugie dokumenty* (Moscow: Mezhdunarodnyi fond "Demokratiia," 2007), 189. Both figures are given in the new ruble introduced in 1961, which redenominated at a rate of 10 old rubles to 1 new.
149. GANISK, f. 5351, op. 1, d. 307, l. 60.
150. RGANI, f. 5, op. 30, d. 419, ll. 151–57
151. GASK, f. 2395, op. 5, d. 560, l. 2
152. David Bronson and Constance Krueger, "The Revolution in Soviet Farm Household Income, 1953–1967," in *Soviet Rural Community*, ed. Millar, 214.
153. Alexander Vucinich, "The Peasants as a Social Class," in *Soviet Rural Community*, ed. Millar, 310.
154. Roy Medvedev and Zhores Medvedev, *Khrushchev: The Years in Power*, trans. Andrew Durkin (New York: Columbia University Press, 1975), 182–83.
155. Nancy Nimitz, "Farm Employment in the Soviet Union, 1928–1963," in *Soviet and East European Agriculture*, ed. Jerzy Karcz (Berkeley: University of California Press, 1967), 204.

CHAPTER 7

1. George Breslauer, *Khrushchev and Brezhnev as Leaders: Building Authority in Soviet Politics* (Boston: Allen and Unwin, 1982), 6.
2. Initially, Sovietologists interpreted these changes as signs of political combat between Khrushchev and Georgii Malenkov. Archival research shows that they resulted neither from that conflict nor from Khrushchev's seemingly "erratic and unpredictable temperament." Yoram Gorlizki, "Anti-Ministerialism and the USSR Ministry of Justice, 1953–1956: A Study in Organizational Decline," *Europe–Asia Studies* 48, no. 8 (1996): 1307. On the 1957 reform, see ibid., 1282.
3. Sociologist Alena Ledeneva sheds light on practices known in Russian as *sviazy* (connections), *znakomstvo* (acquaintance), and *blat* to achieve private ends. *Blat* and the other pervasive but concealed maneuvers "used personal networks and informal contacts to obtain goods and services in short supply and found a way around formal procedures." *Russia's Economy of Favors: Blat, Networking, and Informal Exchange* (New York: Cambridge University Press, 1998), 1. Individuals thus pursued objectives varying according to social position: winter boots, an appointment with a medical specialist, entrance into a prestigious university, elusive theater tickets, or any other necessity.

4. At first, this body maintained its old name and structure from the pre-1953 period, itself an evolution of the Workers' and Peasants' Inspectorate founded in 1920. In 1957, it became the Commission for Government Oversight and, in 1962, the Committee of Public Oversight, or *Komitet narodnogo kontrolia*.
5. GARF, f. R-9477, op. 1, d. 369, ll. 1–14.
6. RGANI, f. 5, op. 30, d. 64, l. 245.
7. RGANI, f. 2, op. 1, d. 124, ll. 48–67.
8. GANISK, f. 1, op. 2, d. 6539, l. 6.
9. RGANI, f. 2, op. 1, d. 124, l. 54.
10. RGASPI, f. 556, op. 22, d. 64, l. 33.
11. On Central Committee personnel who coordinated government policy, see Jerry Hough, *How the Soviet Union Is Governed* (Cambridge, MA: Harvard University Press, 1979), 444–45.
12. GARF, f. R-5446, op. 89, d. 111, l. 259. At the official, overvalued exchange rate, this amounted to over $75 million in 1955, or $667 million if inflation adjusted to 2015 dollars. The union-level government budget for that year, by comparison, was 112 billion rubles.
13. GARF, f. R-5446, op. 89, d. 111, ll. 148–49 and ll. 210–11.
14. GARF, f. R-5446, op. 89, d. 111, ll. 288–89.
15. GARF, f. R-5446, op. 89, d. 111, l. 299.
16. GARF, f. R-8300, op. 24, d. 773, l. 60.
17. GARF, f. R-8300, op. 24, d. 810, l. 5. This total equaled $575 million in 1956, or $5.03 billion adjusted for inflation to 2015.
18. GARF, f. R-8300, op. 24, d. 810, l. 188.
19. GARF, f. R-8300, op. 24, d. 810, l. 93.
20. Donald Filtzer, *Soviet Workers and De-Stalinization: The Consolidation of the Modern System of Soviet Production Relations, 1953–1964* (New York: Cambridge University Press, 1992), 164.
21. RGANI, f. 5, op. 30, d. 107, l. 2.
22. RGANI, f. 5, op. 30, d. 107, l. 8.
23. RGANI, f. 5, op. 30, d. 107, l. 53.
24. Filtzer, *Soviet Workers and De-Stalinization*, 15.
25. RGANI, f. 5, op. 30, d. 107, l. 53.
26. Kendall Bailes, "The American Connection: Ideology and the Transfer of American Technology to the Soviet Union, 1917–1941," *Comparative Studies in Society and History* 23, no. 3 (1981): 428; and Alan Ball, *Imagining America: Influence and Images in Twentieth-Century Russia* (Lanham, MD: Rowman & Littlefield, 2003), 26–30.
27. RGANI, f. 5, op. 30, d. 107, ll. 54–57.
28. RGANI, f. 5, op. 45, d. 168, ll. 55–57 and 61–62.
29. GARF, f. A-259, op. 7, d. 7992, ll. 1–3.

30. Martin McCauley, *Khrushchev and the Development of Soviet Agriculture: The Virgin Land Programme, 1953–1964* (New York: Holmes and Meier, 1976), 137.
31. RGASPI, f. 556, op. 22, d. 381, ll. 120–22.
32. RGASPI, f. 556, op. 22, d. 381, l. 124.
33. RGASPI, f. 556, op. 22, d. 381, l. 141. Similar calls for additional deliveries of tractors, implements, trucks and more continued into 1964 and beyond. RGANI, f. 5, op. 46, d. 479, ll. 1–2, 37, 38–39, 40–41, 42–44, and many others.
34. RGAE, f. 260, op. 2, d. 164, ll. 20–23.
35. GARF, f. R-5446, op. 96, d. 1274, ll. 1–6.
36. Noël Kingsbury, *Hybrid: The History and Science of Plant Breeding* (Chicago: University of Chicago Press, 2009), 217–50.
37. Zhores Medvedev, *The Rise and Fall of T. D. Lysenko*, ed. Lucy Lawrence, trans. I. Michael Lerner (New York: Columbia University Press, 1969); David Joravsky, *The Lysenko Affair* (Cambridge, MA: Harvard University Press, 1970); and the recent archive-based re-evaluation by Ethan Pollock, "From *Partiinost'* to *Nauchnost'* and Not Quite Back Again: Revisiting the Lessons of the Lysenko Affair," *Slavic Review* 68, no. 1 (2009): 95–115. Lysenko has loomed so large in the scholarship that, after 1991, historians conducted archival research on disciplines other than genetics to show that other outcomes were possible. The dominance of an unscrupulous maneuverer was the exception, rather than the rule. Alexei Kojevnikov, *Stalin's Great Science: The Times and Adventures of Soviet Physicists* (River Edge, NJ: Imperial College Press, 2004); and Pollock, *Stalin and the Soviet Science Wars* (Princeton, NJ: Princeton University Press, 2006).
38. Accounting for the harm done to plant breeding by Lysenko and his supporters, the biologist and dissident historian Zhores Medvedev considered hybrid corn a representative case. In his characteristic polemical style, he concluded, "Had there not been the unproved, unfounded, tendentious, and simply ignorant propaganda of the Lysenkoites, [double-cross hybrids] could have been adopted" before World War II, resulting in millions of tons of additional grain annually. Medvedev, *Rise and Fall of T. D. Lysenko*, 181.
39. Joravsky, *Lysenko Affair*, 285–86.
40. RGANI, f. 5, op. 24, d. 574, ll. 20–21 and l. 25
41. RGANI, f. 5, op. 24, d. 574, l. 27.
42. RGANI, f. 5, op. 24, d. 574, l. 29. Other letters from this writer are in RGANI, f. 5, op. 24, d. 574, ll. 30–33; ll. 34–36; and op. 45, d. 83, ll. 23–46.
43. RGANI, f. 5, op. 45, d. 33, ll. 1–5.
44. RGANI, f. 5, op. 45, d. 33, ll. 25–34.
45. RGANI, f. 5, op. 45, d. 33, ll. 35–39; ll. 40–47; and ll. 85–96.
46. RGANI, f. 5, op. 45, d. 33, l. 86.
47. In comparison to wheat and rice, corn hybrids contributed less to the Green Revolution, as traditionally conceived. Small-scale farmers could not annually

reproduce the seeds themselves and frequently lacked the annual cash flow sufficient to purchase new ones, as well as subsidiary chemicals and fertilizers. For one case, see James McCann, *Maize and Grace: Africa's Encounter with a New World Crop, 1500-2000* (Cambridge, MA: Harvard University Press, 2005), especially "Breeding SR-52: The Politics of Science and Race in Southern Africa," 140-73.
48. RGANI, f. 2, op. 1, d. 124, ll. 68-93.
49. RGANI, f. 5, op. 45, d. 33, l. 2.
50. RGAE, f. 7486, op. 22, d. 89, l. 61.
51. Werner Hahn, *The Politics of Soviet Agriculture, 1960-1970* (Baltimore: Johns Hopkins University Press, 1972), 64.
52. RGANI, f. 5, op. 30, d. 107, ll. 130-31.
53. Joravsky, *Lysenko Affair*, 289-91.
54. Jenny Leigh Smith argues that, although a failure as theoretical and experimental science, Lysenko's thought improved management systematizing how to care for livestock that began to remedy some practical failures. *Works in Progress: Plans and Realities on Soviet Farms, 1930-1963* (New Haven, CT: Yale University Press, 2014), 16-17.
55. RGANI, f. 5, op. 30, d. 107, l. 131.
56. This confirms the finding of Zhores Medvedev, who placed it in 1954 rather than 1955. *Rise and Fall of T. D. Lysenko*, 181.
57. RGANI, f. 5, op. 30, d. 107, ll. 151-54.
58. RGANI, f. 5, op. 45, d. 136, ll. 45-58.
59. GARF, f. R-5446, op. 89, d. 997, ll. 144-145. In 1954, the institutes and stations planted 350 hectares of corn, a figure expanded to 3,900 in 1955, as leaders designated 21 institutes, 107 test stations, and 50 seed-selection stations to conduct research. See also Matskevich's April 1955 letter to the USSR Council of Ministers arguing to expand the academy's research into corn. GARF, f. R-5446, op. 89, d. 997, l. 127.
60. Pollock, "From *Partiinost'* to *Nauchnost',*" 109. The formal directive simply declared that Lysenko's "request" to be discharged from the post of president of the Academy had been granted, replacing him with Khrushchev's protégé Pavel Lobanov.
61. For instance, see his remarks at the December 1958 plenum. RGANI, f. 2, op. 1, d. 344, ll. 88-96.
62. Pollock suggests that Lysenko's credentials as a practical operator, in Khrushchev's eyes, ensured he retained support at the top. "From *Partiinost'* to *Nauchnost',*" 110.
63. B. P. Sokolov, "Gibridy kukuruzy v SShA i Kanade," *Sel'skoe khoziaistvo*, February 9, 1956, 4; "Nasushchnye voprosy selektsii i semenovodstva gibridnoi kukuruzy," *Sel'skoe khoziaistvo*, March 14, 1957, 2; "Selektsiia i semenovodstve kukuruzy na novom etape," *Sel'skoe khoziaistvo*, February 16, 1958, 2; and "O prisuzhdenii Leninskikh premii za naibolee vydaiushchiesia raboty v oblasti nauki i tekhniki," *Sel'skaia zhizn'*, April 21, 1963, 3.

64. RGAE, f. 7486, op. 22, d. 89, l. 135.
65. GARF, f. 5446, op. 90, d. 789, l. 174.
66. GARF, f. 5446, op. 90, d. 789, l. 145.
67. GARF, f. R-8300, op. 24, d. 816, ll. 139–40.
68. GARF, f. R-8300, op. 24, d. 817, ll. 31–32.
69. GARF, f. R-8300, op. 24, d. 817, l. 64.
70. GARF, f. R-8300, op. 24, d. 817, l. 87.
71. GARF, f. R-8300, op. 24, d. 817, l. 74.
72. GARF, f. R-8300, op. 24, d. 816, ll. 1–2.
73. GARF, f. R-8300, op. 24, d. 816, ll. 20–21.
74. GARF, f. A-259, op. 7, d. 6736, l. 78.
75. GARF, f. R-8300, op. 24, d. 816, ll. 13–14 and l. 20.
76. GARF, f. A-259, op. 7, d. 6736, ll. 78–80.
77. GARF, f. R-8300, op. 24, d. 828, ll. 4 and 41.
78. GARF, f. R-8300, op. 24, d. 828, l. 5.
79. GARF, f. R-8300, op. 24, d. 828, l. 15.
80. GARF, f. R-8300, op. 24, d. 828, l. 42.
81. GARF, f. R-8300, op. 24, d. 828, l. 7.
82. GARF, f. R-8300, op. 24, d. 828, l. 23.
83. GARF, f. R-8300, op. 24, d. 816, ll. 42–43.
84. GARF, f. R-8300, op. 24, d. 816, ll. 47–48.
85. GARF, f. R-8300, op. 24, d. 816, ll. 53–57; ll. 61–63; and ll. 77–79, respectively.
86. GARF, f. R-8300, op. 24, d. 817, l. 23.
87. For a discussion incentive structures, see Philip Hanson, *The Rise and Fall of the Soviet Economy: An Economic History of the USSR from 1945* (New York: Longman, 2003).
88. TsDAHOU, f. 1, op. 31, d. 404, ll. 148–51.
89. TsDAHOU, f. 1, op. 31, d. 404, ll. 172–74; and RGASPI, f. 556, op. 22, d. 67–69.
90. TsGAMO, f. 2157, op. 1, d. 4579, l. 152.
91. TsGAMO, f. 2157, op. 1, d. 4579, l. 137.
92. TsDAHOU, f. 1, op. 31, d. 411, l. 1. See also, GARF, f. R-8300, op. 24, d. 817, ll. 28–30 and l. 163.
93. TsDAHOU, f. 1, op. 31, d. 411, l. 7.
94. TsDAHOU, f. 1, op. 31, d. 411, ll. 26–27.
95. TsDAHOU, f. 1, op. 31, d. 411, l. 10.
96. GARF, f. R-9477, op. 1, d. 94, l. 337.
97. TsDAHOU, f. 1, op. 31, d. 411, l. 4. The 5 million ruble budget amounted to $1.25 million in 1956 terms at the official exchange rate, or $10.9 million in 2015.
98. TsDAHOU, f. 1, op. 31, d. 411, ll. 32–35.
99. GARF, f. R-8300, op. 24, d. 828, ll. 30–31.
100. TsDAHOU, f. 1, op. 31, d. 411, ll. 38–39.

101. TsDAHOU, f. 1, op. 31, d. 411, ll. 28–29.
102. V. Mints, "Na stroikakh zavodov po obrabotke gibridnykh i sortovykh semian kukuruzy," *Sel'skoe khoziaistvo*, October 31 1956, 4.
103. TsDAHOU, f. 1, op. 31, d. 411, ll. 41 and 43.
104. TsDAHOU, f. 1, op. 31, d. 411, l. 57.
105. TsDAHOU, f. 1, op. 31, d. 411, l. 24.
106. GARF, f. R-9477, op. 1, d. 94, ll. 335–43.
107. GARF, f. R-5446, op. 90, d. 789, ll. 171–74.
108. GARF, f. R-9477, op. 1, d. 94, l. 337.
109. GARF, f. A-340, op. 2, d. 108 ll. 23–30.
110. GARF, f. A-340, op. 2, d. 108 ll. 32–36.
111. GARF, f. A-340, op. 2, d. 108 ll. 58–60
112. GASK, f. 2481, op. 1, d. 399, ll. 129–34.
113. GANISK, f. 1, op. 2, d, 7923, ll. 26–27 and GARF, f. R-9477, op. 1, d. 94, l. 10.
114. Filtzer, *Soviet Workers and De-Stalinization*, 42–47.
115. GANISK, f. 1, op. 2, d. 7568, ll. 15–16.
116. GANISK, f. 1, op. 2, d. 7568, l. 21.
117. GANISK, f. 1, op. 2, d. 7568, l. 22.
118. Thompson, "Industrial Management," in *Nikita Khrushchev*, ed. Sergei Khrushchev, William Taubman, and Abbott Gleason (New Haven, CT: Yale University Press, 2000), 145–46; and Nataliya Kibita, "Moscow-Kyiv Relations and the *Sovnarkhoz* Reform," in *Khrushchev in the Kremlin: Policy and Government in the Soviet Union, 1953–1964*, ed. Jeremy Smith and Melanie Ilič (New York: Routledge, 2011), 94–111.
119. GANISK, f. 1, op. 2, d. 7923, ll. 3–7.
120. GARF, f. R-9477, op. 1, d. 94, l. 268.
121. GARF, f. R-9477, op. 1, d. 94, ll. 259–60.
122. GARF, f. R-9477, op. 1, d. 94, l. 252.
123. GARF, f. R-9477, op. 1, d. 94, ll. 270–71.
124. GARF, f. R-9477, op. 1, d. 94, l. 258.
125. GARF, f. R-9477, op. 1, d. 94, ll. 279–80. For more on these failures, see GARF, f. R-9477, op. 1, d. 94, ll. 344–50.
126. GARF, f. R-9477, op. 1, d. 94, ll. 254–61.
127. GARF, f. R-9477, op. 1, d. 94, ll. 362–70.
128. GASK, f. 2481, op. 1, d. 423, ll. 1–3.
129. GARF, f. R-5446, op. 95, d. 790, ll. 47–48.
130. GANISK, f. 1, op. 2, d. 8863, l. 5.
131. GASK, f. 2481, op. 1, d. 423, ll. 1–3.
132. GANISK, f. 1, op. 2, d. 7482, ll. 74–75.
133. GANISK, f. 1, op. 2, d. 7916, l. 72.
134. GANISK, f. 1, op. 2, d. 7916, l. 70.

135. Samokhval had answered for poor job performance as early as 1953. GARF, f. R-8300, op. 24, d. 532, ll. 28–32.
136. GANISK, f. 1, op. 2, d. 7917, ll. 102–4.
137. GANISK, f. 1, op. 2, d. 7917, ll. 102–4.
138. GARF, f. A-259, op. 7, d. 8050, ll. 18–20.
139. GARF, f. R-5446, op. 94, d. 201, l. 15.
140. Of the 1959 figure of 456,800 tons, 162,440 tons were to be of double-cross hybrids. GARF, f. R-5446, op. 93, d. 659, ll. 18–19. The planned 1960 total reached 696,000 tons, of which only 201,200 tons were to be double-cross hybrids. GARF, f. R-5446, op. 94, d. 201, ll. 5–8, l. 12.
141. GARF, f. R-5446, op. 95, d. 800, l. 1.
142. GARF, f. R-5446, op. 96, d. 871, l. 2.
143. GARF, f. R-5446, op. 96, d. 871, ll. 34–37. At the exchange rate in effect from January 1, 1961, this figure amounted to $169 million in 1961, or $1.34 billion in 2015.
144. GARF, f. R-5446, op. 96, d. 871, l. 2.
145. Some historians have credited the bureaucracy with actually helping the country overcome the effects of Khrushchev's supposedly ill-conceived initiatives. As journalist and historian Anatolii Strelianyi wrote, "If the apparat had delayed in carrying out [Khrushchev's] decisions, the harvests would have been greater." "As far as possible," he continued, "the apparat and in particular its lower reaches adapted [the corn policies] to real conditions, otherwise the results would have been even more deplorable." "Khrushchev and the Countryside," in *Nikita Khrushchev*, ed. Khrushchev, Taubman, and Gleason, 131–32.
146. "The party apparat had been established," historian Aleksandr Nikonov concluded, to make it "capable of overseeing [implementation], and was properly selected and well schooled." Even efforts made in earnest did Khrushchev's corn initiative "more harm than good." Acting "according to formula," authorities pressed "panaceas" on subordinates as a response to the irresistible agitation to plant corn that began in January 1955. A. A. Nikonov, *Spiral' mnogovekovoi dramy: Agrarnaia nauka i politika Rossii, XVIII–XX vv.* (Moscow: Entsiklopediia rossiiskikh dereven', 1995), 302.
147. For more on this lack of concerted resistance, see Strelianyi, "Khrushchev and the Countryside," 132.

CHAPTER 8

1. N. S. Khrushchev, *Stroitel'stvo kommunizma v SSSR i sel'skoe khoziaistvo*, vol. 5 (Moscow: Gosudarstvennoe izdatel'stvo politicheskoi literatury, 1963), 35.
2. Here, I draw on a typology developed by historian Oleg Khlevniuk, which includes corresponding modes of network operation and responses to initiatives from Moscow. He explains how each came into being, interacted with superiors,

and either achieved stability in personnel or fell victim to internal conflicts. "Regional'naia vlast' v SSSR v 1954–kontse 1950-kh gg.: Ustoichivost' i konflikty," *Otechestvennaia istoriia* no. 3 (2007): 33–39.

3. Yoram Gorlizki, "Scandal in Riazan: Networks of Trust and the Social Dynamics of Deception," *Kritika: Explorations in Russian and Eurasian History* 14, no. 2 (2013): 243–78.

4. In the 1960s, political scientist Jerry Hough demonstrated how regional leaders influenced industrial policy in *The Soviet Prefects: The Local Party Organs in Industrial Decision-Making* (Cambridge, MA: Harvard University Press, 1969). Studies of regions typically have privileged individuals, viewing regions as launching pads for the careers of future central leaders. Khlevniuk, "Regional'naia vlast' v SSSR," 31. After the archives opened, foreign and domestic historians have developed local histories, such as that of Aleksandr Agarev, *Tragicheskaia avantiura: Sel'skoe khoziaistvo Riazanskoi oblasti, 1950–1960 gody; A. N. Larionov, N. S. Khrushcheva i drugie; Dokumenty, sobyitiia, fakty* (Riazan': Russkoe slovo, 2005). E. A. Rees has identified "conflicting centripetal and centrifugal forces" in relationships between center and periphery. He has concluded that centralization peaked under Stalin, while the Khrushchev period saw "a relative moderation" enhancing local authority. "Introduction" in *Centre-Local Relations in the Stalinist State, 1928–1941*, ed. E. A. Rees (New York: Palgrave Macmillan, 2002), 1. Viktor Mokhov has argued that after Stalin, local party organizations increasingly accumulated authority particularly over economic activity. *Regional'naia politicheskaia elita Rossii, 1945–1991 gg.* (Perm: Permskoe knizhnoe izdatel'stvo, 2003), 13.

5. Khrushchev demoted the head of the Kazakh government, Zhumabai Shaiakhmetov, ensuring that the republic carried out policy by putting his people in charge. Martin McCauley, *Khrushchev and the Development of Soviet Agriculture: The Virgin Land Programme, 1953–1964* (New York: Holmes & Meier, 1976), 61.

6. Khlevniuk, "Regional'naia vlast' v SSSR," 33–34.

7. M. Abramov, "Pozitsiia 'udobnaia,' no vrednaia," *Sel'skoe khoziaistvo*, June 8, 1954, 2; M. Abramov, "Biurokraticheskii 'posevnoi agregat,'" *Sel'skoe khoziaistvo*, April 25, 1954, 4.

8. V. Ivanov, "Kukuruza kantseliarskaia," *Komsomol'skaia pravda*, June 23, 1954, 2.

9. RGANI, f. 5, op. 31, d. 23, ll. 4–5

10. This was, in the words of two historians of the era, "a system of state edicts, local directives and, at the same time, local bureaucratic administrative tyranny." M. E. Beznin and T. M. Dimoni, "Krest'ianstvo i vlast' v Rossii v kontse 1930-kh–1950-e gg," in *Mentalitet i agrarnoe razvitie Rossiii (XIX–XX vv.): Materialy mezhdunarodnoi konferentsii, 14–15 iiunia 1994 g.*, ed. V. P. Danilov and L. V. Milov (Moscow: ROSSPEN, 1996), 159.

11. RGASPI, f. 82, op. 2, d. 149, ll. 6–9, cited in O. V. Khlevniuk et al. eds., *Regional'naia politika N. S. Khrushcheva: TsK KPSS i mestnye partiinye komitety, 1953–1954* (Moscow: ROSSPEN, 2009), 57.
12. Khlevniuk, *Regional'naia politika N. S. Khrushcheva*, 58.
13. RGASPI, f. 82, op. 2, d. 153, ll. 202–205, in Khlevniuk, *Regional'naia politika N. S. Khrushcheva*, 73. Kalinin Oblast was the Soviet era name of the region that, after 1990, reverted to its earlier name, Tver.
14. RGANI, f. 2, op. 1, d. 121, l. 66.
15. RGANI, f. 2, op. 1, d. 126, l. 119.
16. TsKhDOPIM, f. 3, op. 159, d. 6, l. 7.
17. TsKhDOPIM, f. 3, op. 159, d. 6, l. 39.
18. TsGAMO, f. 191, op. 1. d. 2152, ll. 28–32.
19. TsGAMO, f. 191, op. 1, d. 2305, l. 12.
20. TsGAMO, f. 191, op. 1, d. 2306, ll. 10–11.
21. TsGAMO, f. 191, op. 1, d. 2306, l. 21.
22. Mylarshchikov had served as Khrushchev's subordinate in the Moscow oblast party committee from 1951 to 1953. Anatolii Strelianyi, "The Last Romantic," in *Memoirs of Nikita Khrushchev*, vol. 2, *Reformer, 1945–1964*, ed. Sergei Khrushchev, trans. Stephen Shenfield and George Shriver (University Park: Pennsylvania State University, 2006), 650n115. For more on the Central Committee apparat, its powers, and its evolution under Khrushchev, see Alexander Titov, "The Central Committee Apparatus under Khrushchev," in *Khrushchev in the Kremlin: Policy and Government in the Soviet Union, 1953–1964*, ed. Jeremy Smith and Melanie Ilič (New York: Routledge, 2011), 41–60.
23. RGANI, f. 5, op. 30, d. 157, l. 141.
24. RGANI, f. 5, op. 30, d. 157, ll. 146–47.
25. GARF, f. A-259, op. 7, d. 8050, ll. 44–46.
26. RGANI, f. 5, op. 30, d. 157, ll. 146–47.
27. F. I. Loshchenkov, *Ot Stalina do Gorbacheva: Zhizennye nabliudeniia* (Iaroslavl: LIA, 2000), 29.
28. RGANI, f. 5, op. 30 d. 157, l. 150.
29. GARF, f. R-5446, op. 90, d. 198, ll. 99–101.
30. Khlevniuk, "Regional'naia vlast'" v SSSR," 31.
31. Ibid., 47.
32. Oleg Khlevniuk, "The Economy of Illusions: The Phenomenon of Data-Inflation in the Khrushchev Era," in *Khrushchev in the Kremlin*, ed. Smith and Ilič, 171–89.
33. RGASPI, f. 556, op. 14, d. 106, ll. 59–63.
34. Gorlizki, "Scandal in Riazan," 258.
35. See Agarev, *Tragichestkaia avantiura* and Roy and Zhores Medvedev, *Khrushchev: The Years in Power*, trans. Andrew Durkin (New York: Columbia University Press, 1975), 96–97.

36. Agarev, *Tragicheskaia avantiura*, 25.
37. Taubman, *Khrushchev*, 376–78; and Medvedev and Medvedev, *Khrushchev*, 98–101.
38. Khlevniuk, *Regional'naia politika N. S. Khrushcheva*, 13.
39. Gorlizki, "Scandal in Riazan," 273–77.
40. "Khrushchevskie vremena: Niprinuzhdennye besedy s politicheskimi deiateliami 'velikogo desiatiletiia'; A. N. Shelepin, V. E. Semichastnyi, N. G. Egorychev; Zapisi N. A. Barsukova," in *Neizvestnaia Rossiia XX vek*, vol. 1 (Moscow: Istoricheskoe nasledie, 1992), 275.
41. G. L. Smirnov, "Malenkie sekrety bol'shogo doma: Vospominaniia o rabote v apparate TsK KPSS," in *Neizvestnaia Rossiia XX vek*, vol. 3 (Moscow: Istoricheskoe nasledie, 1993), 372.
42. Khlevniuk, *Regional'naia politika N. S. Khrushcheva*, 13.
43. RGANI, f. 5, op. 30, d. 361, ll. 143–47.
44. RGANI, f. 5, op. 31, d. 144, ll. 73–82, cited in Khlevniuk, *Regional'naia politika N. S. Khrushcheva*, 257.
45. GARF, f. R-5446, op. 95, d. 327, ll. 2–12, cited in Khlevniuk, *Regional'naia politika N. S. Khrushcheva*, 345.
46. RGANI, f. 5, op. 31, d. 144, ll. 73–82, cited in Khlevniuk, *Regional'naia politika N. S. Khrushcheva*, 259.
47. RGANI, f. 5, op. 31, d. 144, l. 183, cited in Khlevniuk, *Regional'naia politika N. S. Khrushcheva*, 260–61n2.
48. RGANI, f. 5, op. 31, d. 168, l. 9.
49. RGANI, f. 5, op. 31, d. 168, l. 19.
50. Khlevniuk, "Regional'naia vlast' v SSSR," 47. On weak secretaries, see Gorlizki, "Scandal in Riazan," 265–73.
51. RGANI, f. 5, op. 31, d. 167, ll. 108–37, cited in Khlevniuk, *Regional'naia politika N. S. Khrushcheva*, 367.
52. RGANI, f. 5, op. 31, d. 167, l. 3.
53. RGANI, f. 5, op. 31, d. 167, l. 18 and l. 74.
54. RGANI, f. 5, op. 31, d. 168, l. 20.
55. GANISK, f. 1, op. 2, d. 8597, ll. 9–11.
56. RGANI, f. 5, op. 31, d. 167, l. 7.
57. RGANI, f. 5, op. 31, d. 13, ll. 48–49.
58. P. E. Shelest, *Da ne sudimye budete: Dnevnikovye zapisi, vospominaniia chlena Politburo TsK KPSS* (Moscow: Edition Q, 1995), 112–28.
59. Ibid., 134–35.
60. A. Kozlov and P. Filimonov, "Berezanskie ochkovtirateli," *Izvestiia*, December 17, 1959, 2.
61. Shelest, *Da ne sudimy budete*, 130 and 137–38.
62. RGANI, f. 5, op., 31, d. 168, l. 42.
63. Shelest, *Da ne sudimy budete*, 144.

64. RGANI, f. 5, op. 31, d. 168, l. 40.
65. RGANI, f. 5, op. 31, d. 168, ll. 41–44.
66. RGASPI, f. 556, op. 22, d. 381, l. 57; and GARF, f. R-8300, op. 24, d. 1363a, ll. 4–5. For more, see Alena Ledeneva, *Russia's Economy of Favors: Blat, Networking, and Informal Exchange* (New York: Cambridge University Press, 1998), 25.
67. GARF, f. 9477, op. 1, d. 358, ll. 16–17.
68. The figure of 4.5 million rubles amounted to $1.125 million in 1960 at the official exchange rate, or $9.04 million in 2015. The sum of 44,000 rubles equaled $11,000, or approximately $88,400 in 2015.
69. RGANI, f. 5, op. 31, d. 168, l. 43.
70. RGANI, f. 5, op. 31, d. 168, l. 43.
71. Shelest, *Da ne sudimy budete*, 145.
72. RGANI, f. 5, op. 31, d. 168, l. 43.
73. Strelianyi, "Khrushchev and the Countryside," 122.
74. Khlevniuk, "Regional'naia vlast' v SSSR," 47–48.
75. Gorlizki, "Too Much Trust," 692.
76. N. S. Khrushchev, *Stroitel'stvo kommunizma v SSSR i sel'skoe khoziaistvo*, vol. 1 (Moscow: Gosudarstvennoe izdatel'stvo politicheskoi literatury, 1962), 165.
77. RGANI, f. 2, op. 1, d. 121, l. 98.
78. RGANI, f. 2, op. 1, d. 121, ll. 109–11.
79. N. S. Khrushchev, *Stroitel'stvo kommunizma v SSSR i sel'skoe khoziaistvo*, vol. 6 (Moscow: Gosudarstvennoe izdatel'stvo politicheskoi literatury, 1963), 216–17.
80. Ibid., 221.
81. "Tak li nado ispol'zovat' kubanskii chernozem?" *Sel'skaia zhizn'*, October 13, 1961, 2.
82. Khrushchev, *Stroitel'stvo kommunizma v SSSR*, 6:60–61.
83. Khrushchev, *Stroitel'stvo kommunizma v SSSR*, 6:69.
84. N. S. Khrushchev, *Stroitel'stvo kommunizma v SSSR i sel'skoe khoziaistvo*, vol. 8 (Moscow: Gosudarstvennoe izdatel'stvo politicheskoi literatury, 1964), 103.
85. GANISK, f. 1, op. 2, d. 8487, l. 29.
86. GANISK, f. 1, op. 2, d. 8487, l. 27.
87. GANISK, f. 1, op. 2, d. 8487, l. 28.
88. GANISK, f. 1, op. 2, d. 8487, l. 32. Typically resulting from dryland irrigation, soil salination is a concentration of salt in the soil that, if unchecked, cuts productivity.
89. GANISK, f. 1, op. 2, d. 8487, ll. 25–26.
90. Strelianyi, "Khrushchev and the Countryside," in *Nikita Khrushchev*, ed. William Taubman, Sergei Khrushchev, and Abbott Gleason (New Haven, CT: Yale University Press, 2000), 124.
91. GANISK, f. 1, op. 2, d. 8597, l. 189.
92. GANISK, f. 5351, op. 1, d. 307, l. 22 and l. 74.

93. For more on the distinctive influence of nationalism on the republic, see Saulius Grybkauskas, "The Role of the Second Party Secretary in the 'Election' of the First: The Political Mechanism for the Appointment of the Head of Soviet Lithuania in 1974," *Kritika: Explorations in Russian and Eurasian History* 14, no. 2 (2013): 344.
94. E. Iu. Zubkova, *Pribaltika i Kreml', 1940–1953 gg.* (Moscow: ROSSPEN, 2008), 320–22.
95. Ibid., 259.
96. Ibid., 322. For more on the earlier periods, see Francine Hirsch, *Empire of Nations: Ethnographic Knowledge and the Making of the Soviet Union* (Ithaca, NY: Cornell University Press, 2005); and Terry Martin, *Affirmative Action Empire: Nations and Nationalism in the Soviet Union, 1923–1939* (Ithaca, NY: Cornell University Press, 2001).
97. Zubkova, *Pribaltika i Kreml'*, 285.
98. Jeremy Smith, "Leadership and Nationalism in the Soviet Republics, 1951–1959," in *Khrushchev in the Kremlin*, ed. Smith and Ilič, 79–93.
99. For an overview, see Ronald Suny and Terry Martin, eds., *A State of Nations: Empire and Nation-Making in the Age of Lenin and Stalin* (New York: Oxford University Press, 2001).
100. Diana Mincyte, "Everyday Environmentalism: The Practice, Politics, and Nature of Subsidiary Farming in Stalin's Lithuania," *Slavic Review* 68, no. 1 (2009): 32–34.
101. LYA, f. 1771, op. 149, d. 302, l. 30.
102. RGANI, f. 2, op. 1, d. 124, l. 12.
103. LYA, f. 1771, op. 149, d. 203, l. 30.
104. RGAE, f. 7486, op. 1, d. 8015, l. 3.
105. RGAE, f. 7486, op. 1, d. 8015, l. 10.
106. LYA, f. 1771, op. 161, d. 13, l. 46.
107. LYA, f. 1771, op. 161, d. 13, ll. 70–71.
108. LYA, f. 1771, op. 161, d. 13, ll. 73–74.
109. LYA, f. 1771, op. 209, d. 24, ll. 1–2.
110. LYA, f. 1771, op. 210, d. 20, ll. 1–4; and LYA, f. 1771, op. 210, d. 24, l. 119 and ll. 131–44.
111. LYA, f. 1771, op. 207, l. 138, ll. 173–200. I have cited a Russian translation of the text. The original Lithuanian text can be found in the same file: LYA, f. 1771, op. 207, d. 138, ll. 150–72.
112. P. Vasinauskas et al., "My protiv neplodorodnykh khleverishch," *Sovetskaia Litva*, November 12, 1961, 1.
113. LYA, f. 177, op. 207, d. 137, l. 6.
114. I. Raudeliunas, "Zemle nuzhny zabotlivye ruki," *Sovetskaia Litva*, December 22, 1961, 2–3.
115. Khrushchev, *Stroitel'stvo kommunizma v SSSR*, 6:322–28.

116. LYA, f. 1771, op. 218, d. 33, ll. 11–12.
117. See, for example V. Figurinas, "Bez travopol'ia, za vysokii urozhai," *Sovetskaia Litva*, January 17, 1962, 1; B. Melamed, "Reshitel'no perekhodit' na propashnuiu sistemu," *Sovetskaia Litva*, January 20, 1962, 2; and P. Vilaishis, "Nashi plany i dela," *Sovetskaia Litva*, March 28, 1962, 2.
118. A. Davidonis, "S travopol'shchikami ne po puti," *Sovetskaia Litva*, January 18, 1962, 1.
119. "Kurs—na propashnuiu sistemu zemledeliia: S mezhdudistrictnogo soveshchaniia rabotnikov sel'skogo khoziaistva v Kaunas," *Sovetskaia Litva*, January 25, 1962, 2.
120. LYA, f. 1771, op. 220, d. 20, l. 20.
121. LYA, f. 1771, op. 218, d. 43, ll. 41–43; and d. 44, ll. 339–40.
122. LYA, f. 1771, op. 220, d. 20, ll. 138–39.
123. "Rech' tovarishcha A. Iu. Snechkusa," *Sovetskaia Litva*, March 8, 1962, 2.
124. LYA, f. 1771, op. 218, d. 8, ll. 2–3.
125. LYA, f. 1771, op. 218, d. 8, ll. 21–24.
126. RGANI, f. 5, op. 31, d. 197, l. 133.
127. LYA, f. 16895, op. 2, d. 92, ll. 4–8.
128. LYA, f. 16895, op. 2, d. 92, ll. 8–16.
129. LYA, f. 16895, op. 2, d. 92, ll. 16–20.
130. LYA, f. 16895, op. 2, d. 92, ll. 16–20.
131. See, for example, "S chest'iu vypolnim zadachi, postavlennye plenumom TsK KPSS: S mezhdistrictnogo soveshchaniia rabotnikov sel'skogo khoziaistva v Dotnuve," *Sovetskaia Litva*, February 4, 1961, 1.
132. LYA, f. 16895, op. 2, d. 92, l. 20.
133. LYA, f. 16895, op. 2, d. 92, l. 17.
134. Vytautas Tininis, *Sniečkus: 33 metai valdžioje; Antano Sniečkus biografinė apybraiža* (Vilnius: Antrasis papildytas pataisytas leidimas, 2000), 171. I am grateful to Saulius Grybkauskas for alerting me to this source, and for translating the relevant section into English.
135. On the second secretary, see Grybkauskas, "Role of the Second Party Secretary," 343–44.
136. LYA, f. 1771, op. 207, d. 28, ll. 394–95.
137. RGANI, f. 2, op. 1, d. 347, l. 24.
138. LYA, f. 16985, op. 2, d. 92, ll. 77–79; and RGANI, f. 2, op. 1, d. 347, l. 21.
139. "Khrushchevskie vremena," 293.
140. LYA, f. 16985, op. 2, d. 92, l. 18.
141. Yuri Slezkine, "The USSR as a Communal Apartment, or How a Socialist State Promoted Ethnic Particularism," *Slavic Review* 53, no. 3 (1994): 414–52.
142. Ronald Suny, *The Revenge of the Past: Nationalism, Revolution, and the Collapse of the Soviet Union* (Stanford, CA: Stanford University Press, 1993), 117.

143. As Donald Raleigh has noted, local studies illuminate larger questions about big events and decisive moments because they bring to light each locale's unique features and interactions with the center. "Introduction," in *Provincial Landscapes: Local Dimensions of Soviet Power, 1917–1953*, ed. Raleigh (Pittsburgh: University of Pittsburgh Press, 2001), 1–5. Peter Holquist has similarly contended that new archival studies of regions and republics must not simply add "local color" to existing accounts of high politics and ideologically conditioned conventions, based only on evidence form the central archives. "A Tocquevillean 'Archival Revolution': Archival Change in the *Longue Durée*," *Jahrbücher für Geschichte Osteuropas* 51 (2003): 77–83.

CONCLUSION

1. Wendell Berry, *Bringing It to the Table: On Farming and Food* (Berkeley, CA: Counterpoint, 2009), 227.
2. Rósa Magnúsdóttir, "Keeping Up Appearances: How the Soviet State Failed to Control Popular Attitudes toward the United States of America, 1945–1959" (PhD diss., University of North Carolina at Chapel Hill, 2006).
3. RGASPI, f. 556, op. 14, d. 187, l. 49.
4. Aleksandr Genis and Peter Vail, *Shestidesiatie: Mir sovetskogo cheloveka* (Ann Arbor, MI: Ardis, 1988), 203.
5. A. Muravlev, "'Potëmkinskaia' shtabka dlia Khrushcheva," *Al'taiskaia pravda*, April 17, 2009; G. Petrov, "V SShA vspominaiut Nikitu Sergeevicha," *Novye Izvestiia*, August 28, 2009, 2; A. Gasiuk, "V Aiove po-russki," *Rossiiskaia gazeta*, August 31, 2009, 5; O. Sul'kin, "Khrushchev, syn Khrushcheva," *Itogi*, June 28, 2010, 32–39; P. Romanov, "Stavka na tsaritsu polei," *Izvestiia*, September 3, 2010, 22: B. Zolotov, "Vot by Khrushchev poradovalsia," *Kubanskie novosti*, August 14, 2013; and O. Nikol'skaia, "Kukuruza vnov' tsaritsa polei," *Vechernaia Moskva*, April 21, 2008.
6. In April 2013, independent pollsters at the Levada Center found that only 6 percent of respondents had a positive view of Khrushchev. A further 39 percent held "more positive than negative" sentiments. This placed Khrushchev ahead only of Mikhail Gorbachev and Boris Yeltsin, and behind Brezhnev, Lenin, Stalin, and Nicholas II. By contrast 35 percent had negative or largely negative views, and 21 percent either did not know of Khrushchev or gave no clear answer to the question. Iurii Levada Analitical Center, "Otnoshenie rossiian k glavam rossiiskogo gosudarstva raznogo vremeni," May 22, 2013, http://www.levada.ru/2013/05/22/otnoshenie-rossiyan-k-glavam-rossijskogo-gosudarstva-raznogo-vremeni/. As historian Oleg Khlevniuk observes, some disdain for Khrushchev stems from attempts to delegitimize the leader who desacralized Stalin, who remains a revered figure. *Stalin: Zhizn' odnogo vozhdia* (Moscow: Corpus, 2015), 445.

7. Iurii Levada Analytical Center, "Sobytiia eopkhi Khrushcheva," March 29, 2016, http://www.levada.ru/2016/03/29/sobytiya-epohi-hrushheva/.
8. RGANI, f. 2, op. 1, d. 780, ll. 104–5.
9. RGANI, f. 5, op. 45, d. 368, ll. 84–92.
10. RGANI, f. 5, op. 45, d. 368, ll. 84–92.
11. RGANI, f. 5, op. 45, d. 368, l. 146.
12. Roy Medvedev and Zhores Medvedev, *Khrushchev: The Years in Power*, trans. Andrew Durkin (New York: Columbia University Press, 1975), 128 and 182.
13. USSR Council of Ministers Central Statistical Department, *Sel'skoe khoziaistvo SSSR: Statisticheskii sbornik* (Moscow: Statistika, 1971), 119–30.
14. USSR Council of Minsters Central Statistical Department, *Sel'skoe khoziaistvo SSSR: Statisticheskii sbornik* (Moscow: Finansy i statistika, 1988), 70 and 192.
15. Andrew Sloin and Oscar Sanchez-Sibony, "Economy and Power in the Soviet Union, 1917–1939," *Kritika: Explorations in Russian and Eurasian History* 15, no. 1 (2014): 9–12. Sloin and Sanchez–Sibony echo Sheila Fitzpatrick, a pioneer of critiques of this first-generation scholarship on Stalin. She commented that scholars had subjected these dictums to a "reversal of signs." "Politics as Practice: Thoughts on a New Soviet Political History," *Kritika: Explorations in Russian and Eurasian History* 5, no. 1 (2004): 36.
16. One prominent contributor, Jerzy Karcz, came to consider Khrushchev's initiatives evidence of "cavalier treatment of stark economic reality." The leader's epochal programs to plant corn, plow up millions of hectares of uncultivated land, and compete with America suggested that he was governed by "the tendency to bend reality to one's own wishes" and an accompanying "propensity to write off economic calculations as deprived of the necessary degree of vision." To Karcz, incessant directives and speeches full of "advice" were evidence not of Khrushchev's limited power, but that his word carried "the force of law." Jerzy Karcz, "Khrushchev's Impact on Soviet Agriculture," *Agricultural History* 40, no. 1 (1966): 31–33. Other scholars argued that Khrushchev substituted slogans for rational financial management. Savaging the leader for imitating the United States by planting corn, Naum Jasny alleged that no one had calculated the economic efficacy of that or any other policy. *Khrushchev's Crop Policy* (Glasgow, UK: Outram, 1965), 11–17.
17. To Karcz, the relationship between "market" and "command" forces in the agrarian economy was one of "fundamental antinomy." Dictating minute details, bureaucracies robbed market-style incentives of effectiveness. This made the Soviet Union fundamentally different from systems conceived of as polar opposites, which were founded on markets, private property, and material incentives. Karcz, "Khrushchev's Impact on Soviet Agriculture," 24–25.
18. Sovietologist Roy Laird wrote that Khrushchev's adherence to peaceful competition and renunciation of political violence was balanced by his inability to enact true reform. Khrushchev "was at best . . . a transitional figure unable to alter the

course charted by Stalin," Laird wrote. "This was certainly the case in agriculture," he continued, where kolkhozes "were finally totally subsumed under the single bureaucratic pyramid that now encompasses the whole Soviet society." Laird, "Khrushchev's Administrative Reforms in Agriculture: An Appraisal," in *Soviet and East European Agriculture*, ed. Jerzy Karcz (Berkeley: University of California Press, 1967), 29–30.

19. In 1970, an interdisciplinary project brought together scholars studying politics, administrative practices, social change, legal codes, technological development, and household economics. James Millar, ed., *The Soviet Rural Community* (Urbana: University of Illinois Press, 1971). This was last in a line of volumes from the United States that included Jerzy Karcz, ed., *Soviet and East European Agriculture* (Berkeley: University of California Press, 1967).

20. In the 1970s, Martin McCauley re-evaluated the Virgin Lands program, qualifying the established verdict that the endeavor had been poor in conception and a fiasco in practice. *Khrushchev and the Development of Soviet Agriculture: The Virgin Land Programme, 1953–1964* (New York: Holmes & Meier, 1976). In dissident publishing circles and then in translation, Roy and Zhores Medvedev offered their nuanced analysis of Khrushchev's policies, praising the successes of his first five years while condemning the shortcomings of the second five. Medvedev and Medvedev, *Khrushchev*.

21. M. L. Bogdenko et al., *Istoriia Krest'ianstva SSSR*, vol. 4, *Krest'ianstvo v gody uprocheniia i razvitiia sovetskogo obshchestva, 1945–konets 50-kh gg.* (Moscow: Nauka, 1988), 5.

22. Textbook explanations of the corn crusade continue to emphasize climatic and technical factors, considering Khrushchev blind to them. "Khrushchev's 1955 scheme to turn vast areas of arable earth into Iowa-like cornfields to feed both livestock and humans turned sour because of unsuitable soil and climate and popular resistance to eating corn," one concludes. "Agriculture remained the weakest link in the system." Catherine Evtuhov and Richard Stites, *A History of Russia since 1800: Peoples, Legends, Events, Forces* (New York: Houghton Mifflin, 2004), 437.

23. Finally able to publish Khrushchev's name, historians renewed inquiry into his era, challenging dictums about "subjectivism" and "volunteerism" handed down by the Brezhnev leadership when they ousted him. Some opponents defended orthodox views. David Nordlander, "Khrushchev's Image in the Light of Glasnost and Perestroika," *Russian Review* 52, no. 2 (1993): 248–64.

24. A. A. Nikonov, *Spiral' mnogovekovoi dramy: Agrarnaia nauka i politika Rossii, XVIII–XX vv.* (Moscow: Entsiklopediia rossiiskikh derreven', 1995); and I. E. Zelenin, *Agrarnaia politika N. S. Khrushcheva i sel'skoe khoziaistvo* (Moscow: Institut istorii Rossiiskoi Akademii Nauk, 2001).

25. "The term 'Khrushchev's reforms' is . . . arbitrary," historian Elena Zubkova concluded, "Reform is a program of consecutive actions directed toward changing

existing political and economic structures or toward their complete replacement." The policies of Khrushchev were "difficult to view ... as coherent and systematic." Although seeking improvement, he did not disavow centralization, state ownership, kolkhozes, and other fundamentals. "The Rivalry with Malenkov," in *Khrushchev*, ed. William Taubman, Sergei Khrushchev, and Abbott Gleason (New Haven, CT: Yale University Press, 2000), 83–84; and *Obshchestvo i reformy, 1945–1964* (Moscow: Rossiia molodaia, 1993), 186.

26. As historian Nick Cullather argues, the Green Revolution was a story not of technology alone, as it is so often told today. Instead, the key variable in India, the Philippines, and Vietnam was incentives, which were shaped by government policies favoring peasant producers prepared to apply technology capable of increasing marketable output of wheat, rice, and other staples. *The Hungry World: America's Cold War Battle against Poverty in Asia* (Cambridge, MA: Harvard University Press, 2010), 232–71.

27. Timothy Mitchell, *Carbon Democracy: Political Power in the Age of Oil* (New York: Verso, 2011), 140–41. On the Soviet Union as a producer of hydrocarbons, see Philip Hanson, *The Rise and Fall of the Soviet Economy: An Economic History of the USSR from 1945* (New York: Longman, 2003).

28. Thomas Piketty, *Capital in the Twenty-First Century*, trans. Arthur Goldhammer (Cambridge, MA: Belknap, 2014), 184.

29. Eric Hobsbawm, *Age of Extremes: The Short Twentieth Century, 1914–1991* (London: Vintage, 1996), 4.

30. Martin McCauley, *Gorbachev* (New York: Longman, 1998), 57.

31. Moshe Lewin, *The Gorbachev Phenomenon: A Historical Interpretation* (Berkeley: University of California Press, 1988), 97–98. McCauley, *Gorbachev*, 66. As political scientist George Breslauer shows, these initiatives were the first stages of what only in 1987 and 1988 became a revolution from above designed to democratize society and dismantle the state-run economy. *Gorbachev and Yeltsin as Leaders* (New York: Cambridge University Press, 2002), 56.

GLOSSARY AND NOTES ON TRANSLITERATION

hectare a metric measurement of area equal to 10,000 square meters—a square with sides 100 meters long—and to 2.47 acres

kolkhoz a nominally independent, democratically governed collective farm (pl. kolkhozes)

kolkhoznik a member of a collective farm (pl. kolkhozniks)

labor-day [Russian: *trudoden*] a unit measuring a kolkhoznik's fulfillment of a daily labor norm, rather than time spent working

metric ton 1,000 kilograms; at 2,200 pounds, 10 percent larger than an English ton

MTS [machine-tractor station] a state enterprise controlling the use of machines, specialists, and other services on the kolkhozes in its locale, all for large in-kind payments of produce; disbanded in 1958

soviet council elected in single-candidate elections for an administrative region from the district to oblast, krai, republic, union; formally a legislative body governing through its executive committee, but in practice possessing little authority

sovkhoz state-owned and -operated farm employing wage laborers

Administrative Divisions

ASSR Autonomous Soviet Socialist Republic; administrative division equal to an oblast in status, but named for a titular ethnonational group

district [Russian: *raion*] in rural areas approximating a county in the United States, but in urban ones closer in size to a ward or borough

krai large administrative unit found within the RSFSR and Kazakh SSR; like an oblast, subordinate to the republic, but also encompassing one or more autonomous oblasts home to a titular ethnonational group

oblast a region comprising a sizable city and its hinterland; a subdivision within some of the fifteen union republics (SSRs)

RSFSR Russian Soviet Federative Socialist Republic, the largest of the constituent SSRs in population, area, and economic output

SSR Soviet Socialist Republic, a constituent union republic of the Soviet Union; during most of the Khrushchev era, the union included fifteen

USSR Union of Soviet Socialist Republics, the Soviet Union

NOTES ON TRANSLITERATION

To avoid confusion, I have used place names common in English: for example Moscow, rather than *Moskva*. I have followed the full Library of Congress system of transliteration in notes and text, except that I have omitted apostrophes used for soft signs from words in the text, as when using oblast instead of *oblast'*. Citations of published works in the Bibliography and Notes sections, however, have strictly followed Library of Congress transliteration conventions. In light of the frequency and familiarity of the family names of Khrushchev and Gorbachev, I have dropped the accent that their final vowel should carry. To avoid privileging Russian in non-Russian contexts, I have, where possible without harming clarity, utilized personal names of non-Russians and place names in non-Russian union republics spelled or transliterated according to the national language. For instance: the capital of Ukraine is rendered as Kyiv, rather than Kiev, and the Latvian party leader is Jānis Kalnbērziņs rather than Ian Kalnberzin. In direct quotes, names remain as transliterated in the original.

BIBLIOGRAPHY

ARCHIVES

Center for Preservation of Documents of the Socio-Political History of Moscow [*Tsentr khraneniia dokumentov obshchestvenno-politicheskoi istorii Moskvy*, TsKhDOPIM], reading room no. 2 of the Central State Archive of the City of Moscow [*Tsentral'nyi gosudarstvennyi arkhiv goroda Moskvy*, TsGAM]
Central State Archive of Moscow Oblast [*Tsentral'nyi gosudarstvennyi arkhiv Moskovskoi oblasti*, TsGAMO]
Central State Archive of Social Organizations of Ukraine [*Tsentral'niy derzhavniy arkhiv hromads'kykh ob'iednan' Ukraïny*, TsDAHOU, Kyiv]
Lithuanian Special Archive [*Lietuvos Ypatingasis Arkhivas*, LYA, Vilnius]
Russian State Archive of Cinema and Photo Documentation [*Rossiiskii gosudarstvennyi arkhiv kinofotodokumentov*, RGAKFD, Moscow]
Russian State Archive of Contemporary History [*Rossiiskii gosudarstvennyi arkhiv noveishei istorii*, RGANI, Moscow]
Russian State Archive of the Economy [*Rossiiskii gosudarstvennyi arkhiv ekonomiki*, RGAE, Moscow]
Russian State Archive of Socio-Political History [*Rossiiskii gosudarstvennyi arkhiv sotsial'no-politicheskoi istorii*, RGASPI, Moscow]
State Archive of the Contemporary History of Stavropol Krai [*Gosudarstvennyi arkhiv noveishei istorii Stavropol'skogo kraia*, GANISK, Stavropol]
State Archive of the Russian Federation [*Gosudarstvennyi arkhiv Rossiiskoi Federatsii*, GARF, Moscow]

State Archive of Stavropol Krai [*Gosudarstvennyi arkhiv Stavropol'skogo kraia*, GASK, Stavropol]

DOCUMENT COLLECTIONS AND REFERENCE MATERIALS

XIII s"ezd Vsesoiuznogo leninskogo kommunisticheskogo soiuza molodëzhi: Stenograficheskii otchet; 14–18 aprelia 1958 g. Moscow: Izdatel'stvo TsK VLKSM "Molodaia gvardiia," 1959.

Artizov, A. N., V. P. Naumov, M. Iu. Prozumenshchikov, Iu. V. Sigachev, N. G. Tomilina, and I. N. Shevchuk, eds. *Nikita Khrushchev, 1964: Stenogrammy Plenuma TsK KPSS i drugie dokumenty*. Moscow: Mezhdunarodnyi fond "Demokratiia," 2007.

Communist Party of China. *The Polemic on the General Line of the International Communist Movement*. Beijing: Foreign Languages Press, 1965. https://www.marxists.org.

Communist Party of the Soviet Union. *Programma kommunisticheskoi partii Sovetskogo Soiuza: Priniata XXII s"ezdom KPSS*. Moscow: Izdatel'stvo politicheskoi literatury, 1964.

Direktivy KPSS i Sovetskogo pravitel'stva po khoziaistvennym voprosam, 1917–1957; vol. 3, *1946–1952 gg.*; vol. 4, *1953–1957 gg.* Moscow: Gosudarstvennoe izdatel'stvo politicheskoi literatury, 1958.

Fursenko, A. A., ed. *Prezidium TsK KPSS, 1954–1964*. Vol. 1, *Chernovye protokol'nye zapisi zasedanii: Stenogrammy*. Moscow: ROSSPEN, 2004.

Garst, Roswell. *Letters from an American Farmer: The Eastern European and Russian Correspondence of Roswell Garst*. Edited by Richard Lowitt and Harold Lee. DeKalb: Northern Illinois University Press, 1987.

Iurii Levada Analytical Center. "Otnoshenie rossiian k glavam rossiiskogo gosudarstva raznogo vremeni." May 22, 2013. http://www.levada.ru/2013/05/22/otnoshenie-rossiyan-k-glavam-rossijskogo-gosudarstva-raznogo-vremeni.

Iurii Levada Analytical Center. "Sobytiia epokhi Khrushcheva." March 29, 2016. http://www.levada.ru/2016/03/29/sobytiya-epohi-hrushheva.

Khlevniuk, O. V., M. Iu. Prozumenshchikov, V. Iu. Vasil'ev, I. Gorlitskii, T. Iu. Zhukova, V. V. Kondrashin, L. P. Kosheleva, R. A. Podkur, and E. V. Sheveleva, eds. *Regional'naia politika N. S. Khrushcheva. TsK KPSS i mestnye partiinye komitety, 1953–1964 gg.* Moscow: ROSSPEN, 2009.

Khrushchev, N. S. *Stroitel'stvo kommunizma v SSSR i sel'skoe khoziaistvo*. 8 vols. Moscow: Gosudarstvennoe izdatel'stvo politicheskoi literatury, 1962–1964.

Kovaleva, N. V., A. V. Korotkov, S. A. Mel'chin, A. I. Stepanov, and Iu. V. Sigachev, eds. *Molotov, Malenkov, Kaganovich, 1957: Stennogramma iiun'skogo plenuma TsK KPSS i drugie dokumenty*. Moscow: Mezhdunarodnyi fond "Demokratiia," 1998.

Lenin, V. I. *Polnoe sobranie sochenenie*. 55 vols. Moscow: Gosudarstvennoe izdatel'stvo politicheskoi literatury, 1958–1965.

Letopis' gazetnykh statei. Moscow: Izdatel'stvo Vsesoiuznoi knizhnoi palaty, 1936–.
Letopis' izobrazitel'nogo isskustva. Moscow: Vsesoiuznaia knizhnaia palata, 1944–.
Letopis' zhurnal'nykh statei. Moscow: Izdatel'stvo Vsesoiuznoi knizhnoi palaty, 1926–.
Tomilina, N. G., A. N. Artizov, L. A. Velichanskaia, I. V. Kazarina, M. Iu. Prozumenshchikov, and S. D. Tavanets, eds. *Nikita Sergeevich Khrushchev: Dva tsveta vremeni; Dokumenty iz lichnogo fonda N. S. Khrushcheva,* vol. 2. Moscow: Mezhdunarodnyi fond "Demokratiia," 2009.
USSR Council of Ministers Central Statistical Department. *Sel'skoe khoziaistvo SSSR: Statisticheskii sbornik.* Moscow: Statistika, 1971.
USSR Council of Minsters Central Statistical Department. *Sel'skoe khoziaistvo SSSR: Statisticheskii sbornik.* Moscow: Finansy i statistika, 1988.

PUBLISHED PRIMARY SOURCES

Emel'ianov, I. E. *Kukuruza: Bibliograficheskii ukazatel' otechestvennoi literatury za 1794–1959 gg.* Moscow: Izdatel'stvo Ministerstva sel'skogo khoziaistva SSSR, 1961.
Jackson, W. A. Douglas. *The Nature and Structure of Soviet Agriculture: A Report for the Use of Specialists in the Field of Agriculture Planning to Visit the Soviet Union.* New York: Institute of International Education, 1963.
Kharlamov, M. A., and O. Vadeev, eds. *Litsom k litsu s Amerikoi: Rasskaz o poezdke N. S. Khrushcheva v SShA, 15–27 sentiabria 1959 g.* Moscow: Gosudarstvennoe izdatel'stvo politicheskoi literatury, 1960.
Ozernyi, M. E. *Kak ia vyrashchivaiu kukuruzu.* Moscow: Ministerstvo sel'skogo khoziaistva SSSR, 1955.
Ozernyi, M. E. *Sovety vyrashchivanie kukuruzy: Otvety M. E. Ozernogo na voprosy kolkhoznikov.* Moscow: Moskovskii rabochii, 1955.
Shevchenko, A. S. *Kukuruza: Dlia obmena opyta dveri shiroko otkryty.* Moscow: Izdatel'stvo sel'skokhoziaistvennoi literatury, zhurnalov, i plakatov, 1961.

MEMOIRS

Khrushchev, Nikita. *Memoirs of Nikita Khrushchev.* Vol. 2, *Reformer (1945–1964);* vol. 3, *Statesman (1953–1964).* Edited by Sergei Khrushchev and translated by Stephen Shenfield and George Shriver. University Park: Pennsylvania State University Press, 2006–2007.
"Khrushchevskie vremena: Niprinuzhdennye besedy s politicheskimi deiateliami 'velikogo desiatiletiia'; A. N. Shelepin, V. E. Semichastnyi, N. G. Egorychev; Zapisi N. A. Barsukova." In *Neizvestnaia Rossiia XX vek,* vol. 1, 270–304. Moscow: Istoricheskoe nasledie, 1992.

Leonov, N. S. *Likholet'e*. Moscow: Mezhdunarodnoe otnosheniia, 1995.
Loshchenkov, F. I. *Ot Stalina do Gorbacheva: Zhizennye nabliudeniia*. Iaroslavl: LIA, 2000.
Mikoian, A. I. *Tak bylo: Razmyshleniia o minuvshem*. Moscow: Vagrius, 1999.
Shelest, P. E. *Da ne sudimye budete: Dnevnikovye zapisi, vospominaniia chlena Politburo TsK KPSS*. Moscow: Edition Q, 1995.
Shepilov, Dmitrii. *The Kremlin's Scholar: A Memoir of Soviet Politics under Stalin and Khrushchev*. Edited by Stephen Bittner. Translated by Anthony Austin. New Haven, CT: Yale University Press, 2007.
Smirnov, G. L. "Malenkie sekrety bol'shogo doma: Vospominaniia o rabote v apparate TsK KPSS." In *Neizvestnaia Rossiia XX vek*, vol. 3, 361–82. Moscow: Istoricheskoe nasledie, 1993.

PERIODICALS

Izvestiia (Moscow)
Komsomol'skaia pravda (Moscow)
Krokodil (Moscow)
Kukuruza (Moscow)
Literaturnaia gazeta (Moscow)
Molodoi leninets (Stavropol)
Moskovskaia pravda (Moscow)
New York Times (New York)
Pravda (Moscow)
Sel'skaia zhizn' (Moscow)
Sel'skoe khoziaistvo (Moscow)
Sovetskaia Litva (Vilnius)
Stavropol'skaia pravda (Stavropol)
Washington Post (Washington, DC)

SECONDARY LITERATURE

Agarev, A. F. *Tragicheskaia avantiura: Sel'skoe khoziaistvo Riazanskoi oblasti, 1950–1960 gg.; A. N. Larionov, N. S. Khrushcheva i drugie; Dokumenty, sobyitiia, fakty*. Riazan: Russkoe slovo, 2005.
Aksiutin, Iu. V., and A. V. Pyzhikov. *Poststalinskoe obshchestvo: Problema liderstva i transformatsiia vlasti*. Moscow: Nauchnaia kniga, 1999.
Allen, Robert C. *Farm to Factory: A Reinterpretation of the Soviet Industrial Revolution*. Princeton, NJ: Princeton University Press, 2003.
Allen, Robert V. *Russia Looks at America: The View to 1917*. Washington, DC: Library of Congress, 1988.

Anderson, Jeremy. "The Soviet Corn Program: A Study in Crop Geography." Ph.D. diss., University of Washington, Seattle, 1964.
Arutiunian, Iu. V. *Sovetskoe krest'ianstvo v gody velikoi otechestvennoi voiny*. 2nd ed. Moscow: Nauka, 1970.
Attwood, Lynne. "Celebrating the 'Frail-Figured Welder': Gender Confusion in Women's Magazines of the Khrushchev Era." *Slavonica* 8, no. 2 (2002): 159–77.
Babiracki, Patryk, and Kenyon Zimmer, eds. *Cold War Crossings: International Travel and Exchange across the Soviet Bloc, 1940s–1960s*. College Station: Texas A&M University Press, 2014.
Bailes, Kendall. "The American Connection: Ideology and the Transfer of American Technology to the Soviet Union, 1917–1941." *Comparative Studies in Society and History* 23, no. 3 (1981): 421–48.
Ball, Alan. *Imagining America: Influence and Images in Twentieth-Century Russia*. Lanham, MD: Rowman & Littlefield, 2003.
Baran, Emily. *Dissent on the Margins: How Soviet Jehovah's Witnesses Defied Communism and Lived to Preach about It*. New York: Oxford University Press, 2014.
Bauer, Raymond, and Alex Inkeles. *The Soviet Citizen: Daily Life in a Totalitarian Society*. Cambridge, MA: Harvard University Press, 1959.
Bauer, Raymond, Alex Inkeles, and Clyde Kluckhohn. *How the Soviet System Works: Cultural, Psychological, and Social Themes*. Cambridge, MA: Harvard University Press, 1956.
Berg, Auri. "Reform in the Time of Stalin: Khrushchev and the Fate of the Russian Peasantry." Ph.D. diss., University of Toronto, 2012.
Berry, Wendell. *Bringing It to the Table: On Farming and Food*. Berkeley, CA: Counterpoint, 2009.
Berry, Wendell. *Citizenship Papers: Essays*. Berkeley, CA: Counterpoint, 2004.
Beznin, M. E., and T. M. Dimoni. "Krest'ianstvo i vlast' v Rossii v kontse 1930-kh–1950-e gody." In *Mentalitet i agrarnoe razvitie Rossii (XIX–XX vv.): Materialy mezhdunarodnoi konferentsii, 14–15 iiunia 1994 g.*, edited by V. P. Danilov and L. V. Milov, 155–167. Moscow: ROSSPEN, 1996.
Bischel, Christine. "'The Drought Does Not Cause Fear': Irrigation History in Central Asia through James C. Scott's Eyes." *Revue d'études comparatives Est-Ouest* 43, no. 1–2 (2012): 73–108.
Bittner, Stephen. "American Roots, French Varietals, Russian Science: A Transnational History of the Great Wine Blight in Late-Tsarist Bessarabia." *Past and Present* no. 227 (2015): 151–77.
Black, Cyril, ed. *The Transformation of Russian Society: Aspects of Social Change since 1861*. Cambridge, MA: Harvard University Press, 1960.
Bogdenko, M. L., O. M. Verbitskaia, I. M. Volkov, M. A. Vyltsan, L. N. Denisova, V. S. Dolgov, A. P. Efremenko, et al. *Istoriia Krest'ianstva SSSR*. Vol. 4, *Krest'ianstvo v gody uprocheniia i razvitiia sovetskogo obshchestva, 1945–konets 50-kh gg*. Moscow: Nauka, 1988.

Bohn, Thomas, Rayk Einax, and Michel Abesser, eds. *De-Stalinization Reconsidered: Persistence and Change in the Soviet Union*. New York: Campus, 2014.
Bonnell, Victoria. "The Peasant Woman in Stalinist Political Art of the 1930s." *American Historical Review* 98, no. 1 (1993): 55–82.
Breslauer, George. *Gorbachev and Yeltsin as Leaders*. New York: Cambridge University Press, 2002.
Breslauer, George. *Khrushchev and Brezhnev as Leaders: Building Authority in Soviet Politics*. Boston: Allen & Unwin, 1982.
Brown, Kate. *Plutopia: Nuclear Families, Atomic Cities, and the Great Soviet and American Plutonium Disasters*. New York: Oxford University Press, 2013.
Brumberg, Abraham, ed. *Russia under Khrushchev: An Anthology from* Problems of Communism. New York: Praeger, 1962.
Buckley, Mary. *Mobilizing Soviet Peasants: Heroines and Heroes of Stalin's Fields*. New York: Rowman and Littlefield, 2006.
Carlson, Peter. *K Blows Top: A Cold War Comic Interlude Starring Nikita Khrushchev, America's Most Unlikely Tourist*. New York: Public Affairs, 2009.
Carstensen, Fred. *American Enterprise in Foreign Markets: Studies of Singer and International Harvester in Imperial Russia*. Chapel Hill: University of North Carolina Press, 1984.
Chandra, Nirmal Kumar. "Relevance of Soviet Economic Model for Non-Socialist Countries." *Economic and Political Weekly* 39, no. 22 (2004): 2287–305.
Chen, Yixin. "Cold War Competition and Food Production in China, 1957–1962." *Agricultural History* 83, no. 1 (2009): 51–78.
Chumachenko, Tatiana. *Church and State in Soviet Russia: Russian Orthodoxy from World War II to the Khrushchev Years*. Edited and translated by Edward Roslof. Armonk, NY: M. E. Sharpe, 2002.
Cohen, Stephen. "The Friends and Foes of Change: Reformism and Conservatism in the Soviet Union." *Slavic Review* 38, no. 2 (1979): 187–202.
Coumel, Laurent. "The Scientist, the Pedagogue, and the Party Official: Interest Groups, Public Opinion, and Decision-making in the 1958 Education Reform." In *Soviet State and Society under Nikita Khrushchev*, edited by Melanie Ilič and Jeremy Smith, 66–85. New York: Routledge, 2009.
Cronon, William. *Nature's Metropolis: Chicago and the Great West*. New York: W. W. Norton, 1991.
Cullather, Nick. *The Hungry World: America's Cold War Battle against Poverty in Asia*. Cambridge, MA: Harvard University Press, 2010.
Dalrymple, Dana. "American Technology and Soviet Agricultural Development." *Agricultural History* 40, no. 3 (1966): 191–214.
Danilov, V. P. *Rural Russia under the New Regime*. Translated and edited by Orlando Figes. Bloomington: Indiana University Press, 1988.

David-Fox, Michael. "Entangled Histories in the Age of Extremes." In *Fascination and Enmity: Russia and Germany as Entangled Histories, 1914–1945*, edited by Michael David-Fox, Peter Holquist, and Alexander Martin, 1–12. Pittsburgh: University of Pittsburgh Press, 2012.

David-Fox, Michael. "The Implications of Transnationalism." *Kritika: Explorations in Russian and Eurasian History* 12, no. 4 (2011): 885–904.

Davies, R. W., Mark Harrison, and Stephen Wheatcroft, eds. *The Economic Transformation of the Soviet Union, 1913–1945*. New York: Cambridge University Press, 1993.

Davies, R. W., and Stephen Wheatcroft. *The Years of Hunger: Soviet Agriculture, 1931–1933*. New York: Palgrave Macmillan, 2004.

Deutsch, Robert. *The Food Revolution in the Soviet Union and Eastern Europe*. Boulder, CO: Westview Press, 1985.

Dobson, Miriam. *Khrushchev's Cold Summer: Gulag Returnees, Crime, and the Fate of Reform after Stalin*. Ithaca, NY: Cornell University Press, 2009.

Draitser, Emil. *Techniques of Satire: The Case of Saltykov-Ščedrin*. New York: Mouton de Gruyter, 1994.

Dronin, Nikolai, and Edward Bellinger. *Climate Dependence and Food Problems in Russia, 1900–1990: The Interaction of Climate and Agricultural Policy and Their Effect on Food Problems*. New York: Central European University Press, 2005.

Dubovskii, Mark. *Istoriia SSSR v anekdotakh, 1917–1992*. Minsk: Smiadyn', 1991.

Dunham, Vera. *In Stalin's Time: Middleclass Values in Soviet Fiction*. New York: Cambridge University Press, 1976.

Edele, Mark. "Soviet Society, Social Structure, and Everyday Life: Major Frameworks Reconsidered." *Kritika: Explorations in Russian and Eurasian History* 8, no. 2 (2007): 349–73.

Edele, Mark. *Stalinist Society, 1928–1953*. New York: Oxford University Press, 2011.

Edele, Mark. "Veterans and the Village: The Impact of Red Army Demobilization on Soviet Urbanization, 1945–1955." *Russian History* 36 (2009): 159–82.

Ellman, Michael. "Against Convergence." *Cambridge Journal of Economics* 4, no. 3 (1980): 199–210.

Engelstein, Laura. "Culture, Culture Everywhere: Interpretations of Modern Russia across the 1991 Divide." *Kritika: Explorations in Russian and Eurasian History* 2, no. 2 (2001): 363–93.

Engerman, David. "Learning from the East: Soviet Experts and India in the Era of Competitive Coexistence." *Comparative Studies of South Asia, Africa, and the Middle East* 33, no. 2 (2013): 227–38.

Engerman, David. "The Second World's Third World." *Kritika: Explorations in Russian and Eurasian History* 12, no. 1 (2011): 183–211.

Engerman, David, Nils Gilman, Mark Haefele, and Michael Latham, eds. *Staging Growth: Modernization, Development, and the Global Cold War*. Amherst: University of Massachusetts Press, 2003.

Evtuhov, Catherine, and Richard Stites. *A History of Russia since 1800: Peoples, Legends, Events, Forces*. New York: Houghton Mifflin, 2004.

Field, Deborah. *Private Life and Communist Morality in Khrushchev's Russia*. New York: Peter Lang, 2007.

Filtzer, Donald. *The Hazards of Urban Life in Late Stalinist Russia: Health, Hygiene, and Living Standards, 1943–1953*. New York: Cambridge University Press, 2010.

Filtzer, Donald. *Soviet Workers and De-Stalinization: The Consolidation of the Modern System of Soviet Production Relations, 1953–1964*. New York: Cambridge University Press, 1992.

Fitzgerald, Deborah. *The Business of Breeding: Hybrid Corn in Illinois, 1890–1940*. Ithaca, NY: Cornell University Press, 1990.

Fitzgerald, Deborah. *Every Farm a Factory: The Industrial Ideal in American Agriculture*. New Haven, CT: Yale University Press, 2003.

Fitzpatrick, Sheila. "Politics as Practice: Thoughts on a New Soviet Political History." *Kritika: Explorations in Russian and Eurasian History* 5, no. 1 (2004): 27–54.

Fitzpatrick, Sheila. *Stalin's Peasants: Resistance and Survival in the Russian Village after Collectivization*. New York: Oxford University Press, 1994.

Frieden, Jeffry. *Global Capitalism: Its Rise and Fall in the Twentieth Century*. New York: W. W. Norton, 2006.

Fritzsche, Peter. "On the Subjects of Resistance." *Kritika: Explorations in Russian and Eurasian History* 1, no. 1 (2000): 147–52.

Fürst, Juliane. "Late Stalinist Society: History, Policies, and People." In *Late Stalinist Russia: Society between Reconstruction and Reinvention*, edited by Juliane Fürst, 1–20. New York: Routledge, 2006.

Fürst, Juliane. *Stalin's Last Generation: Soviet Post-War Youth and the Emergence of Mature Socialism*. New York: Oxford University Press, 2010.

Fürst, Juliane, Polly Jones, and Susan Morrissey. "The Relaunch of the Soviet Project, 1945–1964." *Slavonic and East European Review* 86, no. 2 (2008): 201–7.

Ganson, Nicholas. *The Soviet Famine of 1946–1947 in Global and Historical Perspective*. New York: Palgrave Macmillan, 2009.

Geist, Edward. "Cooking Bolshevik: Anastas Mikoian and the Making of the *Book about Delicious and Healthy Food*." *The Russian Review* 71, no. 1 (2012): 2–20.

Getty, J. Arch, Gábor Rittersporn, and Viktor Zemskov. "Victims of the Soviet Penal System in the Pre-War Years: A First Approach on the Basis of Archival Evidence." *American Historical Review* 98, no. 4 (1993): 1017–49.

Goldman, Wendy and Donald Filtzer, eds. *Hunger and War: Food Provisioning in the Soviet Union during World War II*. Bloomington: Indiana University Press, 2015.

Gorlizki, Yoram. "Anti-Ministerialism and the USSR Ministry of Justice, 1953–1956: A Study in Organizational Decline." *Europe-Asia Studies* 48, no. 8 (1996): 1279–318.

Gorlizki, Yoram. "Party Revivalism and the Death of Stalin." *Slavic Review* 54, no. 1 (1995): 1–22.

Gorlizki, Yoram. "Scandal in Riazan: Networks of Trust and the Social Dynamics of Deception." *Kritika: Explorations in Russian and Eurasian History* 14, no. 2 (2013): 243–78.

Gorlizki, Yoram, and Oleg Khlevniuk. *Cold Peace: Stalin and the Soviet Ruling Circle, 1945–1953.* New York: Oxford University Press, 2004.

Graham, Loren. *The Ghost of the Executed Engineer: Technology and the Fall of the Soviet Union.* Cambridge, MA: Harvard University Press, 1993.

Gronow, Jukka. *Caviar with Champagne: Common Luxury and the Ideals of the Good Life in Stalin's Russia.* New York: Berg, 2003.

Grybkauskas, Saulius. "The Role of the Second Party Secretary in the 'Election' of the First: The Political Mechanism for the Appointment of the Head of Soviet Lithuania in 1974." *Kritika: Explorations in Russian and Eurasian History* 14, no. 2 (2013): 343–66.

Haber, Maya. "Socialist Realist Science: Constructing Knowledge about Rural Life in the Soviet Union, 1943–1958." Ph.D. diss., University of California at Los Angeles, 2013.

Haber, Maya. "The Soviet Ethnographer as Social Engineer: Socialist Realism and the Study of Rural Life, 1945–1958." *Soviet and Post-Soviet Review* 41, no. 2 (2014): 193–219.

Hager, Thomas. *The Alchemy of Air: A Jewish Genius, a Doomed Tycoon, and the Scientific Discovery That Fed the World but Fueled the Rise of Hitler.* New York: Harmony Books, 2008.

Hahn, Werner. *The Politics of Soviet Agriculture, 1960–1970.* Baltimore: Johns Hopkins University Press, 1972.

Hale-Dorrell, Aaron. "Industrial Farming, Industrial Food: Transnational Influences on Soviet Convenience Food in the Khrushchev Era." *Soviet and Post-Soviet Review* 42, no. 2 (2015): 174–96.

Hanson, Philip. *The Rise and Fall of the Soviet Economy: An Economic History of the USSR from 1945.* New York: Longman, 2003.

Harris, Steven. *Communism on Tomorrow Street: Mass Housing and Everyday Life after Stalin.* Baltimore: Johns Hopkins University Press, 2013.

Heinzen, James. *Inventing a Soviet Countryside: Soviet Power and the Transformation of Rural Russia, 1917–1928.* Pittsburgh: University of Pittsburgh Press, 2008.

Heller, Chaia. *Food, Farms, and Solidarity: French Farmers Challenge Industrial Agriculture and Genetically Modified Crops.* Durham, NC: Duke University Press, 2013.

Hirsch, Francine. *Empire of Nations: Ethnographic Knowledge and the Making of the Soviet Union*. Ithaca, NY: Cornell University Press, 2005.
Hixson, Walter. *Parting the Curtain: Propaganda, Culture, and the Cold War, 1945–1961*. New York: St. Martin's Press, 1997.
Hobsbawm, Eric. *Age of Extremes: The Short Twentieth Century, 1914–1991*. New York: Vintage, 1996.
Holquist, Peter. *Making War, Forging Revolution: Russia's Continuum of Crisis, 1914–1921*. Cambridge, MA: Harvard University Press, 2002.
Holquist, Peter. "A Tocquevillean 'Archival Revolution': Archival Change in the *Longue Durée*." *Jahrbücher für Geschichte Osteuropas* 51 (2003): 77–83.
Hornsby, Robert. *Protest, Reform, and Repression in Khrushchev's Soviet Union*. New York: Cambridge University Press, 2013.
Hough, Jerry. *How the Soviet Union Is Governed*. Cambridge, MA: Harvard University Press, 1979.
Hough, Jerry. *The Soviet Prefects: The Local Party Organs in Industrial Decision-Making*. Cambridge, MA: Harvard University Press, 1969.
Hughes, James. *Stalinism in a Russian Province: A Study of Collectivization and Dekulakization in Siberia*. New York: St. Martin's Press, 1996.
Humphrey, Caroline. *Marx Went Away—But Karl Stayed Behind*. Rev. ed. Ann Arbor: University of Michigan Press, 1998.
Isern, Thomas. "Wheat Explorer the World Over: Mark Carleton of Kansas." *Kansas History* 23, no. 1–2 (2000): 12–25.
Jasny, Naum. *Khrushchev's Crop Policy*. Glasgow, UK: Outram, 1965.
Jersild, Austin. "The Soviet State as Imperial Scavenger: 'Catch Up and Surpass' in the Transnational Socialist Bloc, 1950–1960." *American Historical Review* 116, no. 1 (2011): 109–32.
Jones, Jeffrey. *Everyday Life and the "Reconstruction" of Soviet Russia during and after the Great Patriotic War, 1943–1948*. Bloomington, IN: Slavica, 2008.
Joravsky, David. *The Lysenko Affair*. Cambridge, MA: Harvard University Press, 1970.
Joravsky, David. "The Vavilov Brothers." *Slavic Review* 24, no. 3 (1965): 383–85.
Josephson, Paul. *Would Trotsky Wear a Bluetooth? Technological Utopianism under Socialism, 1917–1989*. Baltimore: Johns Hopkins University Press, 2010.
Josephson, Paul, Nikolai Dronin, Ruben Mnatsakanian, Aleh Cherp, Dmitry Efremenko, and Vladislav Larin. *An Environmental History of Russia*. New York: Cambridge University Press, 2013.
Judt, Tony. *Postwar: A History of Europe since 1945*. New York: Penguin, 2005.
Karcz, Jerzy. "Khrushchev's Impact on Soviet Agriculture." *Agricultural History* 40, no. 1 (1966): 19–38.
Karcz, Jerzy, ed. *Soviet and East European Agriculture*. Berkeley: University of California Press, 1967.

Kassof, Allen. *The Soviet Youth Program: Regimentation and Rebellion*. Cambridge, MA: Harvard University Press, 1965.
Kerans, David. "Toward a Wider View of the Agrarian Problem in Russia, 1861–1930." *Kritika: Explorations in Russian and Eurasian History* 1, no. 4 (2000): 657–78.
Khalid, Adeeb. "Backwardness and the Quest for Civilization: Early Soviet Central Asia in Comparative Perspective." *Slavic Review* 65, no. 2 (2006): 231–51.
Kharkhordin, Oleg. *The Collective and the Individual in Russia: A Study of Practices*. Berkeley: University of California Press, 1999.
Khlevniuk, O. V. "Regional'naia vlast' v SSSR v 1954–kontse 1950-kh gg: Ustoichivost' i konflikty." *Otechestvennaia istoriia* no. 3 (2007): 31–49.
Khlevniuk, O. V. *Stalin: Zhizn' odnogo vozhdia*. Moscow: Corpus, 2015.
Khrushchev, Sergei, William Taubman, and Abbott Gleason, eds. *Nikita Khrushchev*. Translated by David Gehrenbeck, Eileen Kane, and Alla Bashenko. New Haven, CT: Yale University Press, 2000.
Kingsbury, Noël. *Hybrid: The History and Science of Plant Breeding*. Chicago: University of Chicago Press, 2009.
Koenker, Diane. *Republic of Labor: Russian Printers and Soviet Socialism, 1917–1930*. Ithaca, NY: Cornell University Press, 2005.
Kojevnikov, Alexei. *Stalin's Great Science: The Times and Adventures of Soviet Physicists*. River Edge, NJ: Imperial College Press, 2004.
Kotkin, Stephen. *Magnetic Mountain: Stalinism as a Civilization*. Berkeley: University of California Press, 1995.
Kotkin, Stephen. "Modern Times: The Soviet Union and the Interwar Conjuncture." *Kritika: Explorations in Russian and Eurasian History* 2, no. 1 (2001): 111–64.
Kozlov, Denis, and Eleonory Gilburd, eds. *The Thaw: Soviet Society and Culture during the 1950s and 1960s*. Toronto: University of Toronto Press, 2013.
Kozlov, Vladimir. *Mass Uprisings in the USSR: Protest and Rebellion in the Post-Stalin Years*. Translated and edited by Elaine McClarnand MacKinnon. Armonk, NY: M. E. Sharpe, 2002.
Krementsov, Nikolai. *The Cure: A Story of Cancer and Politics from the Annals of the Cold War*. Chicago: University of Chicago Press, 2002.
Krylova, Anna. "Soviet Modernity: Stephen Kotkin and the Bolshevik Predicament." *Central European History* 23, no. 2 (2014): 167–92.
Laird, Roy. "The Dilemma of Soviet Agricultural Administration: The Short and Unhappy Life of the TPA." *Agricultural History* 40, no. 1 (1966): 11–18.
Lebina, N. B., and A. N. Chistikov. *Obyvatel' i reformy: Kartiny povsednevnoi zhizni gorozhan v gody NEPa i khrushchevskogo desiatiletiia*. St. Petersburg: Dmitrii Bulanin, 2003.
Ledeneva, Alena. *Russia's Economy of Favors: Blat, Networking, and Informal Exchange*. New York: Cambridge University Press, 1998.
Lee, Harold. *Roswell Garst: A Biography*. Ames: Iowa State University Press, 1984.

Leonard, Carol. *Agrarian Reform in Russia: The Road from Serfdom.* New York: Cambridge University Press, 2010.
Lévesque, Jean. "'Part-Time Peasants': Labour Discipline, Collective Farm Life, and the Fate of Soviet Socialized Agriculture after the Second World War, 1945–1953." Ph.D. diss., University of Toronto, 2003.
Lewin, Moshe. *The Gorbachev Phenomenon: A Historical Interpretation.* Berkeley: University of California Press, 1988.
Lewin, Moshe. *The Making of the Soviet System: Essays in the Social History of Interwar Russia.* New York: Pantheon Books, 1985.
Lewin, Moshe. *Political Undercurrents in Soviet Economic Debates: From Bukharin to the Modern Reformers.* Princeton, NJ: Princeton University Press, 1974.
Lewin, Moshe. *The Soviet Century.* Edited by Gregory Elliott. New York: Verso, 2005.
Linden, Carl. *Khrushchev and the Soviet Leadership, 1957–1964.* Baltimore: Johns Hopkins University Press, 1966.
Linz, Susan, ed. *The Impact of World War II on the Soviet Union.* Totowa, NJ: Rowman and Allanheld, 1985.
Lovell, Stephen. *Summerfolk: A History of the Dacha, 1710–2000.* Ithaca, NY: Cornell University Press, 2003.
Lüthi, Lorenz. *The Sino-Soviet Split: Cold War in the Communist World.* Princeton, NJ: Princeton University Press, 2008.
Magnúsdóttir, Rósa. "Keeping Up Appearances: How the Soviet State Failed to Control Popular Attitudes toward the United States of America, 1945–1959." Ph.D. diss., University of North Carolina at Chapel Hill, 2006.
Martin, Terry. *Affirmative Action Empire: Nations and Nationalism in the Soviet Union, 1923–1939.* Ithaca, NY: Cornell University Press, 2001.
McCann, James. *Maize and Grace: Africa's Encounter with a New World Crop, 1500–2000.* Cambridge, MA: Harvard University Press, 2005.
McCauley, Martin. *Gorbachev.* New York: Longman, 1998.
McCauley, Martin. *Khrushchev and the Development of Soviet Agriculture: The Virgin Land Programme, 1953–1964.* New York: Holmes & Meier, 1976.
Medvedev, Roy, and Zhores Medvedev. *Khrushchev: The Years in Power.* Translated by Andrew Durkin. New York: Columbia University Press, 1975.
Medvedev, Zhores. *The Rise and Fall of T. D. Lysenko.* Edited by Lucy Lawrence, translated by I. Michael Lerner. New York: Columbia University Press, 1969.
Melillo, Edward. "The First Green Revolution: Debt Peonage and the Making of the Nitrogen Fertilizer Trade, 1840–1930." *American Historical Review* 117, no. 4 (2012): 1028–60.
Millar, James, ed. *The Soviet Rural Community.* Urbana: University of Illinois Press, 1971.
Mincyte, Diana. "Everyday Environmentalism: The Practice, Politics, and Nature of Subsidiary Farming in Stalin's Lithuania." *Slavic Review* 68, no. 1 (2009): 31–49.

Mitchell, Timothy. *Carbon Democracy: Political Power in the Age of Oil*. New York: Verso, 2011.

Mokhov, V. P. *Regional'naia politicheskaia elita Rossii, 1945–1991 gg*. Perm: Permskoe knizhnoe izdatel'stvo, 2003.

Moon, David. *The Plough That Broke the Steppes: Agriculture and Environment on Russia's Grasslands, 1700–1914*. Oxford, UK: Oxford University Press, 2013.

Nekrasov, Viacheslav. "Neftekhimicheskii proekt N. S. Khrushcheva (vtoraia polovina 1950-kh–perviaia polovina 1960-kh gg.): Strategiia modernizatsii sovetskoi ekonomiki, eksport nefti i raspredelenie resursnoi renty." *Istoriia* 6, no. 11 (2015). http://history.jes.su/s207987840001371-7-1.

Nikonov, A. A. *Spiral' mnogovekovoi dramy: Agrarnaia nauka i politika Rossii, XVIII–XX vv*. Moscow: Entsiklopediia rossiiskikh dereven', 1995.

Nordlander, David. "Khrushchev's Image in the Light of Glasnost and Perestroika." *Russian Review* 52, no. 2 (1993): 248–64.

Nove, Alec. *An Economic History of the USSR, 1917–1991*. 3rd ed. New York: Penguin, 1993.

Ohayon, Isabelle. *La sédentarisation des Kazakhs dans l'URSS de Staline: Collectivisation et changement social, 1928–1945*. Paris: Maisonneuve et Larose, 2006.

Perkins, John. *Geopolitics and the Green Revolution: Wheat, Genes, and the Cold War*. New York: Oxford University Press, 1997.

Péteri, György. "Nylon Curtain: Transnational and Transsystemic Tendencies in the Cultural Life of State-Socialist Russia and East-Central Europe." *Slavonica* 10, no. 2 (2003): 113–23.

Péteri, György. "The Oblique Coordinate Systems of Modern Identity." In *Imagining the West in Eastern Europe and the Soviet Union*, edited by György Péteri, 1–12. Pittsburgh: University of Pittsburgh Press, 2010.

Piketty, Thomas. *Capital in the Twenty-First Century*. Translated by Arthur Goldhammer. Cambridge, MA: Belknap Press, 2014.

Ploss, Sidney. *Conflict and Decision-Making in Soviet Russia: A Case Study of Agricultural Policy, 1953–1963*. Princeton, NJ: Princeton University Press, 1965.

Pohl, Michaela. "Women and Girls in the Virgin Lands." In *Women in the Khrushchev Era*, edited by Melanie Ilič, Susan Reid, and Lynne Attwood, 52–74. New York: Palgrave Macmillan, 2004.

Pollan, Michael. *The Omnivore's Dilemma: A Natural History of Four Meals*. New York: Penguin, 2006.

Pollock, Ethan. *Stalin and the Soviet Science Wars*. Princeton, NJ: Princeton University Press, 2006.

Raleigh, Donald. *Experiencing Russia's Civil War: Politics, Society, and Revolutionary Culture in Saratov, 1917–1922*. Princeton, NJ: Princeton University Press, 2002.

Raleigh, Donald, ed. *Provincial Landscapes: Local Dimensions of Soviet Power, 1917–1953*. Pittsburgh: University of Pittsburgh Press, 2001.

Raleigh, Donald. *Soviet Baby Boomers: An Oral History of Russia's Cold War Generation*. New York: Oxford University Press, 2011.

Rees, E. A., ed. *Centre-Local Relations in the Stalinist State, 1928–1941*. New York: Palgrave Macmillan, 2002.

Reid, Susan. "Cold War in the Kitchen: Gender and the Destalinization of Consumer Taste in the Soviet Union under Khrushchev." *Slavic Review* 61, no. 2 (2002): 211–52.

Reid, Susan. "Who Will Beat Whom? Soviet Popular Reception of the American National Exhibition in Moscow, 1959." *Kritika: Explorations in Russian and Eurasian History* 9, no. 4 (2008): 855–904.

Richmond, Yale. *Cultural Exchange and the Cold War: Raising the Iron Curtain*. University Park: Pennsylvania State University Press, 2003.

Roth-Ey, Kristen. *Moscow Prime Time: How the Soviet Union Built the Media Empire That Lost the Cultural Cold War*. Ithaca, NY: Cornell University Press, 2011.

Rusinov, I. V. "Agrarnaia politika KPSS v 50-e–pervoi polovine 60-kh gg: Opyt i uroki." *Voprosy istorii KPSS* no. 9 (1988): 35–49.

Ryan, Michael, and Richard Prentice. *Social Trends in the Soviet Union from 1950*. London: Macmillan, 1987.

Saltykov-Shchedrin, M. E. *The History of a Town*. Translated by I. P. Foote. Oxford, UK: W. A. Meeuws, 1980.

Sanchez-Sibony, Oscar. *Red Globalization: The Political Economy of the Soviet Cold War from Stalin to Khrushchev*. New York: Cambridge University Press, 2014.

Sazanova, L. V. *Istoriia rasprostraneniia kukuruzy v nashe strane*. Minsk: Urozhai, 1964.

Scherrer, Jutta. "'To Catch Up and Overtake' the West: Soviet Discourse on Socialist Competition." In *Competition in Socialist Society*, edited by Katalin Miklóssy and Melanie Ilic, 10–23. New York: Routledge, 2014.

Schmid, Sonja. "Celebrating Tomorrow Today: The Peaceful Atom on Display in the Soviet Union." *Social Studies of Science* 36, no. 3 (2006): 331–65.

Scott, James. *Seeing Like a State: How Certain Schemes to Improve the Human Condition Have Failed*. New Haven, CT: Yale University Press, 1998.

Sharma, Shri Ram. *India-USSR Relations, 1947–1971*. Vol. 1, *From Ambivalence to Steadfastness*. New Delhi: Discover Publishing House, 1999.

Shturman, Dora, and Sergei Tiktin. *Sovetskii soiuz v zerkale politicheskogo anekdota*. London: Overseas Publications Interchange, 1985.

Siegelbaum, Lewis. *Cars for Comrades: The Life of the Soviet Automobile*. Ithaca, NY: Cornell University Press, 2008.

Siegelbaum, Lewis. *Stakhanovism and the Politics of Productivity in the USSR, 1935–1941*. New York: Cambridge University Press, 1988.

Slezkine, Yuri. "The USSR as a Communal Apartment, or How a Socialist State Promoted Ethnic Particularism." *Slavic Review* 53, no. 3 (1994): 414–52.

Sloin, Andrew, and Oscar Sanchez-Sibony. "Economy and Power in the Soviet Union, 1917–1939." *Kritika: Explorations in Russian and Eurasian History* 15, no. 1 (2014): 7–22.
Smith, Jenny Leigh. "Empire of Ice Cream: How Life Became Sweeter in the Postwar Soviet Union." In *Food Chains: From Farmyard to Shopping Cart*, edited by Warren Belasco and Roger Horowitz, 142–57. Philadelphia: University of Pennsylvania Press, 2009.
Smith, Jenny Leigh. *Works in Progress: Plans and Realities on Soviet Farms, 1930–1963*. New Haven, CT: Yale University Press, 2014.
Smith, Jeremy, and Melanie Ilič. *Khrushchev in the Kremlin: Policy and Government in the Soviet Union, 1953–1964*. New York: Routledge, 2011.
Smith, Mark. *Property of Communists: The Urban Housing Program from Stalin to Khrushchev*. DeKalb: Northern Illinois University Press, 2013.
Solzhenitsyn, Alexander. *The Solzhenitsyn Reader: New and Essential Writings, 1947–2005*. Edited by Edward Ericson, Jr., and Daniel Mahoney. Wilmington, DE: ISI Books, 2006.
Stites, Richard. *Revolutionary Dreams: Utopian Vision and Experimental Life in the Russian Revolution*. New York: Oxford University Press, 1989.
Suny, Ronald. *The Revenge of the Past: Nationalism, Revolution, and the Collapse of the Soviet Union*. Stanford, CA: Stanford University Press, 1993.
Suny, Ronald, and Terry Martin, eds. *A State of Nations: Empire and Nation-Making in the Age of Lenin and Stalin*. New York: Oxford University Press, 2001.
Sushkov, A. V. *Prezidium TsK KPSS v 1957–1964 gg.: Lichnost' i vlast'*. Ekaterinburg: Ural'skii tsentr akademicheskogo obsluzhivaniia, 2009.
Taubman, William. *Khrushchev: The Man and His Era*. New York: W. W. Norton, 2003.
Taylor, Richard. "Singing on the Steppes for Stalin: Ivan Pyr'ev and the Kolkhoz Musical in Soviet Cinema." *Slavic Review* 58, no. 1 (1999): 143–59.
Thompson, E. P. *The Making of the English Working Class*. New York: Vintage, 1966.
Thompson, E. P. "The Moral Economy of the English Crowd in the Eighteenth Century." *Past and Present* no. 50 (1971): 76–136.
Tininis, Vytautas. *Sniečkus: 33 metai valdžioje; Antano Sniečkus biografinė apybraiža*. Vilnius: Antrasis papildytas pataisytas leidimas, 2000.
Tomilin, V. N. *Nasha krepost': Mashinno-traktornye stantsii Chernozemnogo Tsentra Rossii v poslevoennyi period, 1946–1958 gg*. Moscow: AIRO-XX, 2009.
Tromly, Benjamin. *Making the Soviet Intelligentsia: Universities and Intellectual Life under Stalin and Khrushchev*. New York: Cambridge University Press, 2014.
Tumarkin, Nina. *Lenin Lives! The Lenin Cult in Soviet Russia*. Cambridge, MA: Harvard University Press, 1983.
Vail', Petr, and Alexander Genis. *Shestidesiatye: Mir sovetskogo cheloveka*. Ann Arbor, MI: Ardis, 1986.

van Atta, Don. "The USSR as a 'Weak State': Agrarian Origins of Resistance to Perestroika." *World Politics* 42, no. 1 (1989): 129–49.
Verbitskaia, O. M. *Rossiiskoe krest'ianstvo: Ot Stalina k Khrushchevu, sredina 40-kh–nachalo 60-kh gg.* Moscow: Nauka, 1992.
Verdery, Katherine. *What Was Socialism and What Comes Next?* Princeton, NJ: Princeton University Press, 1996.
Viola, Lynne. *The Best Sons of the Fatherland: Workers in the Vanguard of Soviet Collectivization.* New York: Oxford University Press, 1987.
Viola, Lynne. *Peasant Rebels under Stalin: Collectivization and the Culture of Peasant Resistance.* New York: Oxford University Press, 1996.
Viola, Lynne. *The Unknown Gulag: The Lost World of Stalin's Special Settlements.* New York: Oxford University Press, 2007.
Viola, Lynne, V. P. Danilov, N. A. Ivnitskii, and Denis Kozlov, eds. *The War against the Peasantry, 1927–1930: The Tragedy of the Soviet Countryside.* Translated by Steven Shabad. New Haven, CT: Yale University Press, 2005.
Volin, Lazar. "Khrushchev's Economic Neo-Stalinism." *American Slavic and East European Review* 14, no. 4 (1955): 445–64.
Volin, Lazar. "The Malenkov-Khrushchev New Economic Policy." *Journal of Political Economy* 62, no. 3 (1954): 187–209
Ward, Christopher. *Brezhnev's Folly: The Building of BAM and Late Soviet Socialism.* Pittsburgh: University of Pittsburgh Press, 2009.
Warman, Arturo. *Corn and Capitalism: How a Botanical Bastard Grew to Global Dominance.* Translated by Nancy Westrate. Chapel Hill: University of North Carolina Press, 2003.
Weiner, Amir. "Robust Revolution to Retiring Revolution: The Life Cycle of the Soviet Revolution, 1945–1968." *Slavonic and East European Review* 86, no. 2 (2008): 208–31.
Weiner, Douglas. *Models of Nature: Ecology, Conservation, and Cultural Revolution in Soviet Russia.* Bloomington: Indiana University Press, 1998.
Werner, Michael, and Bénédicte Zimmermann. "Beyond Comparison: *Histoire Croisée* and the Challenge of Reflexivity." *History and Theory* 45, no. 1 (2006): 30–50.
West, Sally. *I Shop in Moscow: Advertising and the Creation of Consumer Culture in Late Tsarist Russia.* DeKalb: Northern Illinois University Press, 2011.
Wolfe, Thomas. *Governing Soviet Journalism: The Press and the Socialist Person after Stalin.* Bloomington: Indiana University Press, 2005.
Yurchak, Alexei. *Everything Was Forever until It Was No More: The Last Soviet Generation.* Princeton, NJ: Princeton University Press, 2006.
Zelenin, I. E. *Agrarnaia politika N. S. Khrushcheva i sel'skoe khoziaistvo.* Moscow: Institut istorii Rossiiskoi Akademii Nauk, 2001.
Zubok, Vladislav. *A Failed Empire: The Soviet Union in the Cold War from Stalin to Gorbachev.* Chapel Hill: University of North Carolina Press, 2007.

Zubkova, E. Iu. [Elena]. *Pribaltika i Kreml', 1940–1953 gg.* Moscow: ROSSPEN, 2008.
Zubkova, Elena. *Russia after the War: Hopes, Illusions, and Disappointments, 1945–1957.* Translated by Hugh Ragsdale. Armonk, NY: M. E. Sharpe, 1998. Russian edition: E. Iu. Zubkova. *Obshchestvo i reformy, 1945–1964.* Moscow: Rossiia molodaia, 1993.

INDEX

Academy of Agricultural Sciences, USSR, 79–80, 176, 179
Afghanistan, 42, 51, 52
Africa, 5, 50, 52, 53
Agnaev, Khadzhiet, 151, 158, 160, 161
agricultural attaché, 40, 43, 49, 50
agrotown, *see* kolkhoz, amalgamation of
Akopov, Stepan, 169, 170–71
Albania, 42
alcohol and alcohol abuse, 120, 149–50, 151, 152, 205, 208, 223
Algeria, 52
All-Union Agricultural Exhibition, 42, 93–95, 118
All-Union Institute for Plant Breeding (VIR), 182, 183, 184
All-Union Research Institute for Agricultural Economics, 140, 143–44, 273n16
Alperovich, (kolkhoz chair), 145–46
Altai Krai, 73, 92, 172
American National Exhibition, 2, 31
Americanism, 33
Angelina, Pasha, 91
Asia, 5, 46, 50, 52, 53

Beatles, the, 9, 32
Belarus, 111, 114, 118, 148, 150, 219
Belgorod Oblast, 78, 107, 117, 126, 184
Belorussian SSR, *see* Belarus
Beliaev, Nikolai, 83
Benediktov, Ivan, 43
Beria, Lavrenti, 10, 216
Berry, Wendell, 226
Black Sea, 58, 81, 132
Bolsheviks, 3, 10–12, 13, 15, 26, 33, 34, 37, 41, 106, 109
Bondarenko, Aleksei, 199, 200
Bonnell, Victoria, 96
Borlaug, Norman, 46
Brezhnev, Leonid, 7, 83, 207, 211
Briansk Oblast, 74, 90, 111, 126, 145, 199
Bukharin, Nikolai, 12
Bulganin, Nikolai, 202, 212
Burma, 42, 52

Cambodia, 52
Canada, 38, 40, 42, 62
capitalism, 5–6, 7–8, 12–13, 27, 29, 31–32, 33, 37, 38–39, 40, 42, 45, 48, 50, 51, 52–53, 64, 101, 137, 143, 213, 226, 231, 232–33

Castro, Fidel, 81
Central Asia, 51, 98, 186
Central Committee of the Communist Party of the Soviet Union, 71, 143, 150, 198, 203, 206, 209
  personnel, 24, 68, 171, 199–200, 201, 205, 279n11
  plenums, 58, 63, 95, 97
    September (1953), 25, 59, 84, 94, 199
    February (1954), 59
    January (1955), 59, 61, 62, 68, 90, 92–93, 112, 169, 177, 200, 212, 217
    December (1958), 75, 168–69, 222–23
    December (1959), 208
    January (1961), 97, 209
    March (1962), 220
    February (1964), 83
    October (1964), 221
    March (1965), 227
  as policy-making body, 81, 93, 94, 140, 169, 171, 176, 188, 193, 199, 205, 212, 229
  Presidium of, 58, 61, 68, 70–71, 81, 167, 179, 198, 201, 221
  secretaries of, 22, 58, 83, 193, 206, 233
Central Statistical Administration, 25, 206
Ceylon, 52
Chachin, Vladimir, 214–15
China, 12
  Communist Party of, 3
  People's Republic of, 37, 41, 42–43, 49
climate, 34, 36, 40, 62, 66, 68, 69, 73, 74, 75, 84, 101, 111, 195, 218, 226, 227, 230
Cold War
  as binary division of the world, 5, 53, 231, 247n14 (*see also* Iron Curtain)
  as binary understanding of the world, 5, 7, 39, 47, 53, 232, 233, 292n17
  effect of on domestic life in the Soviet Union, 18, 90, 247n16
  as military and geopolitical confrontation, 22, 38, 90, 232
  as multifaceted geopolitical competition, 5, 29, 47, 232
  peaceful coexistence policy during, 30, 41
  as peaceful competition, 2, 10, 28–29, 30, 38, 71, 87, 101, 107, 151, 161–62, 203, 211, 213, 226
collective leadership, 58, 71
collectivization of agriculture in the Soviet Union, 3, 7, 8, 10, 13–14, 20, 54, 96, 108, 165–66, 231
Commission for Government Oversight, *see* Ministry of Government Oversight
communism
  abundance as prerequisite to, 1, 28, 29, 30, 37, 44, 56, 64, 84, 85, 95, 100, 101, 107, 225, 226, 232, 257n46
  as alternative to capitalism, 31, 32, 47, 71, 213
  construction of, 29, 64, 99–100, 110, 124, 128, 132, 133–34, 135, 139, 163
  as program of modernization, 4, 37, 99–100, 163, 234, 237n16
Communist Party of Lithuania, Central Committee of, 217, 218–19, 220, 221, 222, 224
Communist Party of the Soviet Union, 8, 12, 134
  Congresses of
    Nineteenth (1952), 58
    Twentieth (1956), 67, 233
    Twenty-first (1959), 75, 121, 134, 168
    Twenty-second (1961), 45

as instrument of governance, 55, 56–57, 58, 88, 109, 111, 124, 127, 135, 168, 172, 196, 197–98
regional networks within, 197–98, 199–200, 203–4, 205–6, 207–8, 211, 217, 221, 222, 223–24, 225, 230, 284n2
structural reform of, 84, 126, 211
as totalitarian party, 2, 56, 138–39, 229, 255n7
Communist Party of Ukraine, Central Committee of, 83, 186, 187, 206, 210
consumption, 22
in the United States, 32, 87, 226
of food, 9, 20, 29, 30, 37, 71, 87, 103, 104–5, 106, 164–65, 230, 232
corn
as component of industrial agriculture, 2, 3, 4, 5, 26, 27–28, 35, 36–37, 38, 41, 42, 44, 54, 75, 84, 127, 153, 167, 176, 195–96, 197, 211, 212, 218, 219, 220, 230, 231
as defense against crop failure, 21, 106
detasseling, 128, 182, 183–184
as feed for livestock, 1, 61, 62, 89, 94, 95, 114, 144, 145, 148, 158, 198, 200, 214–15, 217, 227–28
as food, 102–4, 203, 293n22
as "harebrained scheming," 2, 37, 54, 84, 227, 233
harvesting of, 36–37, 62, 109, 129–30, 141, 144, 145, 146–47, 149, 151, 157, 184–85, 195, 197, 202, 207
harvests of, 35, 44, 50, 55, 59, 60, 64–67, 70, 72, 74–75, 82, 121, 140, 205–6, 210
hybrids, 36, 38, 40, 41, 42, 44, 46, 57, 62, 75, 76, 77, 82, 97, 128, 168, 174, 175–76, 177, 178–87, 190, 193–95, 280n47, 281n59
planting of (*see* square-cluster planting)

as policy initiative, 11, 21, 55, 59, 61–62, 68, 70, 77, 84, 109, 118, 167, 195–96, 215
precedents for in the United States, 62, 176, 177, 178, 260n101
productivity of, 36, 39, 44, 59, 62, 74, 75, 76, 77, 102, 145, 155, 158, 166, 174, 175, 180–81, 213, 220, 227, 228
as source of abundance, 4, 9, 29, 41, 59, 61–62, 64, 76, 77, 82, 87, 89, 92–93, 100, 101–2, 104–5, 139, 158, 211, 226
quantity of plantings of, 1, 59, 62, 64–65, 70, 72, 74, 76, 80–81, 82, 111, 115, 127, 153, 158, 195, 199, 206, 207, 210, 213, 214, 217, 220, 222, 225, 228
silage, 62, 74, 76, 112, 114, 169, 170, 213, 228
as subject of jokes, 1, 107, 226
Corn Belt, 27, 38, 57
"cosmopolitanism," *see* xenophobia
Council of Ministers
as formal government of the Soviet Union, 58, 81, 167, 168, 169, 171, 172, 190, 193, 229 (*see also* Soviet Union, government of)
reforms of, 171, 191–92, 195
(*see also* sovnarkhoz)
Councils of Youth Corngrowers, 123–24, 125–27, 135
crossed history, 6, 28, 238n27
cult of personality, *see* Stalin, cult of personality of
Czechoslovakia, 42

Department of Agriculture, US, 33, 79
Dnieper River, 51
Dnipropetrovsk Oblast, 92, 117, 187
Doliniuk, Evgeniia, 97–98
Donbas, 15–16, 20

Doroshenko, Pëtr, 205
drunkenness, *see* alcohol and abuse
Dygai, Nikolai, 171

East Germany, *see* German Democratic Republic
Eastern Europe, 4, 42
education and upbringing, 79–80, 89, 115, 119–20, 128–36, 184
Egorychev, Nikolai, 223
Egypt, 42, 52
Ehrenburg, Ilya, 201
England, *see* United Kingdom
Enisei River, 72
environment, conceptions of, 3, 34, 53, 100, 108, 176
Estonia, 69–70, 216
Ethiopia, 52
Ezhevskii, Aleksandr, 172–74

famine, 13, 16, 18–19, 20, 21, 22, 41, 138
fertilizer, *see* industrial agriculture, fertilizer as a component of
Finland, 43
Fitzgerald, Deborah, 5
five-year plans. *See also* Seven-Year Plan
  First (1928–1932), 12, 33, 109
  Fifth (1951–1955), 45
  Sixth (1956–1960), 77, 89
Food and Agriculture Organization, United Nations (FAO), 43
food imports, 2, 45
food supply in the Soviet Union, 18–19, 20, 30, 37, 229
Ford Foundation, 46
Ford, Henry, 33
fossil fuels, *see* industrial agriculture, fossil fuels as component of
Foucault, Michel, 88
France, 6, 36, 42, 43

fraud and fraudulent statistics, 66–68, 74, 79, 80, 121–23, 150, 151, 168, 183, 186, 191, 192, 194, 195, 197–98, 202, 203–7, 209, 210, 211, 218, 222, 225

Garst, Roswell, 27, 33, 38, 41–42, 43, 44, 77, 95, 97, 145, 180, 187–88, 213
gender and femininity, 96–97, 98, 100, 103, 119, 120, 142, 159
genetics, 5, 29, 41, 168, 175–78
Genis, Aleksandr, 104, 226
Georgia, 20, 105
Georgian SSR, *see* Georgia
German Democratic Republic (East Germany), 30, 42, 49
Gitalov, Aleksandr, 77, 95, 97, 98, 123, 153
glasnost, 117, 224, 229
Gorbachev, Mikhail, 2, 7, 165, 224, 229–30, 233–34
Gorlova, Mariia, 94
Gosplan, 106, 142, 227–28.
  See also State Planning Agency
grassfield system, 35, 211–15, 218–20, 221–22
Great Depression, 12, 31, 46, 233
Great Leap Forward, 37
Great Patriotic War, *see* World War II
Green Revolution, 42, 43, 45–47, 53, 177, 280n47, 294n26
Gulag, *see* labor camp

Hero of Socialist Labor, 92, 97, 98, 119, 209
high modernism, 3, 25, 26, 108, 232, 254n138
high-yielding varieties, 46, 51.
  See also corn, hybrids
Hobsbawm, Eric, 233
Hungary, 41, 42

ideology, 5, 7, 37, 38–39, 48, 53, 58, 64, 88, 100, 128, 178, 227, 229
Ilchenko, Anna, 113, 119–20
India, 41, 42, 46, 48–50, 51
Indonesia, 52
industrial agriculture
  accounting and financial calculation as component of, 137–38, 144, 153–55, 161, 165–66 (*see also* khozraschet)
  alternatives to, 5, 212, 215, 218, 238n23
  application in the Soviet Union of, 4, 5, 11, 26, 29, 41, 44, 45, 47, 54, 75, 82, 144, 153–54, 163, 165–66, 167, 181–82, 195–96, 197, 211, 218, 225, 226, 231
  as global trend, 3, 4, 6, 8, 34, 36, 37, 46, 53, 177, 195
  as open system of nitrogen fixing, 35, 218, 232, 250n41
  capital investments in, 44, 45, 76, 78, 79, 160, 171, 172, 174, 179, 181, 190, 193, 208
  chemicals as component of, 29, 35, 36, 41, 43, 46, 82, 95, 97, 128, 166, 180, 212, 231
  corn as a component of (*see* corn, as component of industrial agriculture)
  costs of, 5, 54, 232
  early Soviet interest in, 3, 14, 26, 34, 41, 53, 231
  effect on farm labor of, 92, 127–28, 144, 163, 174, 211, 233
  fertilizer as component of, 5, 13, 17, 35, 36, 41, 44, 57, 75, 77, 82, 83, 95, 97, 174, 212, 213, 218, 232, 261n118
  fossil fuels as component of, 4, 35, 37, 232
  origins in the United States of, 2, 5, 31, 34, 36, 38, 77

  productivity and efficiency of, 29, 43, 46, 76, 77, 82, 97, 100, 108, 155–56, 161, 165, 166, 173, 175, 180, 212–13, 229, 230
  regional specialization as component of, 40, 80, 186
  tractors and machines as component of, 33, 35, 36–37, 38, 41, 43, 45, 46, 57, 61, 75, 76, 77–78, 83, 93, 95, 128, 153, 158, 166, 167, 169–75, 180, 212, 228, 231, 232
  transfer mentality as a component of (*see* transfer mentality)
informal practices, 168, 182, 186–87, 195–96, 209–10, 221–22, 225, 278n3
inspection, 69–70, 113–14, 124–26, 145, 147, 156, 162, 168–69, 171, 172, 181, 185–87, 192, 193–94, 206
International Harvester Corporation, 32, 173
Iowa, 27, 35, 37–38, 40, 62, 77, 79, 95
irrigation, 40, 48–49, 50–51, 52, 228
Iron Curtain, 5–6, 27, 32, 231
Italy, 177
Iusupov, Usman, 60
Ivanovo Oblast, 116, 117
Izvestiia, 89, 97, 100, 112, 208

Jewish identity, 18, 210

Kabanets, Ivan, 208–9, 210
Kaganovich, Lazar, 15, 19, 71, 212
Kalinovka, 15, 20, 44
Kalnbērziņs, Jānis, 68, 75–76, 213
Kamensk Oblast, 183, 184
Kapitonov, Ivan, 200
Kaplun, Nikita, 152
Karpenko, Mikhail, 201–2
Kazakh SSR, *see* Kazakhstan
Kazakhstan, 13, 34, 60–61, 83, 92, 97, 113, 135, 198, 205, 221

KGB (Committee for State Security), 44, 111–12, 205, 221
Kharkiv Oblast, 147, 192
Khitrov, Stepan, 197, 198
khozraschet system of financial accounting, 8, 137–38, 144, 154–55, 159–61, 165, 233
Khrushchev, Nikita
   agrarian reforms and agricultural programs of, 1–2, 4, 5, 7, 10, 21–22, 23, 26, 27, 29, 34, 35, 37, 44, 57, 58, 71, 77, 92, 110–11, 113, 137, 139, 140, 145, 147, 171–72, 176, 177, 179, 181, 200, 201, 203–4, 211, 219, 223, 224–25, 229, 230, 246n4, 269n87, 292n18, 293n25
   as head of the Communist Party of the Soviet Union, 56, 58, 81, 198, 199
   as leader in Ukraine, 11, 16, 17, 19, 20, 21–22, 23, 25, 29, 59, 92, 178, 211
   as subordinate of Stalin, 17, 18
   authority of, 2, 29, 31, 55, 56–57, 59, 61, 64, 70, 71–72, 81–85, 167, 168, 171, 172, 195, 211, 220, 221, 224–225, 229, 230, 255n7
   belief in socialism and communism of, 1, 16, 28–29, 30, 31, 32, 37, 45, 64, 71, 85, 100, 101, 124, 213
   cult of, 107, 119
   educational reforms of, 128–32, 135–36
   formative experiences of, 15–16, 20, 21, 29
   leadership style of, 21, 76, 80, 113, 116, 199, 207, 213
   legacy of, 2–3, 30, 56, 57, 84, 85, 87, 107, 164–65, 204, 205, 208, 226–27
   ouster of, 2, 83, 107, 221
   struggle for power by, 10, 20, 25, 29, 55, 56–57, 58–59, 61, 68, 70–71, 78–79, 83–84, 198, 212, 256n13, 278n2
   visits to the United States by, 27, 39, 41, 43, 79
Kitchen Debate, *see* American National Exhibition
Kirichenko, Aleksei, 83, 113, 188–189, 193
kolkhoz, 84
   amalgamation of, 21–22, 58, 72, 275n74
   as a collective enterprise, 69, 141, 149
   economics and operation of, 61, 108, 137, 139, 140–41, 144, 147, 161–62, 169, 229, 272n3
   labor and pay on, 10, 23, 24–25, 77–78, 91, 127–28, 129, 138–40, 141, 142–43, 144, 147, 148–49, 152, 162, 164–65, 174, 184, 211
   low productivity of, 4, 13, 16, 20, 45, 79, 139, 149, 155, 161, 166
Kolkhoznaia zhizn', 151–53, 275n75
kolkhozniks
   as consumers, 108, 143, 150, 163–65, 230–31
   as subjects, 87, 88–89, 90, 138, 140, 164, 166
   living and working conditions of, 22, 25, 26, 49, 51, 88, 98, 107, 108, 127–28, 138, 150, 156–60, 164, 230
Komsomol, 60, 97, 109, 110–15, 116–17, 118, 120–23, 124–27, 129–31, 135, 149, 233
   Central Committee of, 112–13, 114, 116–17, 121, 125–27, 132
   Congresses of, 119–20, 134
Komsomolskaia pravda, 100, 102, 112, 114
Kostroma Oblast, 162, 174

Kozlov, Aleksei, 212
Kozlov, Frol, 193, 206
Krasnodar Krai, 67, 74, 134–35, 145, 184, 187, 188, 189–90, 211, 213–14
Krasnoiarsk Krai, 72–73, 145, 201
Krokodil, 101, 102
Krotkov, Evgenii, 154, 155–56, 157
Kukuruza, 89, 98, 103, 105–6, 128
Kulakov, Fëdor, 154, 163, 164, 214–15
Kurgan Oblast, 59, 69
Kursk Oblast, 15, 44, 126, 184, 211. *See also* Province
Kuzmin, Iosif, 171
Kyiv, 8, 15, 83, 187
Kyiv Oblast, 83, 207, 210
  Red Plowman kolkhoz, 207–9, 210, 225
Kyrgyz SSR, *See* Kyrgyzstan
Kyrgyzstan, 68, 74, 97, 184

labor, 128, 129, 135–36
  conditions, 23, 139–40, 174
  discipline, 25, 160, 191
  market, 8, 24, 139, 146, 166, 191
  pay for (*see* wages and pay)
  unrest, 11, 164
labor camp, 23, 191
labor-day, 10, 17, 23, 24, 138, 140, 141, 142, 143, 146, 148, 154, 155–56, 158–59, 163
labor incentives, 24, 57, 87, 133
  coercion, 24, 25, 87, 88, 107, 109, 127, 129, 135–36, 137, 139, 140, 146–47, 166
  material, 87, 88, 91, 108, 109–12, 117, 118, 121, 129, 135–36, 137, 139–40, 142, 144–45, 146, 154–58, 160–61, 162, 164, 184, 194–95, 229, 269n54 (*see also* wages and pay)

moral, 87, 88–89, 90, 91–92, 100, 107–8, 109–11, 115, 117, 118–19, 121, 123, 125, 129, 135–36, 139, 157, 164, 190
land management, 33, 38, 74, 77, 83, 163, 211–212, 214–15, 216–17, 220, 224–25, 228
Larionov, Aleksei, 203–5, 211, 224
Latin America, 5, 52, 53
Latvia, 68, 70, 75, 92, 111, 116, 216, 217
Lebedev, Ivan, 203
Lenin, Vladimir, 1, 12, 15, 39, 44, 71, 106, 124, 134, 139, 147
Lenin Prize, 180
Leningrad, 33, 100, 115
Leningrad Affair, 58
Leningrad Oblast, 66, 162, 194
Li, Liubov, 98–99, 106
Libya, 52
Lipetsk Oblast, 83, 126
Lithuania, 114, 120, 139, 162, 215–23, 224, 225
Lithuanian SSR, *see* Lithuania
living standards
  as part of Cold War competition, 30, 32, 71, 164, 224
  in the Soviet Union, 1, 29, 30, 32, 43, 64, 230
  loans for development by Soviet Union, 45–46, 48, 51–52, 254n123
Loshchenkov, Fëdor, 202
Lysenko, Trofim, 21, 41, 168, 175–78, 179–80, 280nn37–38, 281n54

machines, agricultural, *see* industrial agriculture, machines as a component of
manufacture of, 78, 167, 169, 170–72, 173–74, 175
Magnitogorsk, 33, 109
Maksimov, Leonid, 213–14

Malenkov, Georgii, 10, 20, 22, 25, 26, 29, 37, 58–59, 61, 70, 71, 78, 83, 176, 191, 212
Maniakin, Sergei, 130–31, 156
Manukovskii, Nikolai, 95, 97, 98, 123, 153
Mao Zedong, 37, 47
Marx, Karl, 8, 15, 71, 250n41
Marxism, *see* ideology
mass media, 23, 56, 63, 67, 87, 88–89, 90, 95, 96, 98, 100, 105, 107, 109, 112–13, 114, 115–16, 117, 124–25, 129, 151, 152, 188, 198, 226
Matskevich, Vladimir, 38, 41, 49, 51, 79, 83, 94, 140, 178, 185, 195
Medvedev, Roy, 228
Medvedev, Zhores, 228
Mendel, Gregor, 176
Metro (Moscow), 16, 48, 104
Mexican Agriculture Program, 46–47
Mexico, 42, 46, 177
Michael, Louis, 33
Mikoyan, Anastas, 20, 48, 59
Ministry of Agriculture, 63, 79, 118, 146, 172, 193
Ministry of Government Oversight, 168, 172, 182, 185, 279n4
Ministry of Grain Procurement, 185, 187, 192, 193
mobilization campaigns, 60, 63, 88, 109, 110, 137, 140, 153, 197, 229. *See also* labor incentives, moral
Moldova, 18, 20, 33, 74–75, 105, 120, 162, 164, 182–83, 184, 186, 199
Moldavian SSR, *see* Moldova
Molodoi leninets, 124, 129–30
Molotov, Viacheslav, 68, 71
Moral Codex of the Builders of Communism, 110
moral economy, 138, 159–60, 164, 166, 272n5
Moroz, V. K., 214–15

Moscow, 8, 13, 15–16, 20, 21, 22, 31, 58, 59, 79, 93, 103, 104, 105, 106, 114, 118, 119, 132, 182, 188, 204, 223, 233
Moscow Oblast, 59, 65, 114, 117, 140, 146, 186, 200–1
MTS (machine-tractor station), 14, 22, 71–72, 110, 140, 143, 147, 150, 169–70, 172, 183, 199, 200–1, 217
    abolition of, 140–41, 144, 158, 161, 231
Muratov, Zinnat, 65, 68
Mylarshchikov, Vladimir, 201–2, 204–5

Nasser, Gemal Abdel, 52
nationalism, 215, 217, 220, 224
Naumenko, Andrei, 206
Nehru, Jawaharlal, 48, 49
Nepal, 52
New Delhi, 49, 50
New Economic Policy (NEP), 12, 14, 25, 26, 155
New Zealand, 43
Nikonov, Aleksandr, 68, 70, 213
Nixon, Richard, 2
North America, 4, 54, 232
North Caucasus, 34, 40, 51, 62, 186
North Ossetia ASSR, 121–23, 190
Novocherkassk, 30
Novosibirsk Report, 234
Nuriev, Zia, 76

Odesa Oblast, 66, 187–88, 192, 207
Omsk Oblast, 69, 111
Order of Lenin, 204
Orël Oblast, 67, 74, 126
Organov, Nikolai, 201
Ozërnyi, Mark, 55, 92, 94

peasants, 10–13, 47, 53, 96, 140, 212, 216, 231
Penza Oblast, 74, 117, 128

perestroika, 229, 233
personal plot, 14, 17, 23, 24–25, 138, 139, 142, 146, 150, 157, 164, 165, 166, 184, 216
petroleum, *see* industrial agriculture, fossil fuels as a component of
Petukhov, Aleksandr, 199
Petukhov, Boris, 202
Philippines, 46
Piketty, Thomas, 232
Pioneers, 112–13, 115, 118
planning in the Soviet Union, 40, 44, 47, 48, 50, 57, 65–66, 69, 79, 140, 173, 229. *See also* five-year plans
  alternatives to, 48, 253n122
  reforms to, 68–69, 70, 80–81, 154, 160
Podgornyi, Nikolai, 83, 149–50, 206, 207, 208
Poland, 30, 42, 49, 149
political economy of Soviet agriculture, 2, 8, 138, 166, 229, 230
popcorn, 93, 103, 104
Pravda, 22, 63, 88, 92, 93, 101, 112, 121, 197, 198, 204
Pravda Ukrainy, 188
press, *see* mass media
private property, 7, 23, 39, 48, 71, 84
procurements and purchases by the state, 58, 76–77, 84, 139, 144, 145, 166, 201–3, 229
Prometheanism, 34, 53
propaganda, 63, 67, 88, 99–100, 107, 128

Radianska Ukraïna, 188
rationing, 17, 19, 20
Red Army, *see* Soviet Army
religion, 14, 88, 98–100
Revolution of 1905, 11
Riazan Oblast, 66, 203–4, 207, 208–9, 210, 215, 221, 223, 224
Riga, 68
Rivne Oblast, 121, 148

Rockefeller Foundation, 46
Romania, 42
Rostov Oblast, 190, 211
RSFSR (Russian Soviet Federative Soviet Republic), *see* Russia
Rudenko, Roman, 206
rural modernization, 4, 5, 6, 10, 26, 47, 52, 53, 104, 158, 230, 231, 237n18
Rusk, Dean, 40
Russia, 13, 15, 17, 40, 66, 74, 123, 125–26, 142, 162, 206, 216, 224, 229
Russian Civil War, 3, 8, 11–12, 15, 64, 106, 109
Russian Revolution, 11, 15, 233

Salisbury, Harrison, 93
Samokhval, (official), 191, 193–94
Sanina, Aleksandra, 140
Saratov, 132, 150
Saratov Oblast, 116, 121
Scientific-Technical Revolution, 7
Selskaia zhizn, *see* Selskoe khoziaistvo
Selskoe khoziaistvo, 63, 67, 88, 97, 98, 101, 102, 103, 112
Semichastnyi, Vladimir, 44, 120
Seven-Year Plan (1959–1965), 44, 75, 76, 84
Sharkov, Boris, 222
Shelepin, Aleksandr, 112, 113, 205
Shelest, Petro, 83, 207–9, 210
Shepilov, Dmitrii, 59, 63, 68, 71
Sheremetev, Aleksandr, 171
Shevchenko, Andrei, 25, 38, 39, 41, 59
Siberia, 34, 60, 69, 72, 177, 201
Slezkine, Yuri, 224
Smirnov, Georgii, 205
Sniečkus, Antanas, 216, 217, 219–24
socialism, 1, 12–13, 17, 29, 31, 33, 64, 71, 88, 110, 118–19, 128, 133, 143, 164, 233

socialist competition, 95, 109, 110–11, 114–17, 118, 120–21, 123–24, 125, 136, 152
socialist realism, 25, 88, 140
Sokolov, Boris, 38, 128, 178–80
Solzhenitsyn, Alexander, 10
Sovetskaia Litva, 218–19, 222
Soviet Army, 15, 16, 17, 102, 215–16, 274n45
Soviet Union
  agrarian crisis in, 2, 10, 17, 25, 26, 54, 55–56
  bureaucracy in, 2, 54, 57, 61, 63, 75, 79, 80–81, 84, 167, 168, 171, 174, 192, 195, 198–200, 230, 231, 284n145–146
  economic growth of, 7, 30, 78–79, 100–1, 229
  government of, 8, 55, 124, 127 (*see also* Council of Ministers)
  industrialization of, 4, 6, 7, 10, 12–13, 22, 25, 31, 47, 54, 107, 109, 230
  rural governance in, 56–58, 79, 85–86, 88, 107–8, 138–39, 198, 199, 216–17, 225, 233–34
sovkhoz
  economics and operation of, 137, 143, 156, 212, 229
  low productivity of, 4, 79, 166, 211–12
  as state-owned enterprise, 48, 53, 68, 79, 229
sovnarkhoz, 191–92
space program, 49. *See also* Sputnik
Sputnik, 7, 100
square-cluster planting, 55, 59, 74, 75, 180, 201
Stakhanovites, 91. *See also* vanguard worker
Stakhurskii, Mikhail, 205–6
Stalin, Iosif, 1, 10, 12–14, 15–16, 18, 19, 25, 27, 48, 58, 63, 85, 98, 124, 127, 128, 168, 176, 198
  agricultural policies of, 55, 71, 88, 137, 138, 139, 167, 212, 230, 231 (*see also* collectivization)
  cult of, 18, 71, 106
Stalinism, 3, 53, 96, 212
Stalinist Plan for the Transformation of Nature, 26, 34
Stalin Prize, 92, 176
Starovskii, Vladimir, 25
State Committee for Foreign Economic Relations, 51
state of all the people, 110, 124, 126, 127, 135
State Planning Agency, *see* Gosplan
Stavropol (city), 8, 130
Stavropol Krai, 23, 68, 74, 90, 115, 124–25, 131–34, 141–42, 151, 153, 155, 157–59, 160, 161–62, 164–65, 169–70, 174–75, 182, 184, 185, 190–92, 193–94, 202–3, 211, 214–15, 233
  Lenin kolkhoz, Goriachevodsk, 151–53, 158, 160, 161, 275n74
Stavropolskaia pravda, 129, 141
Storozhev, Iakov, 199–200
student production brigades, 131–34
Suratgarh, 48–49, 51
Syria, 52

Tambov Oblast, 126, 142, 211, 227
Taubman, William, 2, 3, 38, 81
taxes on agriculture, 10, 12, 13, 17, 22, 23, 24, 25, 58, 139, 143, 150, 155
Taylor, Frederick Winslow, 33
technology
  history of, 6, 32
  transfer from the Soviet Union, 31, 32, 42, 53, 54, 231
  transfer from the United States, 1, 5, 20, 26, 28–29, 31, 32–34, 36, 38, 41, 43, 44, 51, 54, 77, 79, 95, 128, 168, 169, 172–73, 175, 178, 179, 180, 187, 188, 190, 231

transfer to the Soviet Union, 1, 4, 5–6, 29, 32–34, 38, 41, 43, 44, 53, 77, 168, 169, 172–73, 178, 179, 180, 187, 188, 190, 213, 231
  as value-neutral good, 33, 39, 42, 53, 175
Ternopil Oblast, 96–98, 148, 149
territorial production administrations, 80–81, 84, 211
Terror, the, 8, 14, 16
Thaw, the, 6, 29–30, 98, 104, 106, 140, 201
theft, 147–49, 151–52, 184, 205
Third Party Program, 30, 64
Thompson, E. P., 138, 272n5
Timiriazev Agricultural Academy, 79–80
Timiriazev, Kliment, 50
transfer mentality, 5, 29, 33–34, 46, 47, 51, 53
travopole, *see* grass-field system
Trotskii, Lev, 12
Tulupnikov, Aleksandr, 50, 143
Tursunkulov, Khamrakul, 51

Ukraine, 10, 11, 13, 15, 17, 18, 19, 20, 21, 40, 51, 55, 62, 74–75, 93, 97, 105, 119, 141, 146, 147, 148–49, 162, 163, 182, 186, 187, 190, 192, 205, 206–7, 219, 229
United Kingdom, 6, 12, 43
United States
  as international donor, 45–47, 49, 50
  as Soviet Union's main competitor, 6, 71, 224, 226
  visitors to the Soviet Union from, 38, 42, 43, 188, 252nn97–98
urbanization, 4, 8, 14, 22, 29, 104–5, 128, 136, 140, 165, 226, 231
Uzbekistan, 51, 60, 98
Uzbek SSR, *see* Uzbekistan

Vail, Pëtr, 104, 226
vanguard worker, 91, 95, 98–99, 121, 152, 153
Vasinauskas, Petras, 218–19
Vavilov, Nikolai, 33, 50, 176, 177, 178
Venzher, Vladimir, 140
Vietnam, 42
Viliams, Vasilli, 211, 214, 215, 218, 220
Vilnius, 8, 220, 222
Vinnytsia Oblast, 141, 206
Virgin Lands program, 5, 29, 34, 37, 40, 42, 59, 60–61, 68, 83, 84, 107, 109, 113, 114, 117, 132, 135, 198, 221, 227
Volga River, 62, 66
Vologda Oblast, 24, 66, 199
Volovchenko, Ivan, 83
Voronezh Oblast, 93, 94, 95, 114, 126, 184, 197

wage labor, 8, 12, 139, 143–45, 155, 162, 229, 231, 233
wages and pay, 129, 133, 140
  on kolkhozes, 23, 137, 139, 141, 142–44, 146, 148–49, 155–60, 162, 163, 164–65, 201, 230
  on sovkhozes, 139, 141, 156
Wallace, Henry, 46
Washington, DC, 40, 79
Western Europe, 4, 54, 88, 232
women as farm laborers, 22, 96, 98, 141, 142, 146, 151, 156, 159, 162, 269n57. *See also* gender and femininity
World War I, 3–4, 5, 11, 106
World War II, 4, 8, 11, 14, 16–18, 90, 110, 232
  effect on farming in the Soviet Union, 17, 18–19, 26, 54, 230

xenophobia, 18, 31, 176

Young Communist League, *see* Komsomol
youth, 32, 97, 109–15, 128, 133, 135, 267n2
Yugoslavia, 50

Zaslavskaia, Tatiana, 234
Zhebrak, Anton, 177
Zhidkikh, (kolkhoz chair), 148
Zhytomyr Oblast, 205, 225
Zolotoukhin, Grigorii, 211, 227